Wireless
A to Z

Wireless
A to Z

Nathan J. Muller

McGraw-Hill

New York   Chicago   San Francisco   Lisbon   London   Madrid
Mexico City   Milan   New Delhi   San Juan   Seoul
Singapore   Sydney   Toronto

The **McGraw·Hill** Companies

Cataloging-in-Publication Data is on file with the Library of Congress

1 2 3 4 5 6 7 8 9 0   DOC/DOC   0 9 8 7 6 5 4 3 2

ISBN 0-07-141088-0

The sponsoring editor for this book was Steve Chapman, the editing supervisor was Steven Melvin, and the production supervisor was Sherri Souffrance. It was set in New Century Schoolbook following the AZ01 design by Joanne Morbit of McGraw-Hill's Professional's composition unit, Hightstown, N.J.

RR Donnelley was printer and binder.

McGraw-Hill books are available at special quantity discounts to use as premiums and sales promotions, or for use in corporate training programs. For more information, please write to the Director of Special Sales, McGraw-Hill, 2 Penn Plaza, New York, NY 10121-2298. Or contact your local bookstore.

 This book is printed on recycled, acid-free paper containing a minimum of 50% recycled, de-inked fiber.

*To my brother Jim
On his retirement from the
South Burlington, Vermont,
Police Department*

CONTENTS

PREFACE

Of all the communications services available today, wireless services are having the most dramatic impact on our personal and professional lives, enhancing personal productivity, mobility, and security. With every new wireless product and service, the boundary between home and office is blurred further, perhaps to the point that one day they will be indistinguishable. Instead of the flexible work schedule, wireless products and services give us the capability of being "always on." For a growing number of people, a true vacation consists of shutting down communication with the rest of the world.

The wireless industry worldwide is experiencing rapid innovation, increased competition, and diversity in service offerings—all of which have resulted in lower prices for consumers and businesses. Service providers continue to fill in gaps in their national coverage through mergers, acquisitions, license swaps, and joint ventures. Along with this process of footprint building, service providers continue to deploy their networks in an increasing number of markets, expand their digital networks, and develop pricing plans that attract new subscribers and stimulate minutes of usage.

Mobile telephony is a particularly vibrant sector, experiencing strong growth and reaching new levels of competitive development. At year-end 2001, mobile telephony services generated over $65 billion in revenues in the United States, increased subscribers from 109.5 million to 128.5 million, and produced a nationwide penetration rate of over 50 percent.

Broadband PCS carriers and digital SMR providers continue to deploy their networks. According to the Federal Communications Commission (FCC), 268 million people, or 94 percent of the total U.S. population, live in areas served by three or more different operators offering mobile telephone service. Over 229 million people, or 80 percent of the U.S. population, live in counties with five or more mobile telephone operators competing to offer service. And 151 million people, or 53 percent of the population, live in areas in which six different mobile telephone operators are providing service.

Digital technology is now dominant in the mobile telephone sector. At the end of 2001, digital customers made up almost 80 percent of the industry total, up from 72 percent at year-end 2000. In part because of competitive pressures in the marketplace, the average price of residential mobile telephone service declined by 5.5 percent during 2001. The average revenue per minute of mobile telephone use fell 31 percent between 2000 and 2001.

Many mobile telephone carriers are deploying advanced wireless service network technologies such as cdma2000 1xRTT and General Packet Radio Service (GPRS). These deployments have contributed to the further convergence of mobile voice and data. The increased capacity on these digital networks has permitted operators to offer calling plans with large buckets of relatively inexpensive minutes, free enhanced services such as voice mail and caller ID, and wireless data and mobile Internet offerings.

Once solely a business tool, wireless phones are now a mass-market consumer device. By some estimates, 3 to 5 percent of customers use their wireless phones as their only phone. Though relatively few wireless customers have "cut the cord" in the sense of canceling their wireline telephone service, there is growing evidence that consumers are substituting wireless service for traditional phone service. It is also estimated that 20 percent of residential customers have replaced some wireline phone usage with wireless, and that 11 percent have replaced a significant percentage. And almost one in five mobile telephony users regard their wireless phone as their primary phone.

Contributing to these trends is the increasing number of mobile wireless carriers offering service plans designed to compete directly with wireline local telephone service, many with virtually unlimited regional calling plans. For $40 to $50 per month, subscribers get a calling plan that includes 4,000 minutes (usable anytime) and the ability to roam across several states without extra fees.

Several local carriers have attributed declining access-line growth rates in part to substitution by wireless. The number of residential access lines served by BellSouth, SBC, and Verizon dropped by almost 3 percent during 2001, or more than 2.5 million lines. Verizon attributes the decline in the number of access lines in part to the shift to wireless phones. Nationwide, by year-end 2001, wireless had displaced an estimated 10 million access lines, primarily by consumers choosing wireless over installing additional access lines.

Wireless plans are substituting for traditional wireline long distance as well. Many calling plans offered by national wireless carriers include free nationwide long distance. For example, about 20 percent of AT&T's customers, or 5 million people, have replaced some wireline long-distance usage with wireless. AT&T attributes the decline in its long-distance calling volumes and revenues in part to wireless substitution. At least one wireless operator, Cingular Wireless, advertises its nationwide calling plans with the slogan, "Never Pay Long Distance Again."

Because of national advertising and the Internet, consumers all over the country are educated about nationwide rate plans and services enabled by digital technology and the prices of wireless handsets. No matter where they live, customers expect and demand the diversity of services at competitive rates.

Wireless is having an impact in other ways. PDAs, or handheld devices, began as electronic organizers containing personal information management (PIM) functions, such as an address book, calendar, and to-do list that could be "synched" with PIM software on a user's desktop computer. While PDAs still contain these core PIM and software features, handhelds are being repositioned as wireless communication devices instead of simple organizers. All of the PDA models introduced by the major manufacturers since 2001 allow users some method of connecting to the Internet wirelessly. Some combine the features of mobile phones and PDA features into so-called "smartphones."

Compared to traditional mobile handsets, smartphones generally have larger screens, more advanced graphics and processing capabilities, more memory, a more user-friendly operating system, some form of keypad, and the ability to synch data with and download software from a desktop computer. Smartphones also integrate traditional mobile telephone phone number storage and access with a PDA's address book so users are not required to store numbers in two different places.

These smartphones also allow messaging via the Short Message Service (SMS). SMS provides the ability for users to send and receive text messages to and from mobile handsets with maximum message length ranging from 120 to 500 characters. SMS also can be used to deliver a wide range of information to mobile users, including stock prices, sport scores, news headlines, weather reports, and horoscopes. Worldwide, SMS has become increasingly popular, growing to 250 billion messages sent over wireless networks worldwide in 2001.

Using their existing and next-generation networks, major mobile telephone service providers offer text-based wireless web services via mobile telephone handsets at speeds ranging from approximately 14.4 kbps on 2G networks to 60 kbps on advanced 2.5/3G wireless networks. During 2001, mobile telephone providers expanded their data service offerings as they began to transition their networks to higher speed technologies. In addition to offering wireless web service on mobile telephone handsets, several carriers offer wireless Internet connections via wireless modem cards for PDAs and laptop computers as well.

Mobile telephone service providers offer wireless web services that enable customers to surf web sites for news, stock quotes, traffic reports, weather forecasts, movie listings, shopping, and other text-based information. To deliver wireless web content to wireless handset users, carriers currently restrict users to less graphically enriched content. This conserves the resources of the memory-constrained devices. However, customers who connect to the Internet via a wireless modem card attached to a notebook computer are able to access the full content of the web.

Many PDAs have the ability to access almost the entire content of the web. For example, Pocket PC PDAs include a PDA version of Microsoft's Internet Explorer web browser, which can access any web site. While many PDAs have the potential to access web content with their browsing software, they still require a subscription to a wireless Internet Service Provider (WISP) in order to connect to the Internet via wireless links.

As workers become increasingly more mobile and remote, the ability for employees outside the office to access e-mail messages and files stored electronically on corporate servers is likely to become an increasingly more important mobile data application. Analysts claim that giving employees mobile access to e-mail and to data and applications stored on corporate servers are two of the most important uses of PDAs in the enterprise market. Surveys of U.S. firms indicate that mobile access to e-mail is the top priority.

There are short-range data transmission technologies that are gaining in popularity: infrared, Bluetooth, and Wireless Fidelity (Wi-Fi). Infrared is currently used in some PDAs to allow users to transfer data between two devices. Infrared is also the technology commonly used in remote controls and requires line-of-sight transmission. Bluetooth enables multipoint broadcasting applications, and Wi-Fi enables devices to connect to wireless local area networks (WLANs).

Bluetooth is a technology used to establish wireless connectivity between electronic devices that are up to 30 ft (10 m) apart. It allows users to send signals and transfer data among numerous electronic devices, thus creating a personal area network (PAN). Bluetooth uses unlicensed spectrum in the 2.4 GHz band and transmits data at speeds close to 1 Mbps. Bluetooth also uses frequency hopping spread spectrum techniques to provide enhanced communications performance and an initial level of transmission security.

Wi-Fi is another wireless networking technology sharing the 2.4 GHz frequency band with Bluetooth. Also called Wi-Fi, the 802.11b standard is used to connect devices to WLANs, and allows a maximum throughput of 11 Mbps. The technology is being used in a number of WLAN settings, such as college campuses, business parks, office buildings, and even private homes. It is also being implemented by a number of vendors in public places such as airports, hotels, and cafes to give users of notebook computers, handheld devices, and smartphones wireless Internet access anywhere inside those locations.

Wireless technologies and services have become so popular worldwide and sufficiently sophisticated and complex as to merit dozens of books on the topic that are published every year. This encyclopedia is a quick reference that clearly explains the essential concepts of wireless, including services, applications, protocols, network methods, development tools, administration and management, standards, and regulation. It is designed as a companion to other books you may want to read about wireless, providing clarification of concepts that may not be fully covered elsewhere.

The information contained in this book, especially as it relates to specific vendors and products, is believed to be accurate at the time it was written and is, of course, subject to change with continued advancements in technology and shifts in market forces. Mention of specific products and services is for illustration purposes only and does not constitute an endorsement of any kind by either the author or the publisher.

Nathan J. Muller

A

ACCESS POINTS

An access point (AP) provides the connection between one or more wireless client devices and a wired local area network (LAN). The AP is usually connected to the LAN via a Category-5 cable connection to a hub or switch. Client devices communicate with the AP over the wireless link, giving them access to all other devices through the hub or switch, including a router on the other side of the hub, which provides Internet access (Figure A-1).

An AP that adheres to the IEEE 802.11b Standard for operation over the unlicensed 2.4-GHz band supports a wireless link with a data transfer speed of up to 11 Mbps, while an AP that adheres to the IEEE 802.11a Standard for operation over the unlicensed 5-GHz band supports a wireless link with a data transfer speed of up to 54 Mbps. Access points include a number of the following functions and features:

- Radio power control for flexibility and ease of networking setup

- Dynamic rate scaling, mobile Internet Protocol (IP) functionality, and advanced transmit/receive technology to enable multiple access points to serve users on the move

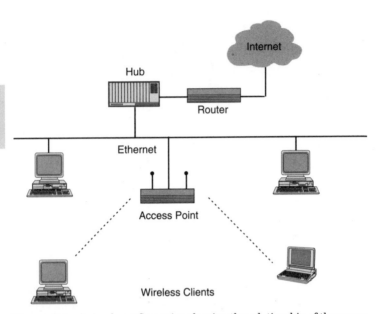

Figure A-1 A simple configuration showing the relationship of the access point to the wired and wireless segments of the network.

- Built-in bridging and repeating features to connect build-ings miles apart (The use of specialty antennas increases range. The AP can support simultaneous bridging and client connections.)
- Wired Equivalent Privacy (WEP), which helps protect data in transit over the wireless link between the client device and the AP, via 64-, 128-, or 256-bit encryption
- Access control list (ACL) and virtual private network (VPN) compatibility to help guard the network from intruders
- Statistics on the quality of the wireless link (Figure A-2)
- Configurability using the embedded Web browser

Consumer-level APs stress ease of setup and use (Figure A-3). Many products are configured with default settings

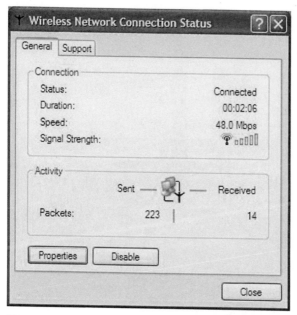

Figure A-2 The 5-GHz DWL-5000 Access Point from D-Link Systems, Inc., keeps the client device notified of the status of the wireless link. In this case, the signal is at maximum strength and is capable of a data transfer rate of 48 Mbps.

that allow the user to plug in the device and use the wireless connection immediately. Later, the user can play with the configuration settings to improve performance and set up security.

Although APs adhere to the IEEE 802.11 Standards, manufacturers can include some proprietary features that improve the data transfer speed of the wireless link. For example, one vendor advertises a "turbo mode" that optionally increases the maximum speed of IEEE 802.11b wireless links from 11 to 22 Mbps. When this turbo feature is applied to IEEE 802.11a wireless links, the maximum speed is increased from 54 to 72 Mbps.

Figure A-3   An example of a consumer AP is this 5-GHz wireless access point (WAP54A) from Linksys, which features antenna with a range of up to 328 feet indoors.

Enterprise-level APs provide more management features, allowing LAN administrators to remotely set up and configure multiple APs and clients from a central location. For monitoring and managing an entire wireless LAN infrastructure consisting of hundreds or even thousands of access points, however, a dedicated management system is usually required. Such systems automatically discover every AP on the network and provide real-time monitoring of an entire wireless network spread out over multiple facilities and subnets. These management systems support the Simple Network Management Protocol (SNMP) and can be tied into higher-level management platforms such as Hewlett-Packard's OpenView.

Among the capabilities of these wireless managers is support of remote reboot, group configuration, or group software uploads for all the wireless infrastructure devices on the network. In addition, the LAN administrator can see how many client devices are connected to each access point, monitor those connections to measure link quality, and monitor all the access points for performance.

Some enterprise APs provide dual-band wireless connections to support both IEEE 802.11a and 802.11b client users at the same time. This is accomplished by equipping the AP with two plug-in radio cards—one that supports the 2.4-GHz frequency specified by the IEEE 802.11b Standard and one that supports the 5-GHz frequency specified by the IEEE 802.11a Standard.

The choice of a dual-band AP provides organizations with a migration path to the higher data transfer speeds available with IEEE 802.11a while continuing to support their existing investment in IEEE 802.11b infrastructure. Depending on manufacturer, these dual-band APs are modular so that they can be upgraded to support future IEEE 802.11 technologies as they become available, which further protects an organization's investment in wireless infrastructure.

Summary

Access points are the devices that connect wireless client devices to the wired network. They are available in consumer and commercial versions, with the latter generally costing more because of more extensive management capabilities and troubleshooting features. They may have more security features as well and support both the 2.4- and 5-GHz frequency bands with separate radio modules that plug into the same unit.

See also

Bluetooth
Wired Equivalent Privacy
Wireless Fidelity

Wireless LANs

Wireless Security

ADVANCED MOBILE PHONE SERVICE

Before the age of digital services, the predominant technology for analog cellular phone services in North America adhered to a set of standards for Advanced Mobile Phone Service (AMPS). Originally, AMPS operated in the 800-MHz frequency band using 30-kHz-wide channels. A variant of AMPS, known as Narrowband AMPS (NAMPS), uses 10-kHz-wide channels and consequently has triple the capacity of AMPS. Although AMPS or a variation of AMPS is still around—chances are that your cellular phone allows you to switch between analog and digital mode—its use is rapidly declining in the face of more sophisticated digital cellular standards.

The mobile telephone service that preceded AMPS was known as Improved Mobile Telephone Service (IMTS), which operated in several frequency ranges: 35 to 44 MHz, 152 to 158 MHz, and 454 to 512 MHz. But IMTS suffered from call setup delay, poor transmission, and limited frequency reuse. AMPS overcame the limitations of IMTS and set the stage for the explosive growth of cellular service, which continues today worldwide. Interestingly, Pacific Bell finally dropped IMTS in 1995.

Proposed by AT&T in 1971, AMPS is still the standard for analog cellular networks. It was tested in 1978, and in the early 1980s cellular systems based on the standards were installed throughout North America. Although AMPS was not the first system for wireless telephony, the existence of a single set of standards enabled the United States to dominate analog cellular throughout the 1980s. Today, Europe dominates cellular primarily because it is a lower-cost alternative to conventional telephone service.

Analog cellular is delivered from a system of cellular hubs and base stations with associated radio towers. A mobile telecommunications switching office (MTSO) authenticates wireless customers before they make calls, switches calls between cells as mobile phone users travel across cell boundaries, and places calls from land-based telephones to wireless customers.

AMPS uses a technique called "frequency reuse" to greatly increase the number of customers that can be served at the same time. Low-powered mobile phones and radio equipment at each cell site permit the same radio frequencies to be reused in different cells, multiplying calling capacity without creating interference. This spectrum-efficient method contrasts sharply with earlier mobile systems that used a high-powered, centrally located transmitter to communicate with high-powered mobile equipment installed in vehicles over a small number of frequencies. Once a channel was occupied with a call, its frequency could not reused over a wide area.

Despite the success of AMPS, this method of transmission has its limitations. Analog signals can be intercepted easily and suffer signal degradation from numerous sources, such as terrain, weather, and traffic volume. Analog systems also could not handle the transmission of data very well. A digital version of AMPS—referred to as DAMPS—solves many of these problems while providing increased capacity and a greater range of services. Both AMPS and DAMPS operate in the 800-MHz band and can coexist with each other. DAMPS is implemented with Time Division Multiple Access (TDMA) as the underlying technology, which provides 10 to 15 times more channel capacity than AMPS and allows the introduction of new feature-rich services such as data communications, voice mail, call waiting, call diversion, voice encryption, and calling-line identification.

A digital control channel available with DAMPS supports such advanced features as a sleep mode, which increases battery life on newer cellular phones by as much as 10 times over the current battery capabilities of analog phones. DAMPS also can be implemented with Code Division Multiple Access (CDMA) technology to increase channel capacity by as much as 20 times and provide a comparable range of services and features. Unlike TDMA, which can be added onto existing AMPS infrastructure, CDMA requires an entirely new network infrastructure.

DAMPS also allows operators to build overlay networks using small micro- and picocells, boosting network capacity still further in high-traffic areas and providing residential and business in-building coverage. Advanced software in the networks' exchanges continuously monitors call quality and makes adjustments, such as handing calls over to different cells or radio channels, when necessary. The network management system provides an early warning to the network operator if the quality of service is deteriorating so that steps can be taken to head off serious problems. Graphical displays of network configuration and performance statistics help ensure maximum service quality for subscribers.

Summary

In 1983, AMPS was approved by the Federal Communications Commission (FCC) and first used in Chicago. In order to encourage competition and keep prices low, the U.S. government required the presence of two carriers in every market, known as A and B carriers. One of the carriers was normally the Local Exchange Carrier (LEC); in other words, the local phone company.

See also

Cellular Data Communications

Cellular Voice Communications

AIR-GROUND RADIOTELEPHONE SERVICE

With the Air-Ground Radiotelephone Service, a commercial mobile radio service (CMRS) provider offers two-way voice, fax, and data service for hire to subscribers in aircraft—in flight or on the ground. Service providers must apply for an FCC license for each and every tower/base site. There are two versions of this service: one for general aviation and one for commercial aviation.

General Aviation Air-Ground Service

Air-Ground Radiotelephone Service has been available to general aviation for more than 30 years. General Aviation Air-Ground systems may operate in the 454.675- to 454.975-MHz and 459.675- to 459.975-MHz bands to provide service to private aircraft, specifically, small single-engine craft and corporate jets.

The service is implemented through general aviation air-ground stations, which comprise a network of independently licensed stations. These stations employ a standardized duplex analog technology called "Air-Ground Radiotelephone Automated Service" (AGRAS) to provide telephone service to subscribers flying over the United States or Canada. Because there are only 12 channels available for this service, it is not available to passengers on commercial airline flights.

Commercial Aviation Air-Ground Systems

Commercial Aviation Air-Ground Systems may operate on 10 channel blocks in the 849- to 851-MHz and 894- to 896-MHz bands. These nationwide systems employ various analog or digital wireless technologies to provide telephone service to passengers flying in commercial aircraft over the United States, Canada, and Mexico. Some systems have satellite-calling capability as well, where the call is sent to an earth station instead of the base station.

Passengers use credit cards or prearranged accounts to make telephone calls from bulkhead-mounted telephones or, in larger jets, from seatback-mounted telephones. This service was available from one company on an experimental basis during the 1980s and began regular competitive operations in the early 1990s. There are currently three operating systems, one of which is GTE Airfone, a subsidiary of Verizon Communications.

When an Airfone call is placed over North America, information is sent from the phone handset to a receiver in the plane's belly and then down to one of the 135 strategically placed ground radio base stations. From there, it is sent to one of three main ground switching stations and then over to the public telephone network to the receiving party's location. When an Airfone call is placed over water, information is sent first to an orbiting satellite. From there, the call transmission path is similar to the North American system, except that calls are sent to a satellite earth station instead of a radio base station. Calls can be placed to any domestic or international location.

To receive calls aboard aircraft, the passenger must have an activation number. In the case of Airfone, an activation number can be obtained by dialing 0 toll-free onboard or 1-800-AIRFONE from the ground. For each flight segment, the activation number will be the same. However, the passenger must activate the phone for each flight segment and include his or her seat number. The person placing the call from the ground dials 1-800-AIR-FONE and follows the voice prompts to enter the passenger's activation number. The passenger is billed for the call on a calling card or credit card but gets to choose whether or not to accept the calls.

The following steps are involved in receiving a call:

- The phone will ring on the plane, and the screen will indicate a call for the seat location.

- The passenger enters a personal identification number (PIN) to ensure that no one else can answer the call.

- The phone number of the calling party will be displayed on the screen.

- If the call is accepted, the passenger is prompted to slide a calling card or credit card to pay for the call.

- Once the call has been accepted, the passenger is automatically connected to the party on the ground.

- If the passenger chooses not to accept the call, he or she follows the screen prompts, and no billing will occur.

Air-to-ground calls are very expensive. The cost to place domestic calls using GTE's Airfone Service, for example, is $2.99 to connect and $3.28 per minute or partial minute, plus applicable tax. By comparison, AT&T's Inflight Calling costs $2.99 to connect plus $2.99 per minute. These rates apply to all data/fax and voice calls. Even calls to 800 and 888 numbers—which are normally toll-free on the ground—are charged at the same rate as regular Airfone and Inflight calls. No billing ever occurs for the ground party. The charges for international calls are higher; both AT&T and GTE charge $5.00 to connect and $5.00 per minute. GTE offers satellite service for use over the ocean and worldwide at $10.00 to connect and $10.00 per minute, but the service is available only on United Airlines.

Summary

FCC rules specifically prohibit the use of cellular transmitters on aircraft, except for aircraft on the ground. This prohibition was not done to protect the aircraft's avionics systems from interference from the cellular transmitter. Rather, this prohibition was made to protect the cellular service on the ground from interference. As the altitude of a cellular handheld transmitter increases, its range also increases and, consequently, its coverage area. At high altitudes, such as would be achieved from an in-flight aircraft, the hand-held unit places its signal over several cellular base stations, preventing other cellular users within range of those base stations from using the same frequency.

This would increase the number of blocked or dropped cellular calls.

See also

   Cellular Data Communications
   Cellular Voice Communications

AMATEUR RADIO SERVICE

Amateur Radio Service is defined by the FCC as "A radio communication service for the purpose of self-training, intercommunication, and technical investigations carried out by amateurs; that is, duly authorized persons interested in radio technique solely with a personal aim and without pecuniary interest."[1]

Amateur radio stations are licensed by the FCC and may engage in domestic and international communications—both two-way and one-way. Applications for new licenses or for a change in operator class are filed through a volunteer examiner-coordinator (VEC). Operators can use their station equipment as soon as they see that information about their amateur operator/primary station license grant appears on the amateur service database. New operators do not need to have the license document in their possession to commence operation of an amateur radio station.

Since amateur stations must share the air waves, each station licensee and each control operator must cooperate in selecting transmitting channels and in making the most effective use of the amateur service frequencies. A specific transmitting channel is not assigned for the exclusive use of any amateur station.

---

[1]There are two exceptions to this rule. A person may accept compensation when in a teaching position and the amateur station is used as a part of classroom instruction at an educational institution. The other exception is when the control operator of a club station is transmitting telegraphy practice or information bulletins.

Types of Communications

With regard to two-way communications, amateur stations are authorized to exchange messages with other stations in the amateur service, except those in any country whose administration has given notice that it objects to such communications.[2] In addition, transmissions to a different country must be made in plain language. Communication is limited to messages of a technical nature relating to tests and to remarks of a personal nature for which, by reason of their unimportance, use of public telecommunications services is not justified.

Amateur radio stations also may engage in one-way communications. For example, they are authorized to transmit auxiliary, beacon, and distress signals. Specifically, an amateur station may transmit the following types of one-way communications:

- Brief transmissions necessary to make adjustments to the station
- Brief transmissions necessary for establishing two-way communications with other stations
- Transmissions necessary to provide emergency communications
- Transmissions necessary for learning or improving proficiency in the use of international Morse code
- Transmissions necessary to disseminate an information bulletin of interest to other amateur radio operators
- Telemetry

Prohibited Communications

Although the FCC does not provide a list of communications that are suitable or unsuitable for the amateur radio service,

---

[2]As of mid-2002, no administration in another country had given notice that they object to communications between the amateur radio stations.

there are several types of amateur-operator communications that are specifically prohibited, including

- Transmissions performed for compensation
- Transmissions done for the commercial benefit of the station control operators
- Transmissions done for the commercial benefit of the station control operator's employer
- Transmissions intended to facilitate a criminal act
- Transmissions that include codes or ciphers intended to obscure the meaning of the message
- Transmissions that include obscene or indecent words or language
- Transmissions that contain false or deceptive messages, signals, or identification
- Transmissions on a regular basis that could reasonably be furnished alternatively through other radio services

Broadcasting information intended for the general public is also prohibited. Amateur stations may not engage in any form of broadcasting or in any activity related to program production or newsgathering for broadcasting purposes. The one exception is when communications directly related to the immediate safety of human life or the protection of property may be provided by amateur stations to broadcasters for dissemination to the public where no other means of communication is reasonably available before or at the time of the event.

Amateur stations are not afforded privacy protection. This means that the content of the communications by amateur stations may be intercepted by other parties and divulged, published, or used for another purpose.

Summary

In August 1999, the FCC's Wireless Telecommunications Bureau (WTB) began the transition to the Universal Licensing

System (ULS) for all application and licensing activity in the Amateur Radio Services. As of February 2000, amateur licensees were required to file using ULS forms, which means that applications using Forms 610 and 610V are no longer accepted by the WTB.[3] The ULS is an interactive licensing database developed by the WTB to consolidate and replace 11 existing licensing systems used to process applications and grant licenses in wireless services. ULS provides numerous benefits, including fast and easy electronic filing, improved data accuracy through automated checking of applications, and enhanced electronic access to licensing information.

See also

Citizens Band Radio Service

Telegraphy

Telemetry

---

[3]For applications that do not need to be filed by a volunteer-examiner coordinator (VEC), such as renewals and administrative updates. Amateur Service licensees may still file FCC Form 605 electronically (interactively) or manually, despite the ULS requirement for other filings.

B

BASIC EXCHANGE TELEPHONE RADIO SERVICE

Developed in the mid-1980s, Basic Exchange Telephone Radio Service (BETRS) is a fixed radio service that uses a multiplexed digital radio link as the last segment of the local loop to provide wireless telephone service to subscribers in remote areas where it would be impractical to provide wireline telephone service. The wireless link allows up to four subscribers to use a single radio channel pair simultaneously without interfering with one another.

Licensed by the Federal Communications Commission (FCC) under the Rural Radiotelephone Service, BETRS may be licensed only to state-certified carriers in the area where the service is provided and is considered a part of the Public Switched Telephone Network (PSTN) by state regulators.

This service operates in the paired 152/158- and 454/459-MHz bands and on 10 channel blocks in the 816- to 820-MHz and 861- to 865-MHz bands. These channels are also allocated for paging services. Since BETRS primarily serves rural areas in the western part of the United States, it typically does not conflict geographically with paging services. When there is a conflict, the FCC provides a remedy.

Rural Radiotelephone Service and BETRS providers obtain site licenses and operate facilities on a secondary basis. This means that if any geographic area licensee subsequently notifies the Rural Radiotelephone Service or BETRS licensee that a facility must be shut down because it may cause interference to the paging licensee's existing or planned facilities, the Rural Radiotelephone Service or BETRS licensee must discontinue use of the particular channel at that site no later than 6 months after such notice.

Summary

BETRS primarily serves rural, mountainous, and sparsely populated areas that might not otherwise receive basic telephone service. Although the industry has raised concerns that auctioning spectrum for BETRS would have the effect of raising the cost of the service, which could deprive these areas of basic telephone service, the FCC does not distinguish BETRS from other services that use radio spectrum to provide commercial communication services.

See also

Rural Radiotelephone Service

BLUETOOTH

Bluetooth is an omnidirectional wireless technology that provides limited-range voice and data transmission over the unlicensed 2.4-GHz frequency band, allowing connections with a wide variety of fixed and portable devices that normally would have to be cabled together. Up to eight devices—one master and seven slaves—can communicate with one another in a so-called piconet at distances of up to 30 feet. Table B-1 summarizes the performance characteristics of Bluetooth products that operate at 1 Mbps in the 2.4-GHz range.

Applications

Among the many things users can do with Bluetooth is swap data and synchronize files merely by having the devices come within range of one another. Images captured with a digital camera, for example, can be dropped off at a personal computer (PC) for editing or a color printer for output on photo-quality paper—all without having to connect cables, load files, open applications, or click buttons.

The technology is a combination of circuit switching and packet switching, making it suitable for voice as well as data. Instead of fumbling with a cell phone while driving, for example, the user can wear a lightweight headset to answer a call and engage in a conversation even if the phone is tucked away in a briefcase or purse.

While useful in minimizing the need for cables, wireless local area networks (LANs) are not intended for interconnecting the range of mobile devices people carry around everyday between home and office. For this, Bluetooth is needed. And in the office, a Bluetooth portable device can be

TABLE B-1 Performance Characteristics of Bluetooth Products

Feature/Function	Performance
Connection type	Spread spectrum (frequency hopping)
Spectrum	2.4-GHz ISM (industrial, scientific, and medical) band
Transmission power	1 milliwatt (mW)
Aggregate data rate	1 Mbps using frequency hopping
Range	Up to 30 feet (9 meters)
Supported stations	Up to eight devices per piconet
Voice channels	Up to three synchronous channels
Data security	For authentication, a 128-bit key; for encryption, the key size is configurable between 8 and 128 bits
Addressing	Each device has a 48-bit Media Access Control (MAC) address that is used to establish a connection with another device

in motion while connected to the LAN access point as long as the user stays within the 30-foot range.

Bluetooth can be combined with other technologies to offer wholly new capabilities, such as automatically lowering the ring volume of cell phones or shutting them off as users enter quiet zones such as churches, restaurants, theaters, and classrooms. On leaving the quiet zone, the cell phones are returned to their original settings.

Topology

The devices within a piconet play one of two roles: that of master or slave. The master is the device in a piconet whose clock and hopping sequence are used to synchronize all other devices (i.e., slaves) in the piconet. The unit that carries out the paging procedure and establishes a connection is by default the master of the connection. The slaves are the units within a piconet that are synchronized to the master via its clock and hopping sequence.

The Bluetooth topology is best described as a multiple-piconet structure. Since Bluetooth supports both point-to-point and point-to-multipoint connections, several piconets can be established and linked together in a topology called a "scatternet" whenever the need arises (Figure B-1).

Piconets are uncoordinated, with frequency hopping occurring independently. Several piconets can be established and linked together ad hoc, where each piconet is identified by a different frequency-hopping sequence. All users participating on the same piconet are synchronized to this hopping sequence. Although synchronization of different piconets is not permitted in the unlicensed ISM band, Bluetooth units may participate in different piconets through Time Division Multiplexing (TDM). This enables a unit to sequentially participate in different piconets by being active in only one piconet at a time.

With its service discovery protocol, Bluetooth enables a much broader vision of networking, including the creation of

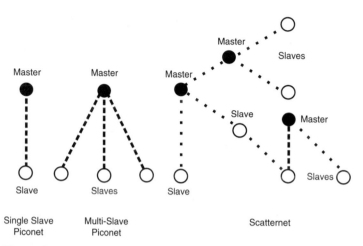

Figure B-1   Possible topologies of networked Bluetooth devices, where each is either a master or slave.

personal area networks, where all the devices in a person's life can communicate and work together. Technical safeguards ensure that a cluster of Bluetooth devices in public places, such as an airport lounge or train terminal, would not suddenly start talking to one another.

Technology

Two types of links have been defined for Bluetooth in support of voice and data applications: an asynchronous connectionless (ACL) link and a synchronous connection-oriented (SCO) link. ACL links support data traffic on a best-effort basis. The information carried can be user data or control data. SCO links support real-time voice and multimedia traffic using reserved bandwidth. Both data and voice are carried in the form of packets, and Bluetooth devices can support active ACL and SCO links at the same time.

ACL links support symmetric or asymmetric packet-switched point-to-multipoint connections, which are typically used for data. For symmetric connections, the maximum data rate is 433.9 kbps in both directions, send and receive. For asymmetric connections, the maximum data rate is 723.2 kbps in one direction and 57.6 kbps in the reverse direction. If errors are detected at the receiving device, a notification is sent in the header of the return packet so that only lost or corrupt packets need to be retransmitted.

SCO links provide symmetric circuit-switched point-to-point connections, which are typically used for voice. Three synchronous channels of 64 kbps each are available for voice. The channels are derived through the use of either Pulse Code Modulation (PCM) or Continuous Variable Slope Delta (CVSD) Modulation. PCM is the standard for encoding speech in analog form into the digital format of ones and zeros. CVSD is another standard for analog-to-digital encoding but offers more immunity to interference and therefore is better suited than PCM for voice communication over a wireless link. Bluetooth supports both PCM and CVSD; the appropriate voice-coding scheme is selected after negotiations between the link managers of each Bluetooth device before the call takes place.

Voice and data are sent as packets. Communication is handled with Time Division Duplexing (TDD), which divides the channel into time slots, each 625 microseconds (µs) in length. The time slots are numbered according to the clock of the piconet master. In the time slots, master and slave can transmit packets. In the TDD scheme, master and slave alternatively transmit (Figure B-2). The master starts its transmission in even-numbered time slots only, and the slave starts its transmission in odd-numbered time slots only. The start of the packet is aligned with the slot start. Packets transmitted by the master or the slave may extend over as many as five time slots.

With TDD, bandwidth can be allocated on an as-needed basis, changing the makeup of the traffic flow as demand warrants. For example, if the user wants to download a large data file, as much bandwidth as is needed will be allocated

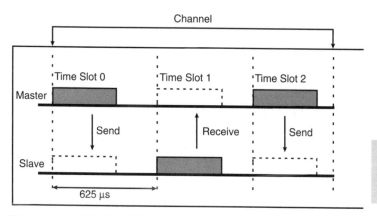

Figure B-2 With the TDD scheme used in Bluetooth, packets are sent over time slots of 625 microseconds (μs) in length between the master and slave units within a piconet.

to the transfer. Then, at the next moment, if a file is being uploaded, that same amount of bandwidth can be allocated to that transfer.

No matter what the application—voice or data—making connections between Bluetooth devices is as easy as powering them up. In fact, one advantage of Bluetooth is that it does not need to be set up—it is always on, running in the background, and looking for other devices that it can communicate with.

When Bluetooth devices come within range of one another, they engage in a service discovery procedure, which entails the exchange of messages to become aware of each other's service and feature capabilities. Having located available services within the vicinity, the user may select from any of them. After that, a connection between two or more Bluetooth devices can be established.

The radio link itself is very robust, using frequency-hopping spread-spectrum technology to overcome interference and fading. Spread spectrum is a digital coding technique in which the signal is taken apart or "spread" so that it sounds more like noise as it is sent through the air. With the addition of frequency hopping—having the signals skip from one frequency to another—wireless transmissions are made even more secure. Bluetooth specifies a rate of 1600 hops per second among 79 frequencies. Since only the sender and receiver know the hopping sequence for coding and decoding the signal, eavesdropping is virtually impossible. For enhanced security, Bluetooth also supports device authentication and encryption.

Other frequency-hopping transmitters in the vicinity will be using different hopping patterns and much slower hop rates than Bluetooth devices. Although the chance of Bluetooth devices interfering with non-Bluetooth devices that share the same 2.4-GHz band is minimal, should non-Bluetooth transmitters and Bluetooth transmitters coincidentally attempt to use the same frequency at the same moment, the data packets transmitted by one or both devices will become garbled in the collision, and a retransmission of the affected data packets will be required. A new data packet will be sent again on the next hopping cycle of each transmitter. Voice packets, because of their sensitivity to delay, are never retransmitted.

Points of Convergence

In some ways, Bluetooth competes with infrared, and in other ways, the two technologies are complementary. With both infrared and Bluetooth, data exchange is considered to be a fundamental function. Data exchange can be as simple as transferring business card information from a mobile phone to a palmtop or as sophisticated as synchronizing personal information between a palmtop and desktop PC. In fact, both technologies can support many of the same applications, raising the question: Why would users need both technologies?

The answer lies in the fact that each technology has its advantages and disadvantages. The very scenarios that leave infrared falling short are the ones where Bluetooth excels, and vice versa. Take the electronic exchange of business card information between two devices. This application usually will take place in a conference room or exhibit floor where a number of other devices may be attempting to do the same thing. This is the situation where infrared excels. The short-range and narrow angle of infrared—30 degrees or less—allow each user to aim his or her device at the intended recipient with point-and-shoot ease. Close proximity to another person is natural in a business card exchange situation, as is pointing one device at another. The limited range and angle of infrared allow other users to perform a similar activity with ample security and no interference.

In the same situation, a Bluetooth device would not perform as well as an infrared device. With its omnidirectional capability, the Bluetooth device must first discover the intended recipient. The user cannot simply point at the intended recipient—a Bluetooth device must perform a discovery operation that probably will reveal several other Bluetooth devices within range, so close proximity offers no advantage here. The user will be forced to select from a list of discovered devices and apply a security mechanism to prevent unauthorized access. All this makes the use of Bluetooth for business card exchange an awkward and needlessly time-consuming process.

However, in other data-exchange situations, Bluetooth might be the preferred choice. Bluetooth's ability to penetrate solid objects and its ability to communicate with other devices in a piconet allow for data-exchange opportunities that are very difficult or impossible with infrared. For example, Bluetooth allows a user to synchronize a mobile phone with a notebook computer without taking the phone out of a jacket pocket or purse. This would allow the user to type a new address at the computer and move it to the mobile phone's directory without unpacking the phone and setting up a cable connection between the two devices. The omnidirectional capability of Bluetooth allows synchronization to occur instantly, assuming that the phone and computer are within 30 feet of each other.

Using Bluetooth for synchronization does not require that the phone remain in a fixed location. If a phone is carried about in a briefcase, the synchronization can occur while the user moves around. This is not possible with infrared because the signal is not able to penetrate solid objects, and the devices must be within a few feet of each other. Furthermore, the use of infrared requires that both devices remain stationary while the synchronization occurs.

When it comes to data transfers, infrared does offer a big speed advantage over Bluetooth. While Bluetooth moves data between devices at an aggregate rate of 1 Mbps, infrared offers 4 Mbps of data throughput. A higher-speed version of infrared is now available that can transmit data between devices at up to 16 Mbps—a four times improvement over the previous version. The higher speed is achieved with the Very Fast Infrared (VFIR) Protocol, which is designed to address the new demands of transferring large image files between digital cameras, scanners, and PCs. Even when Bluetooth is enhanced for higher data rates in the future, infrared is likely to maintain its speed advantage for many years to come.

Bluetooth complements infrared's point-and-shoot ease of use with omnidirectional signaling, longer-distance communications, and capacity to penetrate walls. For some users, having both Bluetooth and infrared will provide the optimal short-range wireless solution. For others, the choice of adding Bluetooth or infrared will be based on the applications and intended usage.

Summary

Communicator platforms of the future will combine a number of technologies and features in one device, including mobile Internet browsing, messaging, imaging, location-based applications and services, mobile telephony, personal information management, and enterprise applications. Bluetooth will be a key component of these platforms. Since Bluetooth radio transceivers operate in the globally available ISM (industrial, scientific, and medical) radio band of 2.4 GHz, products do not require an operator license from a regulatory agency, such as the FCC in the United States. The use of a generally available frequency band means that Bluetooth-enabled devices can be used virtually anywhere in the world and link up with one another for ad hoc networking when they come within range.

See also

Infrared Networking
Spread Spectrum Radio

BRIDGES

Bridges are used to extend or interconnect LAN segments, whether the segments consist of wired or wireless links. At one level, they are used to create an extended network that greatly expands the number of devices and services available to each user. At another level, bridges can be used for segmenting LANs into smaller subnets to improve performance, control access, and facilitate fault isolation and testing without impacting the overall user population.

The bridge does this by monitoring all traffic on the subnets that it links. It reads both the source and destination addresses of all the packets sent through it. If the bridge encounters a source address that is not already contained in its address table, it assumes that a new device has been added to the local network. The bridge then adds the new address to its table.

In examining all packets for their source and destination addresses, bridges build a table containing all local addresses. The table is updated as new packets are encountered and as addresses that have not been used for a specified period of time are deleted. This self-learning capability permits bridges to keep up with changes on the network without requiring that their tables be updated manually.

The bridge isolates traffic by examining the destination address of each packet. If the destination address matches any of the source addresses in its table, the packet is not allowed to pass over the bridge because the traffic is local. If the destination address does not match any of the source addresses in the table, the packet is discarded onto an adjacent network. This filtering process is repeated at each bridge on the internetwork until the packet eventually reaches its destination. Not only does this process prevent unnecessary traffic from leaking onto the internetwork, it acts as a simple security mechanism that can screen unauthorized packets from accessing various corporate resources.

Bridges also can be used to interconnect LANs that use different media, such as twisted-pair, coaxial, and fiberoptic cabling and various types of wireless links. In office environments that use wireless communications technologies such as spread spectrum and infrared, bridges can function as an access point to wired LANs (Figure B-3). On the widea area network (WAN), bridges even switch traffic to a secondary port if the primary port fails. For example, a full-time wireless bridging system can establish a modem connection on the public network if the primary wire or wireless link is lost because of environmental interference.

In reference to the Open Systems Interconnection (OSI) model, a bridge operates at Layer 2; specifically, it operates at the Media Access Control (MAC) sublayer of the Data Link Layer. It routes by means of the Logical Link Control (LLC), the upper sublayer of the Data Link Layer (Figure B-4).

Because the bridge connects LANs at a relatively low level, throughput often exceeds 30,000 packets per second

Figure B-3 For the home or small office network, the Instant Wireless Ethernet Bridge from Linksys extends wireless connectivity to any Ethernet-ready network device, such as a printer, scanner, or desktop or notebook PC.

(pps). Multiprotocol routers and gateways, which provide LAN interconnection over the WAN, operate at higher levels of the OSI model and provide more functionality. In performing more protocol conversions and delivering more functionality, routers and gateways are generally more processing-intensive and, consequently, slower than bridges.

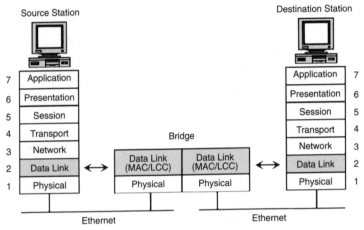

Figure B-4 Bridge functionality in reference to the OSI model

---

See also

Access Points

Repeaters

Routers

C

CELL SITES

A cellular system operates by dividing a large geographic service area into cells (Figure C-1) and assigning the same frequencies to multiple, nonadjacent cells. This is known in the industry as "frequency reuse." As a subscriber travels across the service area, the call is transferred (handed off) from one cell to another without noticeable interruption. All the base stations in a cellular system, including radio towers, are connected to a mobile telephone switching office (MTSO) by landline or microwave links. The MTSO controls the switching between the Public Switched Telephone Network (PSTN) and the cell site for all wireline-to-mobile and mobile-to-wireline calls.

Site Planning

There is a huge investment at stake when determining the location of a cell site. The radio tower alone can cost from $250,000 to $1 million. Thus, before a cell site is installed, a number of studies are performed to justify the cost and calculate the return on investment (ROI). A demographics study, for example, helps forecast the potential subscriber base in the area planned for the cell site. The study begins with the total

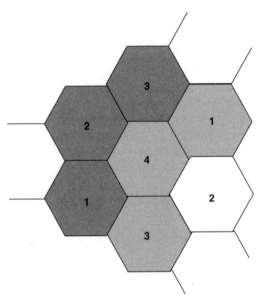

Figure C-1 In a cellular network, the signal coverage of each tower is limited so that the same frequencies can be assigned to multiple nonadjacent cells. This increases the total call-handling capacity of the network while conserving spectrum.

population, broken down by sex, age, race, types of households, occupancy rates, and income levels. Much of this information is gleaned from data compiled by the U.S. Census Bureau.

By generating a topography map, engineers are able to determine if there will be any obvious interference issues. The goal is to discover problems that would impair the performance of the wireless cell site solution or wireless link. Sometimes a 50-foot portable crank-up tower is used to create a temporary cell site. Together with vehicles containing both access points and subscriber units, tests are run to find out what may interfere with the signal and demonstrate a real-time cell site coverage area.

After determining any interference issues and the best strategic location for the cell site, a full site survey is done to establish the final plans for cell site deployment. This includes having all the information about the types of mounts needed as well as having potential interference-filtering measures defined. Certified network engineers then determine the best base station configuration and orientation. If no problems are encountered, a tower can be up and running within 6 weeks.

Regulation

The Telecommunications Act of 1996 specifically leaves in place the authority that local zoning authorities have over the placement of cell towers. It does prohibit the denial of facilities siting based on radio frequency (RF) emissions if the licensee has complied with the Federal Communications Commission's (FCC's) regulations concerning RF emissions. It also requires that denials be based on a reasoned approach and prohibits discrimination and outright bans on construction, placement, and modification of wireless facilities.

The FCC mandates that service providers build out their systems so that adequate service is provided to the public. In addition, all antenna structures used for communications must be approved by the FCC, which determines if there is a reasonable possibility that the structure may constitute a menace to air navigation. The tower height and its proximity to an airport or flight path will be considered when making this determination. If such a determination is made, the FCC will specify appropriate painting and lighting requirements. Thus the FCC does not mandate where towers must be placed, but it may prohibit the placement of a tower in a particular location without adequate lighting and marking.

Summary

Low-powered transmitters are an inherent characteristic of cellular radio and broadband personal communication services (PCS). As these systems mature and more subscribers are added, the effective radiated power of the cell site transmitters is reduced so that frequencies can be reused at closer intervals, thereby increasing subscriber capacity. There are more than 50,000 cell sites operating within the United States and its possessions and territories. Therefore, due to the nature of frequency reuse and the consumer demand for services, cellular and PCS providers must build numerous base sites. The sheer number of towers has caused municipalities to impose new requirements on service providers, such as requiring them to disguise new towers to look like trees, which can add $150,000 to the cost of a tower.

See also

Cellular Data Communications

Cellular Voice Communications

Mobile Telephone Switching Office

Personal Communications Services

CELLULAR DATA COMMUNICATIONS

One of the oldest services for sending data over a cellular communications network is known as "Cellular Digital Packet Data" (CDPD), which provides a way of passing Internet Protocol (IP) data packets over analog cellular voice networks at speeds of up to 19.2 kbps. Although CDPD employs digital modulation and signal processing techniques, the underlying service is still analog. The medium used to transport data consists of the idle radio channels typically used for Advanced Mobile Phone System (AMPS) cellular service.

Channel hopping automatically searches out idle channel times between cellular voice calls. Packets of data select available cellular channels and go out in short bursts without interfering with voice communications. Alternatively, cellular carriers also may dedicate voice channels for CDPD traffic to meet high traffic demand. This situation is common in dense urban environments where cellular traffic is heaviest.

Once the user logs onto the network, the connection stays in place to send or receive data. In accordance with the IP, the data are packaged into discrete packets of information for transmission over the CDPD network, which consists of routers and digital radios installed in current cell sites. In addition to addressing information, each IP packet includes information that allows the data to be reassembled in the proper order via the Transmission Control Protocol (TCP) at the receiving end. The transmissions are encrypted over the air link for security purposes.

Although CDPD piggybacks on top of the cellular voice infrastructure, it does not suffer from the 3-kHz limit on voice transmissions. Instead, it uses the entire 30-kHz RF channel during idle times between voice calls. Using the entire channel contributes to CDPD's faster data transmission rate. Forward error correction ensures a high level of wireless communications accuracy. With encryption and authentication procedures built into the specification, CDPD offers more robust security than any other native wireless data transmission method. As with wireline networks, CDPD users also can customize their own end-to-end security.

To take advantage of CDPD, the user must have an integrated mobile device that operates as a fully functional cellular phone and Internet appliance. For example, the AT&T PocketNet Phone contains both a circuit-switched cellular modem and a CDPD modem to provide users with fast and convenient access to two-way wireless messaging services and Internet information. GTE provides a similar service through its Wireless Data Services. Both companies have negotiated intercarrier agreements that enable their customers to enjoy seamless CDPD service in virtually all markets across the country. AT&T's Wireless IP service, for example, is available in 3000 cities in the United States.

Among the applications for CDPD are access to the Internet for e-mail and to retrieve certain Web-based content. AT&T PocketNet Phone users, for example, have access to two-way messaging, airline flight information, financial information, show times, restaurant reviews, and door-to-door travel directions. AT&T provides unlimited access to featured sites on the wireless Internet, which means that there are no per-minute charges for surfing wireless Web sites.

Companies also can use CDPD to monitor alarms remotely, send/receive faxes, verify credit cards, and dispatch vehicles. Although CDPD services might prove too expensive for heavy database access, the use of intelligent agents can cut costs by minimizing connection time. Intelligent agents gather requested information and report back only the results the next time the user logs onto the network.

Summary

Wireless IP is an appealing method of transporting data over cellular voice networks because it is flexible, fast, widely available, and compatible with a vast installed base of computers and has security features not offered with other wireless data services. One caveat: The carrier's wireless data network is different from its wireless voice network. Therefore, users of AT&T Digital PocketNet service, for example, will not be able to access that service everywhere voice calls can be made. It is important to look at coverage maps and compare service plans before subscribing to this type of service.

See also

Wireless IP

CELLULAR TELEPHONES

Bell Labs built the first cellular telephone in 1924 (Figure C-2). After decades of development, cellular telephones have

Figure C-2 The first cellular telephone, developed by Bell Labs in 1924.

emerged as a "must have" item among mobile professionals and consumers alike, growing in popularity every year since they became commercially available in 1983. Their widespread use for both voice and data communications has resulted from significant progress made in their functionality, portability, the availability of network services, and the declining cost for equipment and services.

System Components

There are several categories of cellular telephone. Mobile units are mounted in a vehicle. Transportable units can be easily moved from one vehicle to another. Pocket phones, weighing in at less than 4 ounces, can be conveniently carried in a jacket pocket or purse. There are even cellular telephones that can be worn. Regardless of how they are packaged, cellular telephones consist of the same basic elements.

Handset/Keypad The handset and keypad provide the interface between the user and the system. This is the only component of the system with which, under normal operation, the user needs to be concerned. Any basic or enhanced system features are accessible via the keypad, and once a connection is established, this component provides similar handset functionality to that of any conventional telephone. Until a connection is established, however, the operation of the handset differs greatly from that of a conventional telephone.

Instead of initiating a call by first obtaining a dial tone from the network switching system, the user enters the dialed number into the unit and presses the "Send" function key. This process conserves the resources of the cellular system, since only a limited number of talk paths are available at any given time. The "Clear" key enables the user to correct misdialed digits.

Once the network has processed the call request, the user will hear conventional call-progress signals such as a busy signal or ringing. From this point on, the handset operates in the customary manner. To disconnect a call, the "End" function key is pressed on the keypad. The handset contains a small illuminated display that shows dialed digits and provides a navigational aid to other features. The keypad enables storage of numbers for future use and provides access to other enhanced features, which may vary according to manufacturer.

Logic/Control The logic/control functions of the phone include the numeric assignment module (NAM) for programmable assignment of the unit's telephone number by the service provider and the electronic serial number of the unit, which is a fixed number unique to each telephone. When a customer signs up for service, the carrier makes a record of both numbers. When the unit is in service, the cellular network interrogates the phone for both of these numbers in order to validate that the calling/called cellular telephone is that of an authentic subscriber.

The logic/control component of the phone also serves to interact with the cellular network protocols. Among other things, these protocols determine what control channel the unit should monitor for paging signals and what voice channels the unit should use for a specific connection. The logic/control component is also used to monitor the control signals of cell sites so that the phone and network can coordinate transitions to adjacent cells as conditions warrant.

Transmitter/Receiver The transmitter/receiver component of the cell phone is under the command of the logic/control unit. Powerful 3-watt telephones are typically of the vehicle-mounted or transportable type, and their transmitters are understandably larger and heavier than those contained within lighter-weight handheld cellular units. These more powerful transmitters require significantly more input wattage than hand-held units that transmit at power levels of only a fraction of a watt, and they use the main battery within a vehicle or a relatively heavy rechargeable battery to do so. Special circuitry within the phone enables the transmitter and receiver to use a single antenna for full-duplex communication.

Antenna The antenna for a cellular telephone can consist of a flexible rubber antenna mounted on a hand-held phone, an extendible antenna on a pocket phone, or the familiar curly stub seen attached to the rear window of many automobiles. Antennas and the cables used to connect them to radio transmitters must have electrical performance characteristics that are matched to the transmitting circuitry, frequency, and power levels. Use of antennas and cables that are not optimized for use by these phones can result in poor performance. Improper cable, damaged cable, or faulty connections can render the cell phone inoperative.

Power Sources Cell phones are powered by a rechargeable battery. Nickel-cadmium (NiCd) batteries are the oldest and cheapest power source available for cellular phones. Newer nickel–metal hydride (NiMH) batteries provide extend talk time compared to lower-cost conventional NiCd units. They provide the same voltage as NiCd batteries but offer at least 30 percent more talk time than NiCd batteries and take approximately 20 percent longer to charge.

Lithium ion batteries offer increased power capacity and are lighter in weight than similar-size NiCd and NiMH batteries. These batteries are optimized for the particular model of cellular phone, which helps ensure maximum charging capability and long life.

Newer cellular phones may operate with optional high-energy AA alkaline batteries that can provide up to 3 hours of talk time or 30 hours of standby time. These batteries take advantage of lithium–iron disulfide technology, which results in 34 percent lighter weight than standard AA 1.5-volt batteries (15 versus 23 grams per battery) and 10-year storage life—double that of standard AA alkaline batteries.

Vehicle-mounted cell phones can be optionally powered via the vehicle's 12-volt dc battery by using a battery eliminator that plugs into the dashboard's cigarette lighter. This saves useful battery life by drawing power from the vehicle's battery and comes in handy when the phone's battery has run down. A battery eliminator will not recharge the phone's battery, however. Recharging the battery can only be done with a special charger.

Lead-acid batteries are used to power transportable cellular phones when the user wishes to operate the phone away from a vehicle. The phone and battery are usually carried in a vinyl pouch.

Features and Options

Cellular telephones offer many features and options, including

- *Voice activation* Sometimes called "hands-free operation," this feature allows the user to establish and answer calls by issuing verbal commands. This safety feature enables a driver to control the unit without becoming visually distracted.

- *Memory functions* These allow storage of frequently called numbers to simplify dialing. Units may offer as few as 10 memory locations or in excess of 100, depending on model and manufacturer.

- *Multimode* This allows the phone to be used with multiple carriers. The phone can be used to access digital service where it is available and then switch to an analog service of another wireless carrier when roaming.

- *Multiband* This allows the phone to be used with multiple networks using different frequency bands. For example, the cell phone can be used to access the 1900-MHz band when it is available and then switch to the 800-MHz band when roaming.

- *Visual status display* This conveys information on numbers dialed, state of battery charge, call duration, roaming indication, and signal strength. Cell phones differ widely in the number of characters and lines of alphanumeric information they can display. The use of icons enhances ease of use by visually identifying the phone's features.

- *Programmable ring tones* Some cellular phones allow the user to select the phone's ring tone. Multiple ring tones can be selected, each assigned to a different caller. A variety of ring tones may be downloaded from the Web.

- *Silent call alert* Features include visual or vibrating notification in lieu of an audible ring tone. This can be particularly useful in locations where the sound of a ringing phone would constitute an annoyance.

- *Security features* These include password access via the keypad to prevent unauthorized use of the cell phone as well as features to help prevent access to the phone's telephone number in the event of theft.

- *Voice messaging* This allows the phone to act as an answering machine. A limited amount of recording time (about 4 minutes) is available on some cell phones. However, carriers also offer voice-messaging services that are not dependent on the phone's memory capacity. While the phone is in standby mode, callers can leave messages on the integral answering device. While the phone is off, callers can leave messages on the carrier's voice-mail system. Users are not billed for airtime charges when retrieving their messages.

- *Call restriction* This enables the user to allow use of the phone by others to call selected numbers, local numbers, or emergency numbers without permitting them to dial the world at large and rack up airtime charges.

- *Call timers* These provide the user with information as to the length of the current call and a running total of airtime for all calls. These features make it easier for users to keep track of call charges.

- *User-defined ring tones* These offer users the option to compose or download ring tones of their choice to replace the standard ring tone that comes with the cell phone.

- *Data transfer kit* For cell phones that are equipped with a serial interface, there is software for the desktop PC that allows users to enter directory information via keyboard rather than the cell phone keypad. The information is transferred via the kit's serial cable. Through the software and cable connection, information can be synchronized between the PC and cell phone, ensuring that both devices have the most recent copy of the same information.

Location-Reporting Technology

Mobile phone companies are under orders from the FCC to incorporate location-reporting technology into cellular phones. Dubbed E-911, or enhanced 911, the initiative is meant to provide law enforcement and emergency services personnel with a way to find people calling 911 from mobile phones when callers do not know where they are or are unable to say. Since no carrier was able to make an October 2001 deadline to fully implement E-911, the FCC issued waivers permitting carriers to add location-detection services to new phones over time so that 95 percent of all mobile phones will be compliant with E-911 rules by 2005.

One way manufacturers can address this requirement is by providing cell phones with a Global Positioning System (GPS) capability in which cell phone towers help GPS satellites fix a cell phone caller's position. Special software installed in the base station hardware serves location information to cell phones, which is picked up at the public safety answering point (PSAP). However, subscribers would need to purchase a new GPS-equipped handset, since this method would not allow legacy handsets to use the location-determination system.

Another location-determination technique is called "Time Difference of Arrival" (TDOA), which works by measuring the exact time of arrival of a handset radio signal at three or more separate cell sites. Because radio waves travel at a fixed known rate (the speed of light), by calculating the difference in arrival time at pairs of cell sites, it is possible to calculate hyperbolas on which the transmitting device is located. The TDOA technique makes use of existing receive antennas at the cell sites. This location technique works with any handset, including legacy units, and only requires modifications to the network.

Internet-Enabled Mobile Phones

Internet-enabled mobile phones potentially represent an important communications milestone, providing users with access to Web content and applications, including the ability to participate in electronic commerce transactions. The Wireless Application Protocol (WAP), an internationally accepted specification, allows wireless devices to retrieve content from the Internet, such as general news, weather, airline schedules, traffic reports, restaurant guides, sports scores, and stock prices.

Users also can personalize these services by creating a profile that might request updated stock quotes every half-hour or specify tastes in music and food. A user also could set up predefined locations, such as home, main office, or transit, so that the information is relevant for that time and location. With access to real-time traffic information, for example, users can obtain route guidance on their cell phone screens via the Internet. Up-to-the-minute road conditions are displayed directly on the cell phone screen. Street-by-street guidance is provided for navigating by car, subway, or simply walking, taking into account traffic congestion to work out the best itinerary. Such services can even locate and guide users to the nearest facilities, such as free parking lots or open gas stations, using either an address entered on the phone keypad or information supplied by an automatic location identification (ALI) service.

One vendor that has been particularly active in developing WAP-compliant Internet-enabled mobile phones is Nokia, the world's biggest maker of mobile phones. The company's Model 7110 works only on GSM 900 and GSM 1800 in Europe and Asia but is indicative of the types of new mobile phones that about 70 other manufacturers are targeting at the world's 200 million cellular subscribers. It displays Internet-based information on the same screen used for voice functions. It also supports Short Messaging Service (SMS) and e-mail and includes a calendar and phonebook as well.

The phone's memory also can save up to 500 messages— SMS or e-mail—sorted in various folders such as the inbox, outbox, or user-defined folders. The phonebook has enough memory for up to 1000 names, with up to five phone and fax numbers and two addresses for each entry. The user can mark each number and name with a different icon to signify home or office phone, fax number, or e-mail address, for example. The phone's built-in calendar can be viewed by day, week, or month, showing details of the user's schedule and calendar notes for the day. The week view shows icons for the jobs the user has to do each day. Up to 660 notes in the calendar can be stored in the phone's memory.

Nokia has developed several innovative features to make it faster and easier to access Internet information using a mobile phone:

- *Large display* The screen has 65 rows of 96 pixels (Figure C-3), allowing it to show large and small fonts, bold or regular, as well as full graphics.

- *Microbrowser* Like a browser on the Internet, the microbrowser feature enables the user to find information by entering a few words to launch a search. When a site of interest is found, its address can be saved in a "favorites" folder or input using the keypad.

- *Navi Roller* This built-in mouse looks like a roller (Figure C-4) that is manipulated up and down with a finger to scroll and select items from an application menu. In each situation, the Navi Roller knows what to do when it is clicked—select, save, or send.

- *Predictive text input* As the user presses various keys to spell words, a built-in dictionary continually compares the word in progress with the words in the database. It selects the most likely word to minimize the need to continue spelling out the word. If there are several word possibilities, the user selects the right one using the Navi

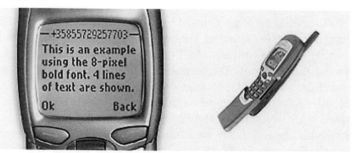

Figure C-3 Display screen of the Nokia 7110.

Figure C-4 Close-up of the Navi Roller on the Nokia 7110.

Roller. New names and words can be input into the phone's dictionary.

However, the Nokia phone cannot be used to access just any Web site. It can access only Web sites that have been developed using WAP-compliant tools. The WAP standard includes its own Wireless Markup Language (WML), which is a simple version of the HyperText Markup Language (HTML) that is used widely for developing Web content. The strength of WAP is that it is supported by multiple airlink standards and, in true Internet tradition, allows content publishers and application developers to be unconcerned about the specific delivery mechanism.

Third-Generation Phones

The world is moving toward third-generation (3G) mobile communications systems that are capable of bringing high-quality mobile multimedia services to a mass market. The International Telecommunication Union (ITU) has put together a 3G framework known as International Mobile Telecommunications-2000 (IMT-2000). This framework encompasses a small number of frequency bands, available on a globally harmonized basis, that make use of existing national and regional mobile and mobile-satellite frequency allocations.

Along the way toward 3G is a 2.5G service known as General Packet Radio Service (GPRS), which offers true packet data connectivity to cell phone users. GPRS leverages Internet Protocol (IP) technologies, adding convenience and immediacy to mobile data services. GPRS is ideal for wireless data applications with bursty data, especially WAP-based information retrieval and database access.

GPRS enables wireless users to have an "always on" data connection, as well as high data transfer speeds. Although GPRS offers potential data transfer rates of up to 115 kbps, subscribers will only really notice faster service at the initial connection. The faster speed is in the connect time. At present, users connect at a maximum of 19.2 kbps.

GPRS packet-based service should cost users less than circuit-switched services, since communication channels are shared rather than dedicated only to one user at a time. It also should be easier to make applications available to mobile users because the faster data rate means that middleware currently needed to adapt applications to the slower speed of wireless systems will no longer be needed. To take advantage of GPRS, however, mobile users will have to buy new cell phones that specifically support the data service.

Summary

Cellular phones are getting more intelligent, as evidenced by the combination of cellular phone, personal digital assistant (PDA), Web browser, and always-on GPRS connection into one unit. These devices not only support data communications, they also support voice messaging, e-mail, fax, and micropayments over the Internet as well. Third-party software provides the operating system and such applications as calendaring, card file, and to-do lists. With more cellular phones supporting data communications, cellular phones are available that provide connectivity to PC desktops and databases via Bluetooth, infrared, or serial RS-232 connections. Information can even be synchronized between cell phones and desktop computers to ensure that the user is always accessing the most up-to-date information.

See also

Cellular Data Communications

Cellular Voice Communications

Global Positioning System

Paging

Personal Communications Services

CELLULAR VOICE COMMUNICATIONS

Cellular telephony provides communications service to automobiles and hand-held portable phones and interconnects with the Public Switched Telephone Network (PSTN) using radio transmissions based on a system of cells and base station antennas.

AT&T's Bell Laboratories developed the cellular concept in 1947, but it was not until 1974 that the FCC set aside radio spectrum between 800 and 900 MHz for cellular radio systems. The first cellular demonstration system was installed in Chicago in 1978, and 3 years later the FCC formally authorized 666 channels for cellular radio signals and established cellular geographic servicing areas (CGSAs) to cover the nation's major metropolitan centers.

At the same time, the FCC created a regulatory scheme for cellular service that specified that two competing cellular companies would be licensed in each market. For each city, one license would be reserved for the local telephone company (a wireline company), and the other license would be granted to another qualified applicant. When the number of applicants became prohibitively large, the FCC amended its licensing rule and specified the use of lotteries to select applicants for all but the top 30 markets.

Cellular service—whether analog or digital—is now available virtually everywhere in the United States and from many more service providers in each market. According the FCC, 259 million people, or almost 91 percent of the total U.S. population, have access to three or more different operators [cellular, broadband personal communications services (PCS), and/or digital specialized mobile radio (SMR) providers] offering mobile telephone service in the counties in which they live. Over 214 million people, or 75 percent of the U.S. population, live in areas with five or more mobile telephone operators competing to offer service. And 133 million people, or 47 percent of the population, can choose from at least six different mobile telephone operators.

If a subscriber uses the cell phone outside the home service area, this is traveling, and an extra charge is applied to the call. When the cell phone is used outside the service provider's network, this is roaming. In this case, if the service provider has agreements with other carriers, the traveling rate is applied to each call. Where digital service is not available and the service provider has agreements with conventional analog service providers, subscribers can use their cell phones in analog mode, in which case airtime and long-distance charges are applied to each call.

Applications

Cellular telephones were targeted originally at mobile professionals, allowing them to optimize their schedules by turning nonproductive driving and out-of-the-office time into productive and often profitable work time. Cellular solutions not only facilitate routine telephone communications, they also increase revenue potential for people in professions that have high-return opportunities as a direct result of being able to respond promptly to important calls.

Today, cellular service is also targeted at consumers, giving them the convenience of anytime, anywhere calling plus the security of instant access to service in times of emergency. As of mid-2001, more than half of U.S. households subscribed to wireless phone service, and a majority of them had two or more mobile phones. Over 15 percent of cellular subscribers use the service for more than half of their long-distance calls, while about 10 percent use it for more than half their local calls.

Developing countries that do not have an advanced communications infrastructure are increasingly turning to cellular technology so that they can take part in the global economy without having to go through the resource-intensive step of installing copper wire or optical fiber. Explosive growth is occurring in India and China. Even among industrialized countries, there is continued high growth in cellular usage. In Japan, the number of cell phones now exceeds the number of analog fixed-line phones.

Technology Components

Cellular networks rely on relatively short-range transmitter/receiver (transceiver) base stations that serve small sections (or cells) of a larger service area. Mobile telephone users communicate by acquiring a frequency or time slot in the cell in which they are located. A master switching center called the "mobile transport serving office" (MTSO) links calls between users in different cells and acts as a gateway to the PSTN. Figure C-5 illustrates the link from the MTSO to the base stations in each cell. The MTSO also has links to local telephone central offices so that cellular users can communicate with users of conventional phones.

Cell Sites Cell boundaries are neither uniform nor constant. The usage density in the area, as well as the landscape, the presence of major sources of interference (e.g., power lines, buildings), and the location of competing carrier cells, contributes to the definition of cell size. Cellular boundaries change continuously, with no limit to the number of frequencies available for transmission of cellular calls in an area. As the density of cellular usage increases, individual

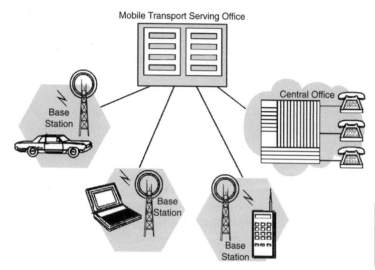

Figure C-5   A typical cellular network configuration.

cells are split to expand capacity. By dividing a service area into small cells with limited-range transceivers, each cellular system can reuse the same frequencies many times. Technologies such as Code Division Multiple Access (CDMA) and Expanded Time Division Multiple Access (E-TDMA) promise further capacity gains in the future.

Master Switching Center In a typical cellular network, the master-switching center operates similar to a telephone central office and provides links to other offices. The switching center supports trunk lines to the base stations that establish the cells in the service area. Each base station supports a specific number of simultaneous calls—from 3 to 15, depending on the underlying technology (i.e., CDMA, TDMA, or some derivative).

Transmission Channels Most cellular systems provide two types of channels: a control channel and a traffic channel.

The base station and mobile station use the control channel to support incoming and outgoing calls, monitor signal quality, and register when a user moves into a new zone. The traffic channel is used only when the station is off-hook and actually involved in a call.

The control and traffic channels are divided into time slots. When the user initiates access to the control channel to place a call, the mobile station randomly selects a subslot in a general-use time slot to reach the system; the system then assigns a time slot to the traffic channel. For an incoming call to a mobile station, the base station initiates conversations on the control channel by addressing the mobile station in a time slot, which at the same time reserves that time slot for the station's reply. If a user's call attempt collides with another user's call attempt, both instruments automatically reselect a subslot and try again. After repeated collisions, if no time slots are available within a predetermined time, the system rejects service requests for incoming and outgoing calls.

When a mobile telephone user places a call, the cell in which the user is traveling allocates a slot for the call. The call slot allows the user access through the base station to the master switching center, essentially providing an extension on which the call can be placed. The master switching center, through an element of the user-to-base-station connection, continuously monitors the quality of the call signal and transfers the call to another base station when the signal quality reaches an unacceptable level due to the distance traveled by the user, obstructions, and/or interference. If the user travels outside the system altogether, the master switching center terminates the call as soon as the signal quality deteriorates to an unacceptable level.

Cellular Telephones Cellular telephones incorporate a combination of multiaccess digital communications technology and traditional telephone technology and are designed to appear to the user as familiar residential or business telephone equipment. Manufacturers use miniaturization and digital signal processing technology to make cellular phones feature rich yet compact and economical.

Cellular instruments consist of a transceiver, an analog/digital converter, and a supervisory/control system that manages calls and coordinates service with both the base station and the master switching center. Cellular telephones can be powered from a variety of sources, including vehicle batteries, ac adapters, and rechargeable battery sets.

Traditional cellular instrument types include hand-held, transportable, and car telephones. However, advances in cellular technology are creating additional types of telephones, including modular and pocket phones. The trend in cellular instruments is toward multipurpose transportable telephones.

There are dual-mode cellular phones that can be used with in-building wireless Private Branch Exchanges (PBXs) as well as with the outside cellular service. The handset registers itself with an in-building base station and takes its commands from the wireless PBX. For out-of-building calling, the handset registers with the nearest cell site transceiver. Aside from convenience, an added benefit of the dual-mode phone is that calls made off the corporate premises can be aggregated with business calls made at home or on the road for the purpose of achieving a discounted rate on all calls.

Network Optimization

Network optimization is a high priority for wireless carriers. A single cell site, including electronics and tower, can cost as much as $600,000 to build. With skyrocketing growth for wireless voice and data access in recent years, wireless service providers want to get the most efficient use out of their current networks and target upgrades appropriately to meet customer demand. The use of network-optimization tools translates into lower wireless service costs and better coverage.

Such tools measure cell site footprints, service areas within those footprints, and frequency assignments—all with the purpose of identifying the disruptive interference that cell sites receive from adjoining sites. By taking steps to limit interference between cells, wireless providers can maximize the bandwidth devoted to moving traffic. In addition to monitoring the cell sites, these network-optimization tools monitor the strength of radio frequency (RF) signals emanating from cell sites as well as how calls are handed off among cell sites.

Regulation

The FCC sets rules, regulations, and policies to, among other things,

- Grant licenses for frequencies and license renewals
- Rule on assignments and transfers of control of licenses
- Govern the interconnection of cellular networks with other wireless and wireline carriers
- Establish access and universal service funding provisions
- Impose fines and forfeitures for violations of any of the FCC rules
- Regulate the technical standards of cellular networks

In addition, the FCC and many states have established universal service programs to ensure affordable, quality telecommunications services for all Americans. Contributions to these programs by cellular/PCS service providers are typically a percentage of end-user revenues.

The FCC currently prohibits a single entity from having a combined attributable interest (20 percent or greater interest in any license) in broadband PCS, cellular, and specialized mobile radio licenses totaling more than 45 MHz in any geographic area, except that in rural service areas no licensee may have an attributable interest in more than 55 MHz of Commercial Mobile Radio Service (CMRS) spectrum.

The FCC must approve any substantial changes in ownership or control of a cellular/PCS license. Noncontrolling interests in an entity that holds a license or operates cellular/PCS networks generally may be bought or sold without prior FCC approval. In addition, the FCC now requires only postconsummation notification of certain pro forma assignments or transfers of control.

All licenses are granted for 10-year terms. Licenses may be revoked if any FCC rules are violated. Licenses may be renewed for additional 10-year terms. Renewal applications are not subject to spectrum auctions. Third parties, however, may oppose renewal applications.

Summary

No other technology has taken the world by storm quite like cellular except, perhaps, the Internet. Cellular systems have expanded beyond providing voice communication to supporting more sophisticated applications, such as Internet access for electronic mail and accessing Web content. New Internet-enabled cellular phones feature larger displays that help make them all-purpose communications appliances. It has reached the point where cellular service is as necessary for the average consumer as for mobile professionals.

See also

Calling Party Pays

Cellular Data Communications

Cellular Telephones

Code Division Multiple Access

Personal Communications Service

Time Division Multiple Access

CITIZENS BAND RADIO SERVICE

Citizens Band (CB) Radio Service is a two-way voice communication service for use in personal and business activities.

The service uses 40 channels in the assigned frequency range of 26.965 to 27.405 MHz, and the effective communication distance is 1 to 5 miles. An FCC license is not required to use this service. CB Rule 3 provides users with all the authority they need to operate a CB unit in places where the FCC regulates radio communications, as long as an unmodified FCC-accepted CB unit is used. An FCC-accepted unit has an identifying label placed on it by the manufacturer. There is no age or citizenship requirement for using this service.[1]

CB users may use an on-the-air pseudonym or "handle" of their own choosing and may operate their CB units within the territorial limits of the 50 states, the District of Columbia, and the Caribbean and Pacific insular areas. Users also may operate their CB units on or over any other area of the world, except within the territorial limits of areas where radio communications are regulated by another agency of the United States or within the territorial limits of any foreign government. In addition, users can use their CB units in Canada, subject to the rules of the Canadian Department of Communications.

The power output of the CB unit may not be raised, since raising the level of radio noise would be unfair to the other users sharing a channel. Users also must not attach a linear amplifier or any other type of power amplifier to their CB unit or modify the unit internally. Doing so cancels its type acceptance, and the user forfeits his or her authorization to use it.

There are no height restrictions for antennas mounted on vehicles or for hand-held units. For structures, the highest point of the antenna must not be more than 20 feet above the highest point of the building or tree on which it is mounted or 60 feet above the ground. There are lower height limits if the antenna structure is located within 2 miles of an airport.

No CB channel is assigned to any specific individual or organization. Any of the 40 CB channels can be used on a

---

[1]The only caveat to CB Rule 3 in this regard is that the user cannot be a foreign government, a representative of a foreign government, or a federal government agency. Of course, if the FCC has issued a cease and desist order, that person cannot be a CB user.

"take turns" basis. Since CB channels are shared, cooperation among users is essential; communications should be
short, with conversations never more than 5 minutes continuously. Users should wait at least 1 minute before starting another communication. Channel 9 should only be used
for emergency communications or for traveler assistance.

Ten-Codes

"Ten-codes" is abbreviations of common questions and answers
used on all types of radio communication. Professional CB
users use these codes to send their message quickly and easily.
Additionally, ten-codes can be readily understood by users
when poor reception or language barriers must be overcome.
Although the FCC authorizes CB operators to use ten-codes, it
does not regulate their meaning. The most commonly used
ten-codes are listed below in Table C-1.

Summary

Initially, users were required to obtain a CB radio license
and call letters from the FCC before they could go on the air.
However, the FCC became so inundated with requests for
CB radio licenses that it finally abandoned formal licensing
and allowed operators to buy CB radio equipment and go on
the air without any license or call letters. Although no
license is required to operate a CB radio, the FCC's rules for
CB radio operation are still in effect and must be followed.
These rules cover CB radio equipment, the ban on linear
amplifiers, and the types of communications permitted on
the air. Manufacturers are required to provide a copy of the
operating rules with each CB set.

See also

Family Radio Service
General Mobile Radio Service

TABLE C-1   Common Ten-Codes Used by CB Operators

10-1 = Receiving poorly
10-2 = Receiving well
10-3 = Stop transmitting
10-4 = Message received
10-5 = Relay message to _____
10-6 = Busy, please stand by
10-7 = Out of service, leaving the air
10-8 = In service, subject to call
10-9 = Repeat message
10-10 = Transmission complete, standing by
10-11 = Talking too rapidly
10-12 = Visitors present
10-13 = Advise weather/ road conditions
10-16 = Make pickup at _____
10-17 = Urgent business
10-18 = Anything for us?
10-19 = Nothing for you, return to base
10-20 = My location is _____
10-21 = Call by telephone
10-22 = Report in person to
10-23 = Stand by
10-24 = Completed last assignment
10-25 = Can you contact _____
10-26 = Disregard last information
10-27 = I am moving to channel _____
10-28 = Identify your station
10-29 = Time is up for contact
10-30 = Does not conform to FCC rules
10-32 = I will give you a radio check
10-33 = Emergency traffic
10-34 = Trouble at this station
10-35 = Confidential information

10-36 = Correct time is
10-37 = Wrecker needed at
10-38 = Ambulance needed at
10-39 = Your message delivered
10-41 = Please turn to channel
10-42 = Traffic accident at
10-43 = Traffic tie-up at
10-44 = I have a message for you
10-45 = All units within range please report
10-50 = Break channel
10-60 = What is next message number?
10-62 = Unable to copy, use phone
10-63 = Net directed to
10-64 = Net clear
10-65 = Awaiting next message/assignment
10-67 = All units comply
10-70 = Fire at _____
10-71 = Proceed with transmission in sequence
10-77 = Negative contact
10-81 = Reserve hotel room for _____
10-82 = Reserve room for _____
10-84 = My telephone number is _____
10-85 = My address is _____
10-91 = Talk closer to the microphone
10-93 = Check my frequency on this channel
10-94 = Please give me a long count (1 to 10)
10-99 = Mission completed, all units secure
10-200 = Police needed at _____

Low-Power Radio Service
Wireless Medical Telemetry Service

CODE DIVISION MULTIPLE ACCESS

Code Division Multiple Access (CDMA) is a spread-spectrum technology that is used for implementing cellular telephone service. Spread spectrum is a family of digital communication techniques originally used in military communications and control applications. Spread spectrum uses carrier waves that consume a much wider bandwidth than that required for simple point-to-point communication at the same data rate. This results in the carrier wave looking more like random noise than real communication between a sender and receiver. Originally, there were two motivations for implementing spread spectrum: to resist enemy efforts to jam vital communications and to hide the fact that communication was even taking place.

For cellular telephony, spread-spectrum technology underlies CDMA, which is a digital multiple access technique specified by the Telecommunications Industry Association (TIA) as IS-95. Commercial applications of CDMA became possible because of two key developments. One was the availability of low-cost, high-density digital integrated circuits, which reduce the size, weight, and cost of the mobile phones. The other was the realization that optimal multiple access communication depends on the ability of all mobile phones to regulate their transmitter power to the lowest level that will achieve adequate signal quality.

CDMA changes the nature of the mobile phone from a predominately analog device to a predominately digital device. CDMA receivers do not eliminate analog processing entirely, but they separate communication channels by means of a pseudorandom modulation that is applied and removed in the digital domain, not on the basis of frequency. This allows multiple users to occupy the same frequency band; this frequency reuse results in high spectral efficiency.

TDMA systems commonly start with a slice of spectrum, referred to as a "carrier." Each carrier is then divided into time slots. Only one subscriber at a time is assigned to each time slot or channel. No other conversations can access this channel until the subscriber's call is finished or until that original call is handed off to a different channel by the system. For example, TDMA systems, designed to coexist with AMPS systems, divide 30 kHz of spectrum into three channels. By comparison, GSM systems create eight time-division channels in 200-kHz-wide carriers.

Wideband Usage

With CDMA systems, multiple conversations simultaneously share the available spectrum in both the time and frequency dimensions. The available spectrum is not "channelized" in frequency or time as in Frequency Division Multiple Access (FDMA) or TDMA systems, respectively. Instead, the individual conversations are distinguished through coding; that is, at the transmitter, each conversation is processed with a unique spreading code that is used to distribute the signal over the available bandwidth. The receiver uses the unique code to accept the energy associated with a particular code. The other signals present are each identified by a different code and simply produce background noise. In this way, many conversations can be carried simultaneously within the same block of spectrum.

The following analogy is used commonly to explain how CDMA technology works. Four speakers are simultaneously giving a presentation, and they each speak a different native language: Spanish, Korean, English, and Chinese (Figure C-6). If English is your native language, you only understand the words of the English speaker and tune out the Spanish, Korean, and Chinese speakers. You hear only what you know and recognize. The rest sounds like background noise. The same is true for CDMA. Each conversation is specially encoded and decoded for a particular user. Multiple users

Figure C-6 In this analogy of CDMA functionality, each conversation is specially encoded and decoded for each particular user. Thus the English-speaking person will only hear another English-speaking person and tune out the other languages, which are heard as background noise.

share the same frequency band at the same time, yet each user hears only the conversation he or she can interpret.

CDMA assigns each subscriber a unique code to put multiple users on the same wideband channel at the same time. These codes are used to distinguish between the various conversations. The result of this access method is increased call-handling capacity.

One of the unique aspects of CDMA is that while there are ultimate limits to the number of phone calls that a system can handle, this is not a fixed number. Rather, the capacity of the system depends on how coverage, quality, and capacity are balanced to arrive at the desired level of system performance. Since these parameters are tightly intertwined, operators cannot have the best of all worlds: 3 times wider coverage, 40 times capacity, and high-quality sound. For example, the 13-kbps vocoder provides better sound quality but reduces system capacity compared with an 8-kbps vocoder. Higher capacity might be achieved through some degree of degradation in coverage and/or quality.

System Features

CDMA has been adapted for use in cellular communications with the addition of several system features that enhance efficiency and lower costs.

Mobile Station Sign-on On power-on, the mobile station already knows the assigned frequency for CDMA service in the local area and will tune to that frequency and search for pilot signals. Multiple pilot signals typically will be found, each with a different time offset. This time offset distinguishes one base station from another. The mobile station will pick the strongest pilot and establish a frequency reference and a time reference from that signal. Once the mobile station becomes synchronized with the base station's system time, it can then register. Registration is the process by which the mobile station tells the system that it is available for calls and notifies the system of its location.

Call Processing The user makes a call by entering the digits on the mobile station keypad and hitting the "Send" button. If multiple mobile stations attempt a link on the access channel at precisely the same moment, a collision occurs. If the base station does not acknowledge the access attempt, the mobile station will wait a random time and try again. On making contact, the base station assigns a traffic channel, whereupon basic information is exchanged, including the mobile station's serial number. At this point, the conversation mode is started.

As a mobile station moves from one cell to the next, another cell's pilot signal will be detected that is strong enough for it to use. The mobile station will then request a "soft handoff," during which it is actually receiving both signals via different correlative elements in the receiver circuitry. Eventually, the signal from the first cell will diminish, and the mobile station will request from the second cell that the soft handoff be terminated. A base station does not hand off the call to another base station until it detects acceptable signal strength.

This soft handoff technique is a significant improvement over the handoff procedure used in analog FM cellular systems, where the communication link with the old cell site is momentarily disconnected before the link to the new site is established. For a short time, the mobile station is not connected to either cell site, during which the subscriber hears background noise or nothing at all. Sometimes the mobile stations Ping-Pong between two cell sites as the links are handed back and forth between the approaching and retreating cell sites. Other times, the calls are simply dropped. Because a mobile station in the CDMA system has more than one modulator, it can communicate with multiple cells simultaneously to implement the soft handoff.

At the end of a call placed over the CDMA system, the channel will be freed and may be reused. When the mobile station is turned off, it will generate a power-down registration signal that tells the system that it is no longer available for incoming calls.

Voice Detection and Encoding With voice activity detection, the transmitter is activated only when the user is speaking. This reduces interference levels—and, consequently, the amount of bandwidth consumed—when the user is not speaking. Through interference averaging, the capacity of the system is increased. This allows systems to be designed for the average rather than the worst interference case. However, the IS-95 CDMA standard requires that no interfering signal be received that is significantly stronger than the desired signal, since it would then jam the weaker signal. This has been called the "near-far problem" and means that high cell capacity does not necessarily translate into high overall system capacity.

The speech coder used in CDMA operates at a variable rate. When the subscriber is talking, the speech coder operates at the full rate; when the subscriber is not talking, the speech coder operates at only one-eighth the full rate. Two intermediate rates are also defined to capture the transitions and eliminate the effect of sudden rate changes. Since the variable-rate operation of the speech coder reduces the average bit rate of the conversations, system capacity is increased.

Privacy Increased privacy is inherent in CDMA technology. CDMA phone calls will be secure from the casual eavesdropper because, unlike a conversation carried over an analog system, a simple radio receiver will not be able to pick out individual digital conversations from the overall RF radiation in a frequency band.

A CDMA call starts with a standard rate of 9.6 kbps. This is then spread to a transmitted rate of about 1.25 Mbps. "Spreading" means that digital codes are applied to the data bits associated with users in a cell. These data bits are transmitted along with the signals of all the other users in that cell. When the signal is received, the codes are removed from the desired signal, separating the users and returning the call to the original rate of 9.6 kbps.

Because of the wide bandwidth of a spread-spectrum signal, it is very difficult to identify individual conversations for eavesdropping. Since a wideband spread-spectrum signal is very hard to detect, it appears as nothing more than a slight rise in the "noise floor" or interference level. With analog technologies, the power of the signal is concentrated in a narrower band, which makes it easier to detect with a radio receiver tuned to that set of frequencies.

The use of wideband spread-spectrum signals also offers more protection against cloning, an illegal practice whereby a mobile phone's electronic serial number is taken over the air and programmed into another phone. All calls made from a cloned phone are "free" because they are billed to the original subscriber.

Power Control CDMA systems rely on strict control of power at the mobile station to overcome the so-called near-far problem. If the signal from a near mobile station is received at the cell site receiver with too much power, the cell site receiver will become overloaded and prevent it from picking up the signals from mobile stations located farther away. The goal of CDMA is to have the signals of all mobile stations arrive at the base station with exactly the same power level. The closer the mobile station is to the cell site receiver, the lower is the power necessary for transmission; the farther away the mobile station, the greater is the power necessary for transmission.

Two forms of adaptive power control are employed in CDMA systems: open loop and closed loop. Open-loop power control is based on the similarity of loss in the forward and reverse paths. The received power at the mobile station is used as a reference. If it is low, the mobile station is assumed to be far from the base station and transmits with high power. If it is high, the mobile station is assumed to be near the base station and transmits with low power. The sum of the two power levels is a constant.

Closed-loop power control is used to force the power from the mobile station to deviate from the open-loop setting. This is achieved by an active feedback system from the base station to the mobile station. Power control bits are sent every 1.25 millisecond (ms) to direct the mobile station to increase or decrease its transmitted power by 1 decibel (dB). Lack of power control to at least this accuracy greatly reduces the capacity of CDMA systems.

With these adaptive power-control techniques, the mobile station transmits only enough power to maintain a link. This results in an average power requirement that is much lower than that for analog systems, which do not usually employ such techniques. CDMA's lower power requirement translates into smaller, lightweight, longer-life batteries—approximately 5 hours of talk time and over 2 days of standby time—and makes possible smaller, lower-cost hand-held computers and hybrid computer-communications devices. CDMA phones can easily weigh in at less than 8 ounces.

Spatial Diversity Among the various forms of diversity is that of spatial diversity, which is employed in CDMA, as well as in other multiple access techniques, including FDMA and TDMA. Spatial diversity helps to maintain the signal during the call handoff process when a user moves from one cell to the next. This process entails antennas in two different cell sites maintaining links with one mobile station. The mobile station has multiple correlative receiver elements that are assigned to each incoming signal and can add these.

CDMA uses at least four of these correlators: three that can be assigned to the link and one that searches for alternate paths. The cell sites send the received data, along with a quality index, to the MTSO, where a choice is made regarding the better of the two signals.

Not all these features are unique to CDMA; some can be exploited by TDMA-based systems as well, such as spatial diversity and power control. These already exist in all TDMA standards today, while soft handoff is implemented in the European Digital Enhanced Cordless Telecommunications (DECT) standard, which is based on TDMA.

Summary

There are still conflicting performance claims for TDMA and CDMA. Since both TDMA and CDMA have become TIA standards—IS-54 and IS-95, respectively—vendors are now aiming their full marketing efforts toward the cellular carriers. Proponents of each technology have the research to back up their claims of superior performance. Of the two, CDMA suffered a credibility problem early on because its advocates made grandiose performance claims for CDMA that could not be verified in the real-world operating environment. In some circles, this credibility problem lingers today. Of note, however, is that both technologies have been successful in the marketplace, each having been selected by many cellular carriers around the world. Both are capable of supporting emerging PCS networks and providing such services as wireless Internet access, Short messaging Service, voice mail, facsimile, paging, and video. Although TDMA-based Global System for Mobile (GSM) telecommunications is the dominant standard in the global wireless market, the use of CDMA is growing rapidly. GSM's head start in the market gives it a much larger presence and practically guarantees that GSM will continue to lead the digital cellular market for the next 5 years.

See also

Digital Enhanced Cordless Telecommunications

Frequency Division Multiple Access

Spread Spectrum Radio

Time Division Multiple Access

COMPETITIVE LOCAL EXCHANGE CARRIERS

Competitive Local Exchange Carriers (CLECs) offer voice, data services, and value-added services at significantly lower prices than the Incumbent Local Exchange Carriers (ILECs), enabling residential and business users to save money on such things as local calls, call-handling features, lines, and Internet access. Typically, CLECs offer service in major cities, where traffic volumes are greatest and, consequently, users are hardest hit with high local exchange charges from the incumbent carrier. Some CLECs call themselves integrated communications providers (ICPs) because their networks are designed from the outset to support voice and data services as well as Internet access. Others call themselves Data Local Exchange Carriers (DLECs) because they specialize in data services such as Digital Subscriber Line (DSL), which is used primarily for Internet access.

As of 2002, the ILECs still controlled 97 percent of the market for local services, according to the FCC, which means that the CLECs are trying to sustain themselves on the remaining 3 percent as they attempt to take market share from the ILECs. To deal with this situation, the CLECs have adopted different strategies based on resale and facilities ownership.

Resale versus Ownership

CLECs may compete in the market for local services by setting up their own networks or by reselling lines and services purchased from the ILEC. They may have hybrid arrangements for a time, which are part resale and part facilities ownership. Most CLECs prefer to have their own networks because the profit margins are higher than for resale. However, many CLECs start out in new markets as resellers. This enables them to establish a local presence, build brand awareness, and begin building a customer base while they assemble their own facilities-based network.

Although this strategy is used by many CLECs, many fail to carry it out properly. They get into financial trouble by using their capital to expand resale arrangements to capture even more market share instead of using that capital to quickly build their own networks and migrate customers to the high-margin facilities. Depending on the service, it could take a carrier 3 to 4 years to break even on a pure resale customer versus only 6 to 9 months on a pure facilities-based customer. With capital markets drying up for telecom companies and customers deferring product and services purchases, prolonged dependence on resale could set the stage for bankruptcy.

CLECs employ different technologies for competing in the local services market. Some set up their own Class 5 central office switches, enabling them to offer "dial tone" and the usual voice services, including Integrated Services Digital Network (ISDN) and features such as caller ID and voice messaging. The larger CLECs build their own fiber rings to serve their metropolitan customers with high-speed data services. Some CLECs have chosen to specialize in broadband data services by leveraging existing copper-based local loops, offering DSL services for Internet access. Others bypass the local loop entirely through the use of broadband wireless technologies, such as Local Multipoint Distribution Service (LMDS), enabling them to feed customer traffic to their nationwide fiber backbone networks without the incumbent carrier's involvement.

Despite the risks, some CLECs view resale as a viable long-term strategy. It not only allows them to enter into new markets more quickly than if they had initially deployed their own network, it also reduces initial capital requirements in each market, allowing them to focus capital resources initially on the critical areas of sales, marketing, and operations support systems (OSS). In addition, the strategy allows them to avoid deployment of conventional circuit switches and maintain design flexibility for the next generation of telecommunications technology.

Unfortunately, the resale strategy also results in lower margins for services than for facilities-based services. This means that the CLEC must pass much of its customer revenues back to the ILEC to pay the monthly fees for access lines. When investors stopped stressing market growth over profits in 2000, these CLECs found that capital was hard to get. By then, many had no money to invest in their own facilities where margins are greater. Most financial analysts doubt that CLECs can rely strictly on resale and survive. Although the ILECs have a vested interest in survival of some resale CLECs in order to receive regulatory approval to provide in-region long distance, once that approval is gained, some analysts believe that the ILECs may have no further interest in cooperating with the CLECs.

Summary

With the Telecommunications Act of 1996, CLECs and other types of carriers are allowed to compete in the offering of local exchange services and must be able to obtain the same service and feature connections as the ILECs have for themselves—and on an unbundled basis. If the ILEC does not meet the requirements of a 14-point checklist to open up its network in this and other ways, it cannot get permission from the FCC to compete in the market for long-distance services.

See also

Federal Communications Commission

Incumbent Local Exchange Carriers

Interexchange Carriers

CORDLESS TELECOMMUNICATIONS

The familiar cordless telephone, introduced in the early 1980s, has become a key factor in reshaping voice communications. Since people cannot be tied to their desks, as much as 70 percent of business calls do not reach the right person on the first attempt. This situation has seen dramatic improvement with cordless technology, which makes phones as mobile as their users. Now almost 30 percent of business calls reach the right person on the first attempt.

Cordless versus Cellular

Although cellular phones and cordless phones are both wireless, they have come to assume quite distinct and separate applications based on their areas of use and the differing technologies developed to meet user requirements. Cellular and cordless are implemented with their own standards-based technologies.

Briefly, cellular telephones are intended for off-site use. The systems are designed for a relatively low density of users. In this environment, macrocellular technology provides wide area coverage and the ability to make calls while traveling at high speeds. Cordless telephones, on the other hand, are designed for users whose movements are within a well-defined area, such as an office building. The cordless user makes calls from a portable handset linked by radio signals to a fixed base station (Figure C-7). The base station is connected either directly or indirectly to the public network.

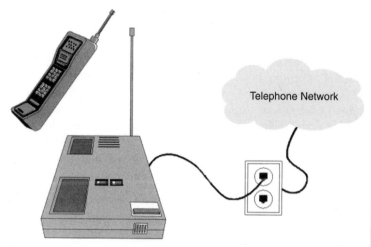

Figure C-7 The familiar cordless telephone found in many homes.

Cordless Standards

The cordless system standards are referred to as CT0, CT1, CT2, CT3, and DECT, with "CT" standing for Cordless Telecommunications. CT0 and CT1 were the technologies for first-generation analog cordless telephones. Comprising base station, charger, and handset and intended primarily for residential use, they had a range of 100 to 200 meters. They used analog radio transmission on two separate channels, one to transmit and one to receive. The potential disadvantage of CT0 and CT1 systems is that the limited number of frequencies can result in interference between handsets, even with the relatively low density of residential subscribers.

Also targeted at the residential user, CT2 represented an improved version of CT0 and CT1. Using Frequency Division Multiple Access (FDMA), the CT2 system splits the available bandwidth into radio channels in the assigned frequency domain. In the initial call setup, the handset scans the available channels and locks onto an unoccupied channel for the duration of the call. Using Time Division Duplexing (TDD), the call is split into time blocks that alternate between transmitting and receiving.

The Digital Enhanced Cordless Telecommunications (DECT) standard started as a European standard for cordless communications, with applications that included residential telephones and wireless Private Branch Exchange (PBX) and wireless local loop (WLL) access to the public network. Primarily, DECT was designed to solve the problem of providing cordless telephones in high-density, high-traffic office and other business environments.

CT3, on the other hand, is a technology developed by Ericsson in advance of the final agreement on the DECT standard and is designed specifically for the wireless PBX application. Since DECT is essentially based on CT3 technology, the two standards are very similar. Both enable the user to make and receive calls when within the range of a base station. Depending on the specific operating conditions, this amounts to a distance of between 164 feet (50 meters) and 820 feet (250 meters) from the base station. To provide service throughout the site, multiple base stations are set up to create a picocellular network. Signal handoff between the cells is supported by one or more radio exchange units, which are ultimately connected to the host PBX.

Both DECT and CT3 have been designed to cope with the highest-density telephone environments, such as city office districts, where user densities can reach 50,000 per block. A feature called Continuous Dynamic Channel Selection (CDCS) ensures seamless handoff between cells, which is particularly important in a picocellular environment where several handoffs may be necessary, even during a short call. The digital radio links are encrypted to provide absolute call privacy.

The two standards, DECT and CT3, are based on multicarrier Time Division Multiple Access/Time Division Duplexing (TMDA/TDD). They do not use the same operating frequencies, though, and consequently have different overall bit rates and call-carrying capacity.

It is the difference in frequencies that governs the commercial availability of DECT and CT3 around the world. Europe is committed to implementing the DECT standard within the frequency range of 1.8 to 1.9 GHz. Other countries, however, have made frequencies in the 800- to 1000-MHz band available for wireless PBXs, thereby paving the way for the introduction of CT3.

Summary

Many of the problems arising from the nonavailability of staff to a wired PBX can be avoided with cordless telephones. They are ideal for people who by the very nature of their work can be difficult to locate (e.g., maintenance engineers, warehouse staff, messengers, etc.) and for places on a company's premises that cannot be effectively covered by a wired PBX (e.g., warehouses, factories, refineries, exhibition halls, dispatch points, etc.).

A key advantage of cordless telecommunications is that it can simply be integrated into the corporate telecom system with add-on products and without the need to replace existing equipment. Another advantage of cordless telecommunications is that the amount of telephone wiring is dramatically reduced. Since companies typically spend between 10 and 20 percent of the original cost of their PBX on wiring the system, the use of cordless technology can have a significant impact on costs. There is also considerable benefit in terms of administration. For example, when moving offices, employees need not change extension numbers, and the PBX does not have to be reprogrammed to reflect the change.

See also

Cellular Voice Communications

Cellular Data Communications

Digital Enhanced Cordless Telecommunications

D

DATA COMPRESSION

Data compression is a standard feature of most bridges and routers, as well as modems, especially those used for transferring bulky files over wireless links. Compression improves throughput by capitalizing on the redundancies found in the data to reduce frame size and thereby allow more data to be transmitted over a link. An algorithm detects repeating characters or strings of characters and represents them as a symbol or token. At the receiving end, the process works in reverse to restore the original data.

There are many different algorithms available to compress data, which are designed for specific types of data sources and the redundancies found in them but do a poor job when applied to other sources of data. For example, the Moving Pictures Experts Group (MPEG) compression standards were designed to take advantage of the relatively small difference from one frame to another in a video stream and so do an excellent job of compressing motion pictures. On the other hand, MPEG would not be effective if applied to still images. For this data source, the Joint Photographic Experts Group (JPEG) compression standards would be applied.

JPEG is "lossy," meaning that the decompressed image is not quite the same as the original compressed image—there is some degradation. JPEG is designed to exploit known limitations of the human eye, notably that small color details are not perceived as well as small details of light and dark. JPEG eliminates the unnecessary details to greatly reduce the size of image files, allowing them to be transmitted faster and take up less space in a storage server.

On wide area network (WAN) links, the compression ratio tends to differ by application. The compression ratio can be as high as 6 to 1 when the traffic consists of heavy-duty file transfers. The compression ratio is less than 4 to 1 when the traffic is mostly database queries. When there are only "keep alive" signals or sporadic query traffic on a T1 line, the compression ratio can dip below 2 to 1.

Encrypted data exhibit little or no compression because the encryption process expands the data and uses more bandwidth. However, if data expansion is detected and compression is withheld until the encrypted data are completely transmitted, the need for more bandwidth can be avoided.

Types of Data Compression

There are several different data-compression methods in use today over WANs—among them are Transmission Control Protocol/Internet Protocol (TCP/IP) header compression, link compression, and multichannel payload compression. Depending on the method used, there can be a significant tradeoff between lower bandwidth consumption and increased packet delay.

TCP/IP Header Compression With TCP/IP header compression, the packet headers are compressed, but the data payload remains unchanged. Since the TCP/IP header must be replaced at each node for IP routing to be possible, this compression method requires hop-by-hop compression and decompression processing. This adds delay to each compressed/decompressed packet and puts an added burden on the router's CPU at each network node.

TCP/IP header compression was designed for use on slow serial links of 32 kbps or less and to produce a significant performance impact. It needs highly interactive traffic with small packet sizes. In such traffic, the ratio of Layer 3 and 4 headers to payload is relatively high, so just shrinking the headers can result in a substantial performance improvement.

Payload Compression Payload compression entails the compression of the payload of a Layer 2 WAN protocol, such as the Point-to-Point Protocol (PPP), Frame Relay, High-Level Data Link Control (HDLC), X.25, and Link Access Procedure–Balanced (LAPB). The Layer 2 packet header is not compressed, but the entire contents of the payload, including higher-layer protocol headers (i.e., TCP/IP), are compressed. They are compressed using the industry standard Lemple-Ziv algorithm or some variation of that algorithm.

Layer 2 payload compression applies the compression algorithm to the entire frame payload, including the TCP/IP headers. This method of compression is used on links operating at speeds from 56 to 1.544 Mbps and is useful on all traffic types as long as the traffic has not been compressed previously by a higher-layer application. TCP/IP header compression and Layer 2 payload compression, however, should not be applied at the same time because it is redundant and wasteful and could result in the link not coming up to not passing IP traffic.

Link Compression With link compression, the entire frame—both protocol header and payload—is compressed. This form of compression is typically used in local area network (LAN)–only or legacy-only environments. However, this method requires error-correction and packet-sequencing software, which adds to the processing overhead already introduced by link compression and results in increased packet delays. Also, like TCP/IP header compression, link compression requires hop-by-hop compression and decompression, so processor loading and packet delays occur at each router node the data traverses.

With link compression, a single data compression vocabulary dictionary or history buffer is maintained for all virtual circuits compressed over the WAN link. This buffer holds a running history about what data have been transmitted to help make future transmissions more efficient. To obtain optimal compression ratios, the history buffer must be large, requiring a significant amount of memory. The vocabulary dictionary resets at the end of each frame. This technique offers lower compression ratios than multichannel, multihistory buffer (vocabulary) data-compression methods. This is particularly true when transmitting mixed LAN and serial protocol traffic over the WAN link and frame sizes are 2 kilobytes or less. This translates into higher costs, but if more memory is added to get better ratios, this increases the upfront cost of the solution.

Mixed-Channel Payload Data Compression By using separate history buffers or vocabularies for each virtual circuit, multichannel payload data compression can yield higher compression ratios that require much less memory than other data-compression methods. This is particularly true in cases where mixed LAN and serial protocol traffic traverses the network. Higher compression ratios translate into lower WAN bandwidth requirements and greater cost savings.

But performance varies because vendors define payload data compression differently. Some consider it to be compression of everything that follows the IP header. However, the IP header can be a significant number of bytes. For overall compression to be effective, header compression must be applied. This adds to the processing burden of the CPU and increases packet delays.

External Data Compression Solutions Bridges and routers can perform data compression with optional software or add-on hardware modules. While compression can be implemented via software, hardware-based compression off-loads the bridge/router's main processor to deliver even higher levels of throughput. With a data-compression module, the compression process can occur without as much processing delay as a software solution.

The use of a separate digital signal processor (DSP) for data compression, instead of the software-only approach, enables the bridge/router to perform all its core functions without any performance penalty. This parallel-processing approach minimizes the packet delay that can occur when the router's CPU is forced to handle all these tasks by itself.

If there is no vacant slot in the bridge/router for the addition of a data-compression module, there are two alternatives: the software-only approach or an external compression device. The software-only approach could bog down the overall performance of the router, since its processor would be used to implement compression in addition to core functions. Although an external data compression device would not bog down the router's core functions, it means that one more device must be provisioned and managed at each remote site.

Summary

Data compression will become increasingly important to most organizations as the volume of data traffic at branch locations begins to exceed the capacity of the wide area links and as wireless services become available in the 2.4- and 5-GHz range. Multichannel payload solutions provide the highest compression ratios and reduce the number of packets transmitted across the network. Reducing packet latency can be effectively achieved via a dedicated processor like a DSP and by employing end-to-end compression techniques rather than node-to-node compression/decompression. All these factors contribute to reducing bandwidth and equipment costs as well as improving the network response time for user applications.

See also

Voice Compression

DECIBEL

Decibel (dB) is a unit of measurement expressing gain or loss. It is used to measure such things as sound, electrical or mechanical power, and voltage. In the telecommunications industry, the decibel is used to conveniently express the gain or loss in transmission systems, whether the medium is copper, optical fiber, or wireless.

The decibel is actually the relationship of some reference point and another point that is above or below the reference point. The base reference point is 0 dB, and subsequent measurements are relative to that reference point. There are a number of decibel notations, each indicating the context of the measurement, such as

- dBi is the antenna gain in dB relative to an isotropic source.
- dBm is the power in dB relative to 1 milliwatt.
- dBW is the power in dB relative to 1 watt.
- dBmV is referenced to 1 millivolt. It is often used as a measure of signal levels (or noise) on a network.

The dB scale related to power is different from the dB scale related to voltage. In power measurements, the power level doubles every 3 dB, instead of every 6 dB as in voltage. Likewise, the dB scale related to audio output is different from the dB scales relating to voltage and power.

Audio Intensity

Since the range of audio intensities that the human ear can detect is so large, the scale frequently used to measure them is a scale based on multiples of 10. This type of scale is sometimes referred to as a "logarithmic scale." The threshold of hearing is assigned a sound level of 0 dB. A sound that is 10 times more intense is assigned a sound level of 10 dB. A sound that is 10 times more intense (10×10) is assigned a sound level of 20 dB. A sound that is 10 times more intense ($10 \times 10 \times 10$) is assigned a sound level of 30 dB. A sound that is 10 times more intense ($10 \times 10 \times 10 \times 10$) is assigned a sound level of 40 dB. Table D-1 lists some common and not so common sounds with an estimate of their intensity and decibel level.

Summary

There are a variety of test instruments available to handle virtually any measurement requirement, including analog impulse meters to measure quick bursts of sound. These devices typically have output jacks for connections to charting devices that plot continuous noise levels across a roll of paper. Digital devices output measurements to light-emitting

TABLE D-1 Decibel Levels of Selected Sounds

Source	No. of Times Greater than Threshold of Hearing	Decibels
Threshold of hearing (TOH)	10^0	0
Rustling leaves	10^1	10
Whisper	10^2	20
Normal conversation	10^6	60
Busy street traffic	10^7	70
Vacuum cleaner	10^8	80
Heavy truck traffic	10^9	90
Walkman at maximum level	10^{10}	100
Power tools	10^{11}	110
Threshold of pain to the ear	10^{12}	120
Airport runway	10^{13}	130
Sonic boom	10^{14}	140
Perforation of eardrum	10^{16}	160
12 feet from a battleship cannon muzzle	10^{20}	220

diode (LED) screens. Band filters allow selection of narrow frequency ranges to isolate specific noises for measurement. Optional calibrators are available for in-field adjustments.

See also

Hertz

DIGITALLY ENHANCED CORDLESS TELECOMMUNICATIONS

The Digital Enhanced Cordless Telecommunications (DECT) standard defines a protocol for secure digital telecommunications and is intended to offer an economical alternative to existing cordless and wireless solutions. DECT uses Time Division Multiple Access (TDMA) technology to provide ten 1.75-MHz channels in the frequency band between 1.88 and 1.90 GHz. Each channel can carry up to 12 simultaneous two-way conversations. Speech quality is comparable to conventional land-based phone lines. Frequency bands have been made available for DECT in more than 100 countries.

Whereas conventional analog cordless phones have a range of about 100 meters, the DECT version can operate reliably up to 300 meters. What started out as a European standard for replacing analog cordless phones has been continually refined by the European Telecommunication Standards Institute (ETSI) to become a worldwide standard that provides a platform for wireless local loops (WLL), wireless LANs, and more recently, wireless Internet access. In addition, DECT services are compatible with GSM and ISDN, and dual-mode DECT/GSM handsets are available.

Advantages

A key advantage of DECT is dynamic reconfiguration, which means that implementation does not require advance load, frequency, or cell planning. Other wireless architectures require a predetermined frequency-allocation plan. Conventional analog cellular networks, for example, are organized as cells in honeycomb fashion. To avoid conflict from adjacent cells, each base station is allotted only a fraction of the allowable frequencies. Changing a particular station's frequency band to accommodate the addition of more base stations to increase network capacity entails an often difficult and expensive hardware upgrade. However sparsely the base stations are constructed at the start of an installation, all possible base stations must be assigned frequencies before any physical systems are put into place.

In a DECT system, planning for uncertain future growth is unnecessary. This is because a DECT base station can dynamically assign a call to any available frequency channel in its band. The 12 conversations occurring at any one time can take place on any of the 10 channels in any combination. The handset initiating a call identifies an open frequency and time slot on the nearest base station and grabs it. DECT systems also can reconfigure themselves on the fly to cope with changing traffic patterns. Therefore, adding a base station requires no modification of existing base stations and no prior planning of channel allocations.

Compared to conventional analog systems, DECT systems do not suffer from interference or cross-talk. Neither different mobile units nor adjacent DECT cells can pose interference problems because DECT manages the availability of frequencies and time slots dynamically. This dynamic reconfiguration capability makes DECT useful also as a platform for WLLs. DECT allows the deployment of a few base stations to meet initial service demand, with the easy addition of more base stations as traffic levels grow.

Voice compression [i.e., Adaptive Differential Pulse Code Modulation (ADPCM)] and the higher levels of the DECT protocol are not implemented at the base stations but are handled separately by a concentrator. The concentrator routes calls between the WLL network and the public switch telephone network (PSTN). This distributed architecture frees up base station processing power so that it can better handle the up to 12 concurrent transmission and reception activities.

For high-end residential and small-business users, DECT permits wireless versions of conventional PBX equipment, supporting standard functions such as incoming and outgoing calls, call hold, call forwarding, and voice mail without having to install new wiring. In this application, DECT dynamic reconfiguration means that implementation does not require advance load, frequency, or cell planning. Users can begin with a small system and then simply add components as needs change.

The DECT/GSM Interworking Profile allows a single handset to address both DECT systems and conventional cellular networks. This allows users to take advantage of the virtually free wireless PBX service within a corporate facility and then seamlessly switch over to GSM when the handset passes out of range of the PBX base station. When the call is handled by GSM, appropriate cellular charges accrue to the user. If the call cannot get through on either type of network, it is diverted to a voice mailbox.

Wireless Local Loops

Although residential cordless communication represents the largest current market for DECT-based products, other applications look promising for the future. In developing countries, where lack of a universal wired telecommunications infrastructure can limit economic growth, DECT permits the creation of a wireless local loop (WLL), thereby avoiding the considerable time and expense required to lay wire lines. WLLs can be implemented in several ways, which are summarized in Figure D-1.

In a small cell installation in densely populated urban or downtown areas, the existing telephone network can be used as a backbone that connects the base stations for each DECT

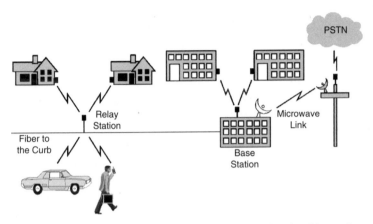

Figure D-1 DECT supports the deployment of wireless local loops that offer a high degree of configuration flexibility and cost savings over conventional wireless and wireline solutions.

cell. These DECT base stations may be installed on telephone poles or other facilities. Customer boxes (i.e., transceivers) installed on the outsides of houses and office buildings connect common phone, fax, and modem jacks inside. Through the transceivers, customers use their telephone, fax, and modem equipment to communicate with the base stations outside. In addition, customers can use DECT-compliant mobile phones, which can receive and transmit calls to the same base station.

In larger cell installations, such as suburban or rural areas, fiberoptic lines may provide the backbone that connects local relay stations to the nearest base station. These relay stations transmit and receive data to and from customer boxes. In these installations, the customer box must have a direct line of sight to the relay station.

Network feeds over long distances may be accomplished via microwave links, which is more economical than having to install new copper or fiber lines. Large cells can be converted easily into smaller cells by installing additional base units or relay stations. Since DECT system has a self-organizing air interface, no top-down frequency planning is necessary, as is the case with other wireless connection techniques such as GSM or its derivative Digital Cellular System 1800 (DCS 1800).

While most WLL installations focus on regular telephone and fax services, DECT paves the way for enhanced services. Multiple channels can be bundled to provide wider bandwidth, which can be tailored for each customer and billed accordingly. Among other things, this allows the mapping of ISDN services all the way through the network to the mobile unit.

Wireless LANs

In many data applications with low bit rate requirements, DECT can be a cost-effective solution. One example is remote wireless access to corporate LANs. By bundling channels, full-duplex transmission of up to 480 kbps per frequency carrier is theoretically possible. For multiple data links, a DECT base station can be complemented by additional DECT base stations controlled by a DECT server. This forms a multicell system for higher traffic requirements. With a transparent interface to ISDN, data access and videoconferencing through wireless links can be realized. Such installations also may include such services as voice mail, automatic call back, answering and messaging services, data on demand, and Internet access.

Summary

DECT is a radio access technology. As such, it has been designed and specified to work with many other types of networks, including the PSTN, ISDN, GSM, and the Internet, as well as LANs and telephone systems in office buildings and homes. DECT modules incorporated into building control and security systems provide intelligent systems that allow automatic control and alerting to augment or replace today's customized telemetry and wired systems. DECT also may find its way into the home, providing automatic security alerting in the event of unauthorized entry, fire, or flood; remote telephone control of appliances; and return channels for interactive television. While DECT is an international standard, it has been adapted only recently for use in North America, where it operates in the unlicensed 2.4-GHz ISM (industrial, scientific, and medical) band. The standard in North America is known as Worldwide Digital Cordless Telephone (WDCT), which is based on DECT.

See also

Global System for Mobile (GSM) Telecommunications

Wireless LANs

Wireless Local Loops

DIRECT BROADCAST SATELLITE

Direct broadcast satellite (DBS) operators use satellites to transmit video programming to subscribers, who must buy or rent a small parabolic dish antenna and pay a subscription fee to receive the programming service. DBS meets consumer demand for entertainment programming, Internet connectivity, and multimedia applications. DBS offers more programming choices for consumers and a platform for the development of new services, including video on demand, interactive TV, Internet messaging services, and personalized on-demand stock quotes. Much of the growing popularity of DBS is attributable to the programming choices available to consumers as well as the picture quality provided by digital technology. And like cable television systems, DBS offers programming in the high-definition television (HDTV) format.

One of the most popular DBS services is DirecTV, a unit of Hughes Electronics, which markets the service worldwide. First introduced in the United States in 1994, DirecTV offers over 225 channels and has over 10 million customers. The satellite service requires the user to have an 18-inch dish, a digital set-top decoder box, and a remote control. The system features an on-screen guide that lets users scan and select programming choices using the remote. Customers also can use the remote control to instantly order pay-per-view movies, as well as set parental controls and spending limits.

The DirecTV installation includes an access card, which provides security and encryption information and allows customers to control the use of the system. The access card also enables DirecTV to capture billing information. A standard telephone connection is also used to download billing information from the decoder box to the DirecTV billing center. This telephone line link enables DirecTV subscribers to order pay-per-view transmission as desired.

DirecTV allows users to integrate local broadcast channels with satellite-based transmissions. In markets where broadcast or cable systems are in place, users can maintain a basic cable subscription or connect a broadcast antenna to the DirecTV digital receiver to receive local and network broadcasts. A switch built into the remote control enables consumers to instantly switch between DirecTV and local stations.

HDTV programming from DirecTV is delivered from its 119° west longitude orbital slot location. To receive HDTV programming, consumers must have an HDTV set with a built-in DirecTV receiver or a DirecTV-enabled HDTV set-top converter box. A small elliptical satellite dish is needed to receive HDTV programming from the 119° orbital slot location, as well as core DirecTV programming from the 101° orbital location.

Internet access is provided via two services. The older service is DirecPC, a product that uses DirecTV technology in conjunction with a PC to deliver high-bandwidth, satellite-based access to the Internet. The DirecPC package includes a satellite dish and an expansion card designed for a PC's input-output (I/O) bus. This receiver card transmits data from the Internet to the computer at 400 kbps, a rate 14 times faster than that of a 28.8-kbps modem connection.

Users connect to the Internet service provider (ISP) through a modem connection, but the ISP is responsible for routing data through the satellite uplink and transmitting the data to the receiver card and into the computer (Figure D-2). The service also provides users with the option to "narrowcast" software from the head end of a network to branch users during off-peak hours. Additionally, DirecPC transmits television broadcasts from major networks, such as CNN and ESPN, to the user's computer system.

The company's newer service, DirecWAY, offers a two-way broadband connection that offers 400 kbps on the downlink and about 150 kbps on the uplink, which eliminates the need for a modem and separate phone line. A new dish antenna provides access to the Internet and cable programming. A business-class DirecWAY service is also available. Multiple-seat account options (2 seats is the entry-level service; 5-, 10-, and 20-seat options are available), LAN software routing, and firewall security are offered as part of the business class service.

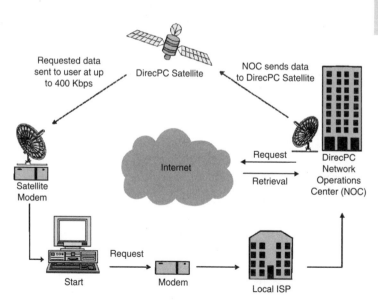

Figure D-2   Typical DBS configuration for Internet access.

Operation

DBS operates in the K_u band, the group of frequencies from 12 to 18 GHz. TV shows and movies are stored on tape or in digital form at a video server, while live events are broadcast directly to a satellite (Figure D-3). Stored programs are sent to the uplink (ground-to-satellite) center manually via tape or electronically from the video server over fiberoptic cable. Live events also pass through the uplink center. There, all programs—whether live or stored—are digitized (or redigitized) and compressed before they are uplinked to the satellites. All DBS systems use the MPEG-2 compression scheme because it supports a wide range of compression ratios and data rates. It is capable of delivering a clean, high-resolution video signal and CD-quality sound. The satellites broadcast over 200 channels simultaneously via the downlink. The home satellite dish picks up all the channels and sends them via a cable to a set-top decoder. The set-top decoder tunes one channel, decodes the video, and sends an analog signal to the TV.

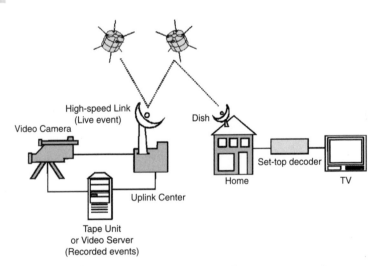

Figure D-3   Typical DBS configuration for television programming.

Service Providers

More than 1 million U.S. residents have installed small TV satellite dishes to receive programming via satellite services. At this writing, there are four direct broadcast satellite systems in operation: PrimeStar, EchoStar, Digital Satellite Service (DSS), and AlphStar. DirectTV uses DSS and PrimeStar.

Ordering PrimeStar service is similar to ordering cable: After the order is placed, a technician installs the dish and activates programming. DSS, EchoStar, and AlphaStar services also give users the option of installing the dish themselves. The dish must be placed so that it can capture a clean signal from the nearest satellite—usually on the roof, facing south. To activate service, the user calls the programming provider to obtain a unique satellite dish address.

Equipment

The key component of the DBS system is the dish antenna, which comes in various sizes. Dish size depends on the strength of the satellite signal; the stronger the signal, the smaller the dish can be. Users select the dish according to their geographic proximity to the satellite source. This also explains why it is necessary to install the dish so that it points in a specific direction. If the satellite sits on the southern horizon, the dish must be pointed south.

The user also needs a receiver-decoder unit, which tunes in one channel from the multitude of channels it receives from the dish. The decoder then decompresses and decodes the video signal in real time so that the programs can be watched on the television set. These set-top units also may include a phone-line connection for pay-per-view ordering and Internet access. Taping DBS programs requires the set-top unit to be tuned to the correct channel. To make recording easier, some receiver-decoders include an event scheduler and an on-screen programming guide.

As with most audio-video components, DBS units come with a remote control. Some manufacturers offer a universal remote that also can be used to operate the TV and VCR.

The accessories available for DBS systems deal with secondary and tertiary installations. Users can buy additional receiver-decoder units or multiroom distribution kits, which use either cable or radio frequencies to transmit the signals from the original set-top unit to other rooms. Some kits enable the VCR to be plugged into the distributor.

Programming

Each of the four DBS systems currently available provides similar core services. The differences lie mainly in the availability of premium movie channels, audio channels, pay-per-view events, Internet services, and custom features.

With more than 200 channels to choose from, the on-screen programming guide can become an important factor when selecting a service. Most guides enable users to sort the available programming based on content area—such as sports, movies, comedies—or list favorite channels at the top of the menu. Depending on the equipment selected, users can even store the favorite-channel profiles of multiple family members.

Parental lockout enables adults to block specific channels or programming with a specific content rating or to set a maximum pay-per-view spending limit. Channel-blocking options are protected by passwords; with multiprofile units, parents can customize the system for each child.

The addition of digital television recording systems such as TiVo allows viewers to easily find and schedule their favorite television shows automatically and digitally record or store up to 35 hours of video content without the use of videotape. Such systems provide the ability to pause, rewind, replay, and slow motion live shows. An advanced programming guide allows viewers to check program listings up to 14 days in advance.

Regulation

Despite increases in the number of subscribers to DBS systems in recent years, CATV systems remain the dominant supplier in what is called the "multichannel video program distribution" (MVPD) market. The FCC has regulatory authority over DBS and is charged with implementing the Satellite Home Viewer Improvement Act of 1999 (SHVIA).

This law provides that after December 31, 2001, each satellite carrier providing television broadcast signals to subscribers within the local market of a television broadcast station of a primary transmission made by that station shall carry on request the signals of all television broadcast stations located within that local market.

Until January 1, 2002, satellite carriers were granted a royalty-free copyright license to retransmit broadcast signals on a station-by-station basis, subject to obtaining a broadcaster's retransmission consent. This transition period was intended to provide the satellite industry with time to begin providing local signals into local markets—in effect, providing local-into-local satellite service.

Summary

While DBS competes well against cable television in terms of television programming, it may not be able to compete with cable on the data front. In contrast with the finite bandwidth available to wireless and satellite systems, the terrestrial broadband pipe technologies available to cable systems offer bandwidth that is virtually limitless for almost all current practical purposes. Duplication of this pipe requires an investment of tens of billions of dollars and therefore would be impractical. Realizing this, DBS services limit downlink throughput per subscriber at about 400 kbps and reserve the right to limit bandwidth-hogging activities, such as audio and video streaming, and automatic file exchange applications. These restrictions are justified as being necessary to preserve an adequate level of service for all subscribers.

See also

Satellite Communications

E

ENHANCED DATA RATES FOR GLOBAL EVOLUTION

It is expected that packet data will dominate circuit-switched data in the future, primarily to give users high-speed Internet access from mobile phones and other handheld devices. One of the key enabling technologies that will allow this to happen is known as Enhanced Data Rates for Global Evolution (EDGE), which combines multiple 30-kHz time slots available under Time Division Multiple Access (TDMA) to provide data rates of up to 384 kbps.

An interim technology is known as General Packet Radio Service (GPRS), which combines TDMA time slots to provide data rates of up to 115 kbps. EDGE technology builds on GPRS, offering enhanced modulation that adapts to radio circumstances, thereby offering the highest data rates in good propagation conditions while ensuring wider area coverage at lower data speeds per time slot. Typical applications for this type of service include multimedia messaging, Web browsing, enhanced short messages, wireless imaging with instant pictures, video services, document and information sharing, surveillance, voice over the Internet, and broadcasting.

Europe is ahead of the United States in the deployment of EDGE technology on their GSM networks. Instead of the advertised speed of 384 kbps, however, the actual speed may not even reach half that. Where EDGE is already deployed in the United Kingdom, for example, the top speed is 160 kbps.

Nevertheless, EDGE (and GPRS) offers a 2.5G migration path to the global standard Universal Mobile Telecommunications System (UMTS), which is considered a third-generation (3G) wireless communications platform that will be capable of supporting speeds of up to 2.4 Mbps.

Summary

The introduction of GPRS and then EDGE as an overlay to existing TDMA networks builds on the operator's existing investment in infrastructure. EDGE provides a boost to data speeds using existing TDMA networks, allowing the operator to offer personal multimedia applications before the introduction of UMTS. As wireless data become available to all subscribers and they demand a full set of high-speed services and shorter response times, EDGE will provide an operator with a competitive advantage. EDGE also enables data capacity to be deployed when and where demand warrants, minimizing the investment required.

See also

General Packet Radio Service

Global System for Mobile Telecommunications

Time Division Multiple Access

Universal Mobile Telecommunications System

F

FAMILY RADIO SERVICE

Family Radio Service (FRS) is one of the Citizens Band Radio Services. It is for family, friends, and associates to communicate among themselves within their neighborhood and while on group outings. Users may select any of the 14 FRS channels on a "take turns" basis. No FRS channel is assigned to any specific individual or organization.

Although manufacturers advertise a range of up to 2 miles, users can expect a communication range of less than 1 mile. Although FRS may be used for business-related communications, it cannot be connected to the public switched telephone network (PSTN) and used for telephone calls.

License documents are neither needed nor issued. FRS Rule 1 provides all the authority necessary to operate an FRS unit (Figure F-1) in places where the Federal Communications Commission (FCC) regulates radio communications as long as an unmodified FCC-certified FRS unit is only used. An FCC-certified FRS unit has an identifying label placed on it by the manufacturer. There is no age or citizenship requirement.

FRS units may be operated within the territorial limits of the 50 United States, the District of Columbia, and the

Figure F-1   Motorola's TalkAbout 280 SLK is a palm-size FRS unit that is small enough to carry in a shirt pocket. The 6-ounce radio runs on three AA batteries.

Caribbean and Pacific insular areas. Such units also may be operated on or over any other area of the world, except within the territorial limits of areas where radio communications are regulated by another agency of the United States or within the territorial limits of any foreign government.

Users cannot make any internal modification to an FRS unit. Any internal modification cancels the FCC certification and voids the user's authority to operate the unit over the FRS. In addition, users may not attach any antenna, power amplifier, or other apparatus to an FRS unit that has not been FCC certified as part of that FRS unit. There are no exceptions to this rule, and attaching any such apparatus to an FRS unit cancels the FCC certification and voids everyone's authority to operate the unit over the FRS.

Summary

Family Radio Service is used for conducting two-way voice communications with another person. One-way transmission may be used only to establish communications with another person, send an emergency message, provide traveler assistance, make a voice page, or conduct a brief test. Operators must, at all times and on all channels, give priority to emergency communication messages concerning the immediate safety of life or the immediate protection of property.

See also

Citizens Band Radio Service
General Mobile Radio Service
Low-Power Radio Service

FEDERAL COMMUNICATIONS COMMISSION

The Federal Communications Commission (FCC) is an independent federal agency in the United States that is responsible directly to Congress. Established by the Communications Act of 1934, the FCC is charged with regulating interstate and international communications by radio, television, wire, satellite, and cable. Its jurisdiction covers the 50 states and territories, the District of Columbia, and U.S. possessions.

The FCC is directed by five commissioners appointed by the President and confirmed by the Senate for 5-year terms, except when filling an unexpired term. The President designates one of the commissioners to serve as chairman, who presides over all FCC meetings. The commissioners hold regular open and closed agenda meetings and special meetings. By law, the FCC must hold at least one open meeting per month. It also may act between meetings by "circulation," a procedure whereby a document is submitted to each commissioner individually for consideration and official action.

Certain other functions are delegated to staff units and bureaus and to committees of the commissioners. The chairman coordinates and organizes the work of the FCC and represents the agency in legislative matters and in relations with other government departments and agencies.

Operating Bureaus

At the staff level, the FCC comprises seven operating bureaus and 10 staff offices. Most issues considered by the FCC are developed by the bureaus and offices, which are organized by substantive area (Figure F-2):

- The Common Carrier Bureau handles domestic wireline telephony.
- The Wireless Telecommunications Bureau oversees wireless services such as private radio, cellular telephone, personal communications service (PCS), and pagers.
- The Cable Services Bureau regulates cable television and related services.
- The International Bureau regulates international and satellite communications.
- The Enforcement Bureau enforces the Communications Act, as well as the FCC's rules, orders, and authorizations.

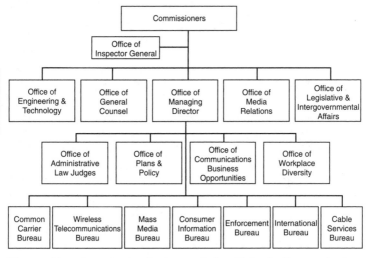

Figure F-2 Organizational chart of the Federal Communications Commission.

- The Mass Media Bureau regulates AM and FM radio and television broadcast stations, as well as multipoint distribution (i.e., cable and satellite) and instructional television fixed services.
- The Consumer Information Bureau communicates information to the public regarding FCC policies, programs, and activities. This bureau is also charged with overseeing disability mandates.

Other Offices

In addition, the FCC includes the following other offices:

- The Office of Plans and Policy serves as the FCC's chief economic policy advisor.
- The Office of the General Counsel reviews legal issues and defends FCC actions in court.
- The Office of the Managing Director manages the internal administration of the FCC.
- The Office of Legislative and Intergovernmental Affairs coordinates FCC activities with other branches of government.
- The Office of the Inspector General reviews FCC activities.
- The Office of Communications Business Opportunities provides assistance to small businesses in the communications industry.
- The Office of Engineering and Technology allocates spectrum for nongovernment use and provides expert advice on technical issues before the FCC.
- The Office of Administrative Law Judges adjudicates disputes.
- The Office of Workplace Diversity ensures equal employment opportunities within the FCC.
- The Office of Media Relations informs the news media of FCC decisions and serves as the FCC's main point of contact with the media.

Reorganization Plan

In mid-1999, the FCC unveiled a 5-year restructuring plan that will enable it to better meet the fast-changing needs of the communications business. The plan eliminates unnecessary rules in areas where competition has emerged and reorganizes the FCC along functional rather than technological lines. The plan's success hinges on thriving competition, enough to reduce the need for the FCC to regulate directly. Under the plan, the FCC is making the transition from an industry regulator to a market facilitator.

The plan addresses the disintegration of boundaries between wire, wireless, satellite, broadcast, and cable communications—categories around which the FCC is currently grouped.

The internal bureaus are now grouped around functions such as licensing and competition rather than by technology. The plan consolidates enforcement and consumer information into two separate bureaus of the FCC rather than having those functions spread across the agency.

The plan has streamlined and speeded up the FCC's services, for example, by instituting agency-wide electronic filing and automated licensing systems. Other goals include reducing backlog and making greater use of alternative dispute-resolution mechanisms. The plan also makes it faster and easier for consumers to interact with the agency. Under the reorganization plan, the FCC keeps many of its existing priorities, such as protecting consumers from fraud and keeping phone rates affordable for the poor.

Summary

The top-down regulatory model of the industrial age is as out of place in today's digital economy as the rotary telephone. As competition and convergence develop, the FCC must streamline its operations and continue to eliminate regulatory burdens. However, despite the FCC's planned reorganization from "regulator" to "facilitator," the commission will still have to contend with a set of core functions that are not normally addressed by market forces. These core functions include universal service, consumer protection and information, enforcement and promotion of procompetition goals domestically and worldwide, and spectrum management.

See also

Wireless Telecommunications Bureau

FIXED WIRELESS ACCESS

Fixed wireless access technology provides a wireless link to the public switched telephone network (PSTN) as an alternative to traditional wire-based local telephone service. Since calls and other information (e.g., data, images) are transmitted through the air rather than through conventional cables and wires, the cost of providing and maintaining telephone poles and cables is avoided. Unlike cellular technologies, which provide services to mobile users, fixed wireless services require a rooftop antenna to an office building or home that is lined up with a service provider's hub antenna.

Fixed wireless access systems come in two varieties: narrowband and broadband. A narrowband fixed wireless access service can provide bandwidth up to 128 kbps, which can support one voice conversation and a data session such as Internet access or fax transmission. A broadband fixed wireless access service can provide bandwidth in the multi-megabit-per-second range, which is enough to support telephone calls, television programming, and broadband Internet access.

A narrowband fixed wireless service requires a wireless access unit that is installed on the exterior of a home or business (Figure F-3) to allow customers to originate and receive calls with no change to their existing analog telephones.

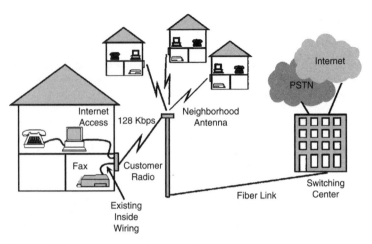

Figure F-3 Fixed wireless access configuration.

This transceiver is positioned to provide an unobstructed view to the nearest base station receiver. Voice and data calls are transmitted from the transceiver at the customer's location to the base station equipment, which relays the call through the carrier's existing network facilities to the appropriate destination. No investment in special phones or facsimile machines is required; customers use all their existing equipment.

Narrowband fixed wireless systems use the licensed 3.5-GHz radio band with 100-MHz spacing between uplink and downlink frequencies. Subscribers receive network access over a radio link within a range of 200 meters (600 feet) to 40 kilometers (25 miles) of the carrier's hub antenna. About 2000 subscribers can be supported per cell site.

Broadband fixed wireless access systems are based on microwave technology. Multichannel Multipoint Distribution Service (MMDS) operates in the licensed 2- to 3-GHz frequency range, while Local Multipoint Distribution Service (LMDS) operates in the licensed 28- to 31-GHz frequency range. Both services are used by Competitive Local Exchange

Carriers (CLECs) primarily to offer broadband Internet access. These technologies are used to bring data traffic to the fiberoptic networks of Interexchange Carriers (IXCs) and nationwide CLECs, bypassing the local loops of the Incumbent Local Exchange Carriers (ILECs).

Summary

Fixed wireless access technology originated out of the need to contain carriers' operating costs in rural areas, where pole and cable installation and maintenance are more expensive than in urban and suburban areas. However, wireless access technology also can be used in urban areas to bypass the local exchange carrier for long-distance calls. Since the IXC or CLEC avoids having to pay the ILEC's local loop interconnection charges, the savings can be passed back to the customer. This arrangement is also referred to as a "wireless local loop."

See also

Cellular Voice Communications

Local Multipoint Distribution Service

Multichannel Multipoint Distribution Service

FRAUD MANAGEMENT SYSTEMS

Despite advances in technology, mobile phone fraud continues to be an ongoing problem. Fraud management systems that incorporate mechanisms to integrate service usage data are becoming a necessary weapon among wireless carriers. The challenge is to monitor and profile the activity using hard data and to be alert to the ever-changing nature of fraud. To combat cell phone fraud, a number of systems have been implemented that can discourage cell phone cloning and stop thieves from obtaining free access to cellular networks.

Personal Identification Numbers

Some service providers offer their subscribers a free fraud protection feature (FPF) to help protect against unauthorized use of their cell phones. Like an ATM bank card, FPF uses a private combination, or personal identification number (PIN), that only the subscriber knows. The PIN code does not interfere with regular phone usage. Even if pirates capture the phone's signal, they would not be able to use it without the PIN code.

For example, the subscriber locks the phone account by pressing *56 + PIN + SEND. This blocks all outgoing calls, except for 911 (emergency) and the carrier's customer care number. While the account is locked, calls can still be received, and voice mail will continue to take messages. When the subscriber wants to make a call, the account is unlocked by pressing *56 + 0 + PIN + SEND. Both the lock and unlock sequences can be programmed directly into the mobile phone's speed-dial option.

There are database applications available that help service providers make better use of PINs to detect potentially fraudulent calls and prevent their completion. One of these products comes from Sanders Telecommunications Systems, a Lockheed Martin company. The company's MicroProfile software enables the carrier to determine when a call is being placed to a number outside the subscriber's normal calling pattern. Combined with another product, an intelligent network-based application called Intelligent PIN, the MicroProfile software requires that the caller provide a PIN only for out-of-profile calls. By establishing and maintaining calling profiles of subscribers, out-of-profile call requests can be identified in real time, whereupon PINs can be requested to ensure that only legitimate calls are allowed to be placed.

With this system, unauthorized calls and lost revenues resulting from phone cloning or other illegal means can be reduced significantly. MicroProfile also provides the ability to monitor and analyze call histories to detect patterns of fraud and identify their origins. Since more than 90 percent of mobile calls will be within an established profile and, therefore, will not require routine PIN use, this method of fraud prevention poses little or no inconvenience to users.

Other fraud detection schemes are able to detect when more than one phone with the same personality is attempting to access the system. On detection, the subscriber is requested to enter a PIN prior to all new call attempts until further investigation.

There are also fraud detection schemes that can be controlled by subscribers. For example, the subscriber can enter numbers on a targeted list comprised of 900 numbers or international area codes that require a PIN when dialed. If the PIN entry fails, the subscriber record is marked as fraudulent and requires a PIN for all future calls until further investigation.

Call Pattern Analysis

Cloned phones can be identified with a technology called "call pattern analysis." When a subscriber's phone deviates from its normal activity, it trips an alarm at the service provider's fraud management system. There it is put into a queue, where a fraud analyst ascertains whether the customer has been victimized and then remedies the situation by dropping the connection.

Coral Systems offers an innovative software solution designed to help wireless carriers worldwide to fight the ongoing fraud problem. The company's FraudBuster system is designed specifically to detect and manage fraud in multiple wireless systems. Through its support of wireless standards including Advanced Mobile Phone Service (AMPS), Code Division Multiple Access (CDMA), and Global System for Mobile (GSM), FraudBuster targets the most pervasive types of fraud, including cloned phones and subscription fraud. Through the development of personalized customer profiles based on the subscriber's typical calling patterns, FraudBuster can immediately identify suspicious usage.

After primary detection, a member of the carrier's fraud investigation team is immediately alerted, supplied with detailed information on the calls in question, and provided with recommended actions to address the alleged fraud (Figure F-4).

Authentication

Authentication works by automatically sending a series of encoded passwords over the airwaves between the cellular telephone and the cellular network to validate a customer each time a call is placed or received. Authentication uses a complex security feature that contains a secret code and special number based on an algorithm shared only by the cellular telephone and the wireless network. Whenever a customer places or receives a call, the wireless system asks the cellular telephone to prove its identity through a question-and-answer process. This process occurs without delaying the time it takes to connect a legitimate cellular call. There is no charge for this service.

Authentication uses advanced encryption technology that makes it almost impossible for a subscriber's mobile phone number to be cloned. It involves the exchange of a secret code based on an intricate algorithm between the phone and the

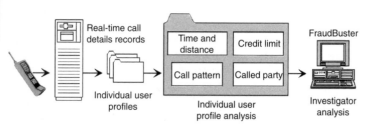

Figure F-4   Coral System's FraudBuster software creates individual user profiles on a real-time basis to allow for per-subscriber monitoring. The profiles are updated, referenced, and analyzed on a call-by-call basis. When a suspicious deviation is detected, a fraud investigator is immediately alerted and provided with a detailed accounting of events that triggered the alert.

switch. These algorithms are so complex that service providers say that they will remain impenetrable for at least 20 years.

Authentication technology identifies cloned phone numbers immediately, before costly communications can take place. The digital network and the authentication-ready phones operating on it carry matching information. When a user initiates a call, the network challenges the phone to verify itself by performing a mathematical equation only that specific phone can solve. An authenticatable phone will match the challenge, confirming that it and the corresponding phone number are being used by a legitimate customer. If it does not match, the network determines that the phone number is being used illegally, and service to that phone is terminated. All this takes place in a fraction of a second.

Prior to authentication, a PIN had to be entered before the call was connected. All that the user needs now is an authentication-ready phone to take advantage of the service, which is usually offered free. Where authentication service is available, subscribers no longer need to use a PIN to make calls, except when roaming in areas where authentication is not yet available.

PINs have become an effective fraud prevention tool. If a subscriber's phone gets cloned anyway, a simple call to the service provider to obtain a new PIN is the extent of the customer's inconvenience—there is no need to change phone numbers, which also obviates the need for new business cards and letterheads. The subscriber does not even need to come into the service center. But PINs are far from bulletproof, and cloners have proven particularly adept at cracking most security systems carriers have deployed. Still, PINs have decreased cellular phone fraud by as much as 70 percent.

Radio frequency Fingerprinting

Radio frequency fingerprinting is a step up from the use of PINs. With digital analysis technology that recognizes the unique characteristics of radio signals emitted by mobile phones, a fingerprint can be made that can distinguish individual phones within a fraction of a second after an attempt to place a call is made. Once the fraudulent call is detected, it is immediately disconnected. The technology works so well that it has cut down on fraudulent calls by as much as 85 percent in certain high-fraud markets, including Los Angeles and New York.

Voice Verification

Most fraud prevention technologies—such as PINs, call pattern analysis, authentication, and radio frequency fingerprinting—are only partially effective. Rather than verifying the caller, they merely authenticate a piece of information (PIN, ESN/MIN), a piece of equipment (RF fingerprinting, authentication), or the subscriber's call patterns (call pattern analysis).

Voice verification systems are based on the uniqueness of each person's voice and the reliability of the technology that can distinguish one voice from another by comparing a digitized sample of a person's voice with a stored model or "voice print."

One of the most advanced voice verification systems comes from T-NETIX, Inc. The company uses a combination of decision-tree and neural network technologies to implement what it calls a "neural tree network". The neural tree is comprised of nodes, or neurons, that are discriminantly trained through multiple repeated utterances of a subscriber-selected password or a small sample of speech. Discriminant training contrasts the acoustic features of the speaker being enrolled to features of the speakers already enrolled in the service.

During the verification process, each neuron must decide whether the acoustic features of the spoken input are more like those of the person whose identity is claimed or more like those of other speakers in the system. The neural tree network technology permits this complex decision-making, or discriminant, process to be completed in a relatively short period of time in contrast to other technologies. In effect, yes/no decisions are reached at each neuron of the neural tree, and a conclusion is reached after moving through five or six branches of the tree. The relative simplicity of the neural network decision-path design facilitates rapid analysis of spoken input with no upward limit on the number of enrollees.

The technology is also robust in its ability to determine and isolate channel environmental conditions. The front-end analysis recognizes and normalizes conditions such as background noise, channel differences, and microphone variances.

The mobile service subscriber goes through an enrollment process consisting of the following steps:

- To access the enrollment system, the subscriber inputs his or her identity using a PIN.

- The voice response unit prompts the subscriber to speak the password a few times (typically three or four). The speaker verification technology averages the voice samples to obtain a more robust voice model for the subscriber.

- The technology then analyzes the characteristics of the subscriber's statement of the password and characterizes its tonal aspects. The process also results in characterization and isolation of the channel environment (i.e., line type, hand-set type).

- The system segments the voice utterance into its subword units in order to examine the utterance in greater detail.

- Models for the voice segments are created and compared with other samples stored in the database to train the system to distinguish between individuals with similar voice characteristics.

- Finally, the system loads the subscriber's voice model into the voice identification database, indexing it to the subscriber's numeric identifier.

The voice verification system can reside on a public or private network as an intelligent peripheral or can be placed as an adjunct serving a Private Branch Exchange (PBX) or

Automatic Call Distributor (ACD). In a mobile environment, the system can be an adjunct to a Mobile Switching Center (MSC).

Data Mining

Another method of fraud detection entails the use of data-mining software that examines billing records and picks up patterns that reveal the behavior of cloning fraud.

Two major patterns are associated with cloning fraud. One is called the "time overlap pattern," which means that a phone is involved in two or more independent calls simultaneously. The other is called the "velocity pattern," which originates from the assumption that a handset cannot initiate or receive another call from a location far away from its previous call in a short time. If this happens, there is a high probability that a clone phone is being used somewhere.

The cloning information is passed to the service provider's billing and management system, and countermeasures will be executed to prevent further damages. Cloning history data are also kept in the database and can be queried by service representatives via a wide area network (WAN).

Real-Time Usage Reporting

Several software vendors offer products that provide real-time collection of data from cellular switches that can be used for identifying fraudulent use. Subscriber Computing's FraudWatch Pro software, for example, provides workable cases of fraud rather than just alarms. With this system, analysts have the ability to select the types of fraud that they work, such as subscription fraud or cloning. Based on an analysis of call detail records (CDR) received from home switches and roaming CDR data feeds, the system detects fraudulent activities and provides prioritized cases to the analyst for action. The system provides investigative tools that allow the analyst to ascertain whether fraudulent activity has indeed occurred. The cases are continuously prioritized according to fraud certainty factors or the probability of fraud occurring.

FraudWatch Pro receives CDR records directly from the switches and from roaming CDR exchanges as quickly as the data become available to the service provider. Profiles are created that contain details about the specific daily activity of a given subscriber. The subscriber profile contains about 40 single and multidimensional data "buckets" that are updated in real time. Analysts have a Graphical User Interface (GUI) that allows them to query the database to search for a wide range of behavior patterns as well as address queries about specific individuals. The system stores up to 12 months of summary statistics as well as up to 6 months of daily call detail records.

The software also provides support for the investigative analysis of additional dialed digit relationships. This allows the relationships among clones and their dialed digits to be graphically displayed as a means to help build prosecutable cases.

Smart Cards

While it is easy to intercept information from mobile phones used on analog networks, it is much harder to do so on such digital networks as the North American Personal Communication Services (PCS) and the European Global System for Mobile (GSM) telecommunications. This is so because the signals are encrypted, making them much more difficult to intercept without expensive equipment and a higher degree of technical expertise.

Although signal encryption during airtime is a standard feature of GSM networks, a network operator can choose not to implement it. In this case, when a handset is turned on to access the host base station for services, the subscriber is vulnerable to the same eavesdropping attacks as with analog systems.

GSM signal encryption is done via a programmable smart card—the Subscriber Identification Module (SIM), which slips into a slot built into the handset. Each customer has a personal smart card holding personal details (short codes, frequently called numbers, etc.) as well as an international mobile subscriber identity (IMSI)—equivalent to Mobile Identification Number (MIN) for analog systems—and an authentication key on the microprocessor. Plugging the smart card into another phone will allow that phone to be used as if it were the customer's own. This is convenient in that the subscriber needs only to carry the SIM while traveling and plug it into a rental phone at the destination location where the difference in frequency would preclude use of the owner's phone.

However, it is still possible for a technically savvy fraudster to access a microprocessor's firmware for identification details and to reprogram them into other SIM cards. Counterfeiting SIM cards for GSM phones can be accomplished by programming computer chips using a laptop computer and other peripheral equipment. Although cloning fraud is possible, the technical expertise required is such that fewer people will be able to engage in this activity. And the nature of the process is such that it cannot be done on a massive scale cost-effectively.

While cloning may have hit a higher technological barrier on GSM and PCS networks, other types of fraud, such as technical fraud in international roaming markets and subscription fraud, are on the rise. With GSM networks, there is little protection from the theft of authentication keys. In fact, the use of mobile communications services by a growing subscriber base across an expanding network of roaming partners has created opportunities to defraud the digital networks to a degree not envisioned by participants in the early design phases of the technology.

The weak link in GSM networks is the challenge and response technique incorporated into the authentication process that allows the SIM to verify its IMSI by demonstrating knowledge of the authentication algorithm and the unique key, referred to as the Ki. The home system sends a random challenge to the handset, and only that handset can encrypt the challenge using both the algorithm and the Ki resident within the SIM assigned to that subscriber. Using the stored algorithm, the SIM generates the correct response back to the home system. In this scenario, the single point of failure is the authentication center in the home system.

Authentication should prevent fraudulent access to the wireless service, and the authentication center itself can be secured against internal and external theft of the IMSI and Ki sets. However, even this solution has its share of problems. For example, when high call volumes and nonsignaling traffic threaten to overburden the system—as when subscribers roam to international locations where the potential of intersystem bottlenecks is greatest—network administrators can reconfigure their systems to reuse the results of previous authentications or bypass the authentication process entirely to reduce the traffic back to the home system. As traffic increases, network administrators come under increasing pressure to alter or dispense with authentication out of the need to keep congestion down, customers happy, and revenues up. Unfortunately, tampering with authentication creates opportunities for fraud, which also can result in customer churn and lost revenues. The solution is to increase network capacity, which also costs money and can lead to customer churn if prices are increased.

Fraud Management Procedures

In addition to implementing new technologies to combat the fraudulent use of cellular phones, carriers and retail agents have implemented new fraud management procedures to help prevent the opportunity for abuse.

Service Providers Revenue losses to mobile phone fraud are so great that many service providers have an internal unit that is responsible for advising the company on ways to minimize opportunities for fraud, investigating instances of fraud and working with law enforcement authorities to prosecute criminals, and to educate corporate mobile phone users about ways to prevent fraud.

The following procedures, when adopted by service providers, can minimize opportunities for fraud:

- *Check clearance.* Allow a new customer to access the network only after the check has been cleared. In the United States, the Federal Reserve requires three business days to clear personal checks.

- *Credit card verification.* Allow a new customer to access the network only after the credit card has been verified. This can be done in a few minutes via card readers connected to the card issuer's regional database.

- *Welcome call.* Verify a new customer's identity before paying out a bonus or commission to the salesperson or licensed dealer. Call the customer on a fixed line to verify name and address, and confirm other details on the application form—such as mother's maiden name or spouse's initials. Ask for the name and address of the employer; if in doubt, call the workplace to verify employment information.

- *Welcome letter.* Send a letter to the customer in an envelope that does not give the appearance of being junk mail. In order to activate service, the recipient must call the customer service toll-free phone number specified in the letter.

- *Commission withholding.* Implement a 3-month withholding period for connection commissions to dealers. If a new subscriber fails to make the first payment within that period, the sale should be declared invalid and the connection commission canceled.

- *Premium services bar.* Only provide those services which a new subscriber has asked for. International calling and premium rate services would be automatically barred by default when the phone is given to the customer.

- *Early invoicing.* Send the first phone bill out early to minimize airtime fraud or expose commission fraud.

- *Credit alert.* Generate a report of phones with a high-volume of calls. A complex algorithm is used to detect fraudulent activity based on customer profiling, how long the phone has been connected, average monthly spend, and period of inactivity followed by hectic usage.

- *Nonphone alert.* This is a commission fraud involving a bogus phone, or the phone was never switched on. If the phone was switched on only two or three times over a period, the customer should be contacted as a precaution.

Subscribers Businesses and government agencies are very susceptible to cellular fraud. Not only do employees have no personal stake in taking basic antitheft precautions, but the phone bills of many organizations are so massive that fraudulent calls are difficult, if not impossible, identify. Most fraudulent phone calls are paid for by unknowing victims, often traveling executives who pass the invoices on to their companies without inspecting them. Nevertheless, organizations (and consumers) can minimize mobile phone fraud by taking the following precautions:

- If available, purchase phones equipped with authentication technology, which uses secret codes that are never transmitted across the airwaves.

- Ask the mobile service for a PIN that must be entered before a call can go through. Given the infrequency of its transmission, this code is not easily intercepted, and without it, cell phones cannot be cloned.

- If not needed, ask the service provider to shut off access to international service. This would prevent anyone from making illegal calls to other countries, where many fraudulent calls are directed.

- If possible, use a beeper to screen incoming calls. With the phone is turned off, it will not be transmitting its identifying numbers, thereby minimizing vulnerability to fraud. Since call retrieval usually entails air time charges, it is more economical to use the beeper as a screening device for calls that do not demand immediate attention.

- Do not lend the cell phone to anyone, and put it in a safe place when it is not in use. Even if the cell phone is insured against theft, most subscriber agreements limit coverage to one incident, after which the user must pay for a replacement phone.

- When leaving cars unattended, remove the handset and antenna to prevent mobile phone thieves from targeting the vehicle.

- Never leave the subscriber agreement or contract out in plain view; it usually contains such sensitive information as Social Security Number, drivers license number, credit card number, mobile number, and electronic serial number.

- Report all problems to the service provider immediately, especially if there is trouble placing calls. This may indicate that someone may have cloned the phone and is using it at that time. Other warning signs include difficulty retrieving voice-mail messages, excessive hangups, and callers receiving busy signals or wrong numbers.

While there is great progress in cracking down on mobile phone fraud in the United States and Europe, other countries are experiencing an increase in this kind of criminal activity. According to some experts, the international arena looms as the next frontier for mobile phone fraud, particularly in locations where U.S.-based multinationals are setting up shop and buying this kind of service. Foreign governments have just not been aggressive in finding and prosecuting this kind of criminal, they note. In some countries such as China, there are even operations dedicated to building cell phones that get illegally programmed and then sold on the black market.

Scanning the airwaves for cell phone identification numbers and programming them into clones has been made more difficult with sophisticated authentication processes and digital networks that support encryption. GSM will change the nature of fraud in the future. The authentication mechanism implemented with the SIM is forcing many would-be criminals to turn their attention to subscription fraud, which is still time-consuming to track down, even for the largest service providers.

Summary

Mobile phone fraud is an extension of "phone phreaking," the name given to a method of payphone fraud that originated in the 1960s and employed an electronic box held over the speaker. When a user is asked to insert money, the electronic box plays a rapid sequence of tones that fool the billing computer into thinking that money has been inserted. Since then, criminals and hackers have devoted time and money to develop and refine their techniques, applying them to mobile phones as well. The growing popularity and spread of the Internet to distribute tips and tricks to defraud carriers makes mobile phone fraud a billion-dollar international activity. Not only is mobile phone fraud lucrative, the stolen handsets have also provided anonymity to callers engaged in criminal activities. Apart from the called numbers, often no other evidence is left behind. The calls are charged to the legitimate subscribers' accounts.

See also

Wireless Security

FREQUENCY DIVISION MULTIPLE ACCESS

Three multiple access schemes are in use today, providing the foundation for mobile communications systems (Figure F-5):

- Frequency Division Multiple Access (FDMA), which serves the calls with different frequency channels
- Time Division Multiple Access (TDMA), which serves the calls with different time slots
- Code Division Multiple Access (CDMA), which serves the calls with different code sequences

All three technologies are widely used in cellular networks. FDMA is still used on some first-generation cellular analog networks, such as Advanced Mobile Phone Service (AMPS)[1] and TACS (Total Access Communications System).[1] TDMA is used on second-generation digital cellular networks, such as North American Digital Cellular and Global System for Mobile (GSM) communications. CDMA is also used on second-generation digital cellular networks, such as PCS 1900. Both TDMA and CDMA have been enhanced to support emerging third-generation networks.

Of the three, FDMA is the simplest and still the most widespread technology in use today for mobile communications. For example, FDMA is used in the CT2 system for

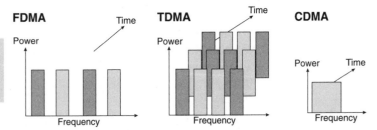

Figure F-5 Simple comparison of Frequency Division Multiple Access (FDMA), Time Division Multiple Access (TDMA), and Code Division Multiple Access (CDMA).

---

[1]There is a digital version of AMPS called D-AMPS (Digital-Advanced Mobile Phone Service). D-AMPS adds Time Division Multiple Access (TDMA) to AMPS to get three subchannels for each AMPS channel, tripling the number of calls that can be handled on a channel.

cordless telecommunications. The familiar cordless phone used in the home is representative of this type of system. It creates capacity by splitting bandwidth into radio channels in the frequency domain. In the initial call setup, the handset scans the available channels and locks onto an unoccupied channel for the duration of the call.

The traditional analog cellular systems, such as those based on AMPS, also use FDMA to derive the channel. In the case of AMPS, the channel is a 30-kHz "slice" of spectrum. Only one subscriber at a time is assigned to the channel. No other conversations can access the channel until the subscriber's call is finished, or until the call is handed off to a different channel in an adjacent cell.

The analog operating environment poses several problems. One is that the wireless devices are often in motion. Current analog technology does not deal with call handoffs very well, as evidenced by the high incidence of dropped calls. This environment is particularly harsh for data, which is less tolerant of transmission problems than voice. Whereas momentary signal fade, for instance, is a nuisance in voice communications, it may cause a data connection to drop.

Another problem with analog systems is their limited capacity. To increase the capacity of analog cellular systems, the 30-kHz channel can be divided into three narrower channels of 10 kHz each. This is the basis of the narrowband AMPS (N-AMPS) standard. However, this band-splitting technique incurs significant base station costs, and its limited growth potential makes it suitable only as a short-term solution.

While cell subdivision often is used to increase capacity, this solution has its limits. Since adjacent cells cannot use the same frequencies without risking interference, a limited number of frequencies are being reused at closer distances, which makes it increasingly difficult to maintain the quality of communications. Subdividing cells also increases the amount of overhead signaling that must be used to set up and manage the calls, which can overburden switch resources. In addition, property or rights of way for cell sites are difficult to obtain in metropolitan areas where traffic volume is highest and future substantial growth is anticipated.

These and other limitations of analog FM radio technology have led to the development of second-generation cellular systems based on digital radio technology and advanced networking principles. Providing reliable service in this dynamic environment requires digital radio systems that employ advanced signal processing technologies for modulation, error correction, and diversity. These capabilities are provided by TDMA and CDMA.

Summary

FM systems have supported cellular service for nearly 20 years, during which demand has finally caught up with the available capacity. Now first-generation cellular systems based on analog FM radio technology are rapidly being phased out in favor of digital systems that offer higher capacity, better voice quality, and advanced call handling features. TDMA- and CDMA-based systems are contending for acceptance among analog cellular carriers worldwide.

See also

   Code Division Multiple Access
   Cordless Telecommunications
   Time Division Multiple Access

G

GLOBAL MARITIME DISTRESS AND SAFETY SYSTEM

The Global Maritime Distress and Safety System (GMDSS) is an internationally recognized distress and radio communication safety system for ships that replaced the previous ship-to-ship safety system that relied on manual Morse code operating on 500 kHz and voice radiotelephony on Channel 16 and on 2182 kHz.

The GMDSS is an automated ship-to-shore system using satellites and digital selective calling technology. The GMDSS is mandated for ships internationally by the International Maritime Organization (IMO) Safety of Life at Sea Convention (SOLAS), 1974, as amended in 1988, and carries the force of an international treaty. The procedures governing use are contained in the International Telecommunications Union (ITU) recommendations and in the International Radio Regulations and also carry the force of an international treaty.

There are many advantages of the GMDSS over the previous system, including

- Provides worldwide ship-to-shore alerting; it is not dependent on passing ships.

- Simplifies radio operations; alerts may be sent by "two simple actions."

- Ensures redundancy of communications; it requires two separate systems for alerting.

- Enhances search and rescue; operations are coordinated from shore centers.

- Minimizes unanticipated emergencies at sea; maritime safety broadcasts are included.

- Eliminates reliance on a single person for communications; it requires at least two licensed GMDSS radio operators and typically two maintenance methods to ensure distress communications capability at all times.

Radio officers (trained in manual Morse code) are not part of the GMDSS regulations or system. In lieu of a single radio officer, the GMDSS regulations require at least two GMDSS radio operators and a GMDSS maintainer if the ship elects at-sea repair as one of its maintenance options.

The Federal Communications Commission (FCC) requires two licensed radio operators to be aboard all GMDSS-certified ships, one of which must be dedicated to communications during a distress situation. The radio operators must be holders of a GMDSS Radio Operator's License. The GMDSS radio operator is an individual licensed to handle radio communications aboard ships in compliance with the GMDSS regulations, including basic equipment and antenna adjustments. The GMDSS radio operator need not be a radio officer.

Another IMO convention requires all masters and mates to hold the GMDSS Radio Operator's License and attend a two-week training course and demonstrate competency with operation of the GMDSS equipment. These requirements also would carry to any person employed specifically to act as a dedicated radio operator if the ship elected to carry such a position.

The international GMDSS regulations apply to "compulsory" ships, including

- Cargo ships of 300 gross tons and over when traveling on international voyages or in the open sea
- All passenger ships carrying more than 12 passengers when traveling on international voyages or in the open sea

Fishing vessels to which GMDSS previously applied are under a waiver until a date to be announced in the future. The waiver is conditioned on the requirement that, for the duration of the waiver, fishing vessels of 300 gross tons or greater continue to carry a 406-MHz receiver and survival craft equipment including at least three portable VHF radiotelephones and two 9-GHz radar transponders. The GMDSS regulations do not apply to vessels operating exclusively on the Great Lakes.

Summary

GMDSS was mandated by international treaty obligations. In 1988, the International Maritime Organization (IMO), an organization of the United Nations, amended the Safety of Life at Sea (SOLAS) Convention to implement the GMDSS worldwide. The United States has been a strong advocate of the GMDSS internationally. In January 1992, the FCC adopted the GMDSS regulations for U.S. compulsory vessels. The only way to ensure compliance with GMDSS requirements is to use equipment that has been specifically approved by the FCC for GMDSS use.

See also

Wireless E911

GENERAL MOBILE RADIO SERVICE

General Mobile Radio Service (GMRS) is one of several personal radio services; specifically, it is a personal two-way ultra-high-frequency (UHF) voice communication service that can be used to facilitate the activities of an individual's immediate family members. This FCC-licensed service has a communications range of 5 to 25 miles and cannot be used to make telephone calls.

A GMRS system consists of station operators, a mobile station (often composed of several mobile units), and sometimes one or more land stations. The classes of land stations are base station, mobile relay station (also known as a "repeater"), and control stations. A small base station is one that has an antenna no more than 20 feet above the ground or above the tree on which it is mounted and transmits with no more than 5 watts.

Users communicate with a GMRS radio unit (Figure G-1) over the general area of their residence in an urban or rural

Figure G-1 Motorola's TalkAbout Distance DPS is a hand-held GMRS radio that weighs 11.7 ounces and runs on a rechargeable NiCad battery or 6 AA batteries. A GMRS license, issued by the FCC, and a fee are required for use of this radio.

area. This area must be within the territorial limits of the 50 United States, the District of Columbia, and the Caribbean and Pacific insular areas. In transient use, mobile station units from one GMRS system may communicate through a mobile relay station in another GMRS system with the permission of its licensee.

There are 23 GMRS channels. None of the GMRS channels are assigned for the exclusive use of any system. License applicants and licensees must cooperate in the selection and use of the channels in order to make the most effective use of them and to reduce the possibility of interference.

Any mobile station or small base station in a GMRS system operating in the simplex mode may transmit voice-type emissions with no more than 5 watts on the following 462-MHz channels: 462.5625, 462.5875, 462.6125, 462.6375, 462.6625, 462.6875, and 462.7125 MHz. These channels are shared with the Family Radio Service (FRS).

Any mobile station in a GMRS system may transmit on the 467.675-MHz channel to communicate through a mobile relay station transmitting on the 462.675-MHz channel. The communications must be for the purpose of soliciting or rendering assistance to a traveler or for communicating in an emergency pertaining to the immediate safety of life or the immediate protection of property.

Each GMRS system license assigns one or two of eight possible channels or channel pairs (one 462-MHz channel and one 467-MHz channel spaced 5 MHz apart) as requested by the applicant. Applicants for GMRS system licenses are advised to investigate or monitor to determine the best available channel(s) before making their selection. Each applicant must select the channel(s) or channel pair(s) for the stations in the proposed system from the following list:

- For a base station, mobile relay station, fixed station, or mobile station: 462.550, 462.575, 462.600, 462.625, 462.650, 462.675, 462.700, and 462.725 MHz.

- For a mobile station, control station, or fixed station in a duplex system: 467.550, 467.575, 467.600, 467.625, 467.650, 467.675, 467.700, and 467.725 MHz.

GMRS system station operators must cooperate in sharing the assigned channel with station operators in other GMRS systems by monitoring the channel before initiating transmissions, waiting until communications in progress are completed before initiating transmissions, engaging in only permissible communications, and limiting transmissions to the minimum practical transmission time.

Summary

Any individual 18 years of age or older who is not a representative of a foreign government is eligible to apply for a GMRS system license. There is a filing fee for new licenses and license renewals. For general information regarding the fee and filing requirements, contact the FCC's Consumer Center toll free at 1-888-225-5322.

See also

Citizens Band Radio Service

Family Radio Service

Low-Power Radio Service

GENERAL PACKET RADIO SERVICE

Many carriers will take an interim step to third generation (3G), referred to as 2.5G, that uses the Internet Protocol (IP) to provide fast access to data networks via General Packet Radio Service (GPRS) technology. Compared to circuit-switched data (CSD), which operates at up to 14.4 kbps and high-speed circuit-switched data (HSCSD), which operates at up to 43.2 kbps, GPRS uses packet-switching technology to transmit short bursts of data over an IP-based network to deliver speeds of up to 144 kbps over an "always on" wireless connection.

True 3G networks based on enhanced data rates for GSM evolution (EDGE) technology deliver data at speeds of up to 384 kbps. EDGE is a step beyond GPRS that will allow up to three times higher throughput compared to GSM/GPRS using the same bandwidth. Carriers in the United States have been moving toward 3G for several years by overlaying various technologies onto their existing networks to enhance their data-handling capabilities.

For carriers with TDMA-based networks, the first step to offering true 3G services is to deploy Global Systems for Mobile communications (GSM) and then GPRS. The GPRS enhancement to GSM can support peak network speeds of wireless data transmissions between 64 and 170 kbps depending on the various claims of hardware vendors. The new GSM/GPRS network does not replace existing TDMA networks; carriers will continue supporting these networks long into the future to service their voice customers. Eventually, all TDMA customers will be migrated to GSM/GPRS.

Once the GSM/GPRS overlay is in place in a market, the carriers can upgrade their networks with EDGE-compliant software to boost data transmission rates to as much as 384 kbps and begin the availability of true 3G services.

Carriers whose wireless networks are based on CDMA will take a different technology path to 3G, going through CDMA2000, before eventually arriving at Wideband Code Division Multiple Access (W-CDMA). Both EDGE and W-CDMA offer a migration path to the global standard Universal Mobile Telecommunications System (UMTS).

Coverage for 2.5/3G services is still ramping up, despite the impressive figures thrown out by individual carriers. The next step is for service providers to engage in more roaming arrangements, which is a way to save costs, reduce time to market, and add value to attract more customers.

The data speed of 2.5/3G services is determined by many factors, including the equipment and software in the wireless network, the distance of the user from the nearest base station, and how fast the user may be moving. The claimed speed of the service is rarely, if ever, achieved in the real-world operating environment.

The pricing plans and price points differ by carrier, from a simple add-on to existing digital voice plans for a basic data service, to tiered pricing plans based on actual data usage.

Depending on the applications, users can opt for 2.5/3G cell phones with multimedia messaging capabilities. Alternatively, users with heavy messaging and file transfer requirements may opt for PC cards for notebooks and personal digital assistants (PDAs).

The choice of 2.5G platform depends on whether the carrier has a TDMA- or CDMA-based network. Both technologies are capable of eventual migration to full 3G and at a higher level will be able to interoperate in compliance with the global IMT-2000 initiative.

Some service providers intend to support Wireless Fidelity (Wi-Fi), as well as GPRS. Wi-Fi networks are based on the IEEE 802.11b Standard for Ethernet, which operates in the unlicensed 2.4-GHz band to provide a maximum stated speed of 11 Mbps. In actual operation, however, Wi-Fi offers between 5 and 6 Mbps. Some carriers are looking at ways to offer both Wi-Fi and 3G from the same device to meet the diverse needs of customers.

The existing GPRS and upcoming EDGE networks provide wide area coverage for applications where customers want constant access to such as e-mail and calendar, whereas Wi-Fi networks will be available in convenient locations where customers are likely to spend time accessing larger data files.

Summary

The United States lags behind the rest of the world when it comes to wireless technologies for a number of reasons. The telecommunications infrastructure in the United States is more developed than in many European and Asian countries. As a result, the demand for wireless devices has been lower in the United States because consumers have other low-cost options. Also, the United States has a number of competing technical standards for digital services, while European and Asian countries are predominately centered on GSM. In the United States, carriers have only recently adopted GSM. Now that carriers in the United States have mapped out their migration strategies, they have been busy positioning their networks to 2.5G, investing billions of dollars for infrastructure upgrades, with billions more committed to go the rest of the way to 3G. In the process, they are betting they can attract millions of new customers who will want high-speed wireless data on their mobile phones, notebooks, and PDAs.

See also

Code Division Multiple Access

Enhanced Data Rates for Global Evolution

Global System for Mobile Telecommunications

Time Division Multiple Access

Universal Mobile Telecommunications System

GLOBAL POSITIONING SYSTEM

The Global Positioning System (GPS) is a network of 24 Navstar[1] satellites orbiting Earth at 11,000 miles up. Established by the U.S. Defense Department for military applications, access to GPS is now free to all users, including those in other countries. The system's positioning and timing data are used for a variety of applications, including air, land,

---

[1]Originally, NAVSTAR was an acronym for Navigation System with Timing and Ranging.

and sea navigation; vehicle and vessel tracking; surveying and mapping; and asset and natural resource management. With military accuracy restrictions lifted in May 2000, the GPS can now pinpoint the exact location of people as they move about with their receivers powered on. This development has ushered in a wave of new commercial applications for GPS.

GPS Components

The first GPS satellite was launched in 1978. The first 10 satellites were developmental satellites. From 1989 to 1993, 23 production satellites were launched. The launch of the twenty-fourth satellite in 1994 completed the $13 billion constellation. The satellites are positioned so that signals from 6 of them can be received nearly 100 percent of the time at any point on earth.

The GPS consists of satellites, receivers, and ground control systems. The satellites transmit signals (1575.42 MHz) that can be detected by GPS receivers on the ground. These receivers can be portable or mounted in ships, planes, or cars to provide exact position information, regardless of weather conditions. They detect, decode, and process GPS satellite signals to give the precise position of the user.

The GPS control or ground segment consists of five unmanned monitor stations located in Hawaii, Kwajalein in the Pacific Ocean, Diego Garcia in the Indian Ocean, Ascension Island in the Atlantic Ocean, and Colorado Springs, Colorado. There is also a master ground station at Falcon Air Force Base in Colorado Springs, Colorado, and four large ground antenna stations that broadcast signals to the satellites. The stations also track and monitor the GPS satellites.

System Operation

With GPS, signals from several satellites are triangulated to identify the exact position of the user. To triangulate, GPS measures distance using the travel time of a radio message from the satellite to a ground receiver. To measure travel time, GPS uses very accurate clocks in the satellites. Once the distance to a satellite is known, knowledge of the satellite's location in space is used to complete the calculation. GPS receivers on the ground have an "almanac" stored in their computer memory that indicates where each satellite will be in the sky at any given time. GPS receivers calculate for ionosphere and atmosphere delays to further tune the position measurement.

To make sure both satellite and receiver are synchronized, each satellite has four atomic clocks that keep time to within 3 nanoseconds, or 3 billionths of a second. For cost savings, the clocks in the ground receivers are not that accurate. To compensate, an extra satellite range measurement is taken. Trigonometry says that if three perfect measurements locate a point in three-dimensional space, then a fourth measurement can eliminate any timing offset. This fourth measurement compensates for the receiver's imperfect synchronization.

The ground unit receives the satellite signals, which travel at the speed of light. Even at this speed, the signals take a measurable amount of time to reach the receiver. The difference between when the signals are sent and the time they are received, multiplied by the speed of light, enables the receiver to calculate the distance to the satellite. To measure precise latitude, longitude, and altitude, the receiver measures the time it took for the signals from several satellites to get to the receiver (Figure G-2).

GPS uses a system of coordinates called the Worldwide Geodetic System 1984 (WGS-84). This is similar to the latitude and longitude lines that are commonly seen on large wall maps used in schools. The WGS-84 system provides a built-in, standardized frame of reference, enabling receivers from any vendor to provide exactly the same positioning information.

GPS Applications

The GPS system has amply proven itself in military applications, most notably in Operation Desert Storm where U.S.

Navstar Satellites

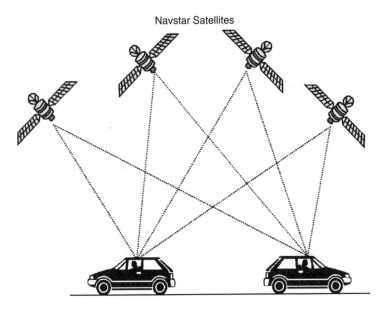

GPS-equipped Vehicles

Figure G-2   Signals from four satellites, captured by a vehicle's onboard GPS receiver, are used to determine precise location information.

and allied troops faced a vast, featureless desert. Without a reliable navigation system, sophisticated troop maneuvers could not have been performed. This could have prolonged the operation well beyond the 100 hours it actually took.

With GPS, troops were able to go places and maneuver in sandstorms or at night when even the troops who were native to the area could not. Initially, more than 1000 portable commercial receivers were purchased for their use. The demand was so great that before the end of the conflict, more than 9000 commercial receivers were in use in the Gulf region. They were carried by ground troops and attached to vehicles, helicopters, and aircraft instrument panels. GPS receivers were used in several aircraft, including F-16 fighters, KC-135 tankers, and B-52s. Navy ships used GPS receivers for rendezvous, minesweeping, and aircraft operations.

While the GPS system was developed originally to meet the needs of the military community, new ways to use its capabilities are continually being found, from the exotic to the mundane. Among the former is the use of GPS for wildlife management. Endangered species such as Montana elk and Mojave Desert tortoises have been fitted with tiny GPS receivers to help determine population distribution patterns and possible sources of disease. In Africa, GPS receivers are used to monitor the migration patterns of large herds for a variety of research purposes.

Handheld GPS receivers are now used routinely in field applications that require precise information gathering, including field surveying by utility companies, mapping by oil and gas explorers, and resource planning by timber companies.

GPS-equipped balloons are used to monitor holes in the ozone layer over the polar ice caps. Air quality is being monitored using GPS receivers. Buoys tracking major oil spills transmit data using GPS. Archaeologists and explorers are using the system to mark remote land and ocean sites until they can return with proper equipment and funding.

Vehicle tracking is one of the fastest-growing GPS applications. GPS-equipped fleet vehicles, public transportation systems, delivery trucks, and courier services use receivers to monitor their locations at all times.

GPS data are especially useful to consumers when they are linked with digital mapping. Accordingly, some automobile manufacturers are offering moving-map displays guided by GPS receivers as an option on new vehicles. The displays can even be removed and taken into a home to plan a trip. Some GPS-equipped vehicles give directions to drivers on display screens and through synthesized voice instructions. These features enable drivers to get where they want to go more rapidly and safely than has ever been possible before. GPS receivers are also included in newer mobile phones, and add-on receivers are available for hand-held computers, such as the Palm III (Figure G-3).

Figure G-3 Palm III users can clip on Rand McNally's StreetFinder GPS receiver, turning the unit into a portable navigational tool. Customized maps from the StreetFinder software can be downloaded to the Palm III, along with address-to-address directions accessible via the Internet. With trip information stored on the Palm III, the GPS receiver enables the user to track travel progress and manage itinerary changes en route.

GPS is also helping save lives. Many police, fire, and emergency medical service units are using GPS receivers to determine the police car, fire truck, or ambulance nearest to an emergency, enabling the quickest possible response in life-or-death situations.

When GPS data are used in conjunction with geographic data collection systems, it is possible to instantaneously arrive at submeter positions together with feature descriptions to compile highly accurate geographic information systems (GIS). When used by cities and towns, for example, GPS can help in the management of the geographic assets summarized in Table G-1.

Some government agencies, academic institutions, and private companies are using GPS to determine the location of a multitude of features, including point features such as pollutant discharges and water supply wells, line features such as roads and streams, and area features such as waste lagoons and property boundaries. Before GPS, such features had to be located with surveying equipment, aerial photographs, or satellite imagery. With GPS, the precise location of these and other features can be determined with a hand-held GPS receiver.

GPS and Cellular

GPS technology is even being used in conjunction with cellular technology to provide value-added services. With the push of a button on a cellular telephone, automobile drivers and operators of commercial vehicles in some areas can talk to a service provider and simultaneously signal their position, emergency status, or equipment failure information to auto clubs, security services, or central dispatch services.

TABLE G-1 Types of Geographic Assets That Can Be Managed with the Aid of the Global Positioning System (GPS)

Point Features	Line Features	Area Features
Signs	Streets	Parks
Manhole covers	Sidewalks	Landfills
Fire hydrants	Fitness trails	Wetlands
Light poles	Sewer lines	Planning zones
Storm drains	Water lines	Subdivisions
Driveways	Bus routes	Recycling centers

This is possible with Motorola's Cellular Positioning and Emergency Messaging Unit, for example, which offers mobile security and tracking to those who drive automobiles and/or operate fleets. The system is designed for sale to systems integrators that configure consumer and commercial systems that operate via cellular telephony. The Cellular Positioning and Emergency Messaging Unit communicates GPS-determined vehicle position and status, making it suited for use in systems that support roadside assistance providers, home security monitoring firms, cellular carriers, rental car companies, commercial fleet operators, and auto manufacturers seeking a competitive advantage.

As an option, the OnStar system is available for select vehicles manufactured by General Motors, which uses a GPS receiver in conjunction with analog cellular phone technology to provide a variety of travel assistance services, including emergency response. At the push of a button, a cellular call is placed to an OnStar operator. Although digital technology is more advanced, OnStar uses analog cellular because it has the broadest geographic coverage in the United States. Over 90 percent of the country is covered by the analog system, whereas digital coverage is less than 30 percent. OnStar has worked to "clear" the OnStar emergency button call through all analog cellular phone companies so that it will go through no matter which carrier is used locally. GPS comes into play by providing the OnStar operator with the precise location of the vehicle.

Summary

Because of its accuracy, GPS is rapidly becoming the location data-collection method of choice for a variety of commercial, government, and military applications. GPS certainly has become an important and cost-effective method for locating terrestrial features too numerous or too dynamic to be mapped by traditional methods. Although originally funded by the U.S. Defense Department, access to the GPS network is free to all users in any country. This has encouraged applications development and created an entirely new consumer market, particularly in the area of vehicular location and highway navigation.

See also

Satellite Communications

GLOBAL SYSTEM FOR MOBILE (GSM) TELECOMMUNICATIONS

Global System for Mobile (GSM) telecommunications—formerly known as Groupe Spéciale Mobile, for the group that started developing the standard in 1982—was designed from the beginning as an international digital cellular service. It was intended that GSM subscribers should be able to cross national borders and find that their mobile services crossed with them. Today, GSM is well established in most countries, with the highest concentration of service providers and users in Europe.

Originally, the 900-MHz band was reserved for GSM services. Since GSM first entered commercial service in 1992, it has been adapted to work at 1800 MHz for the Personal Communications Networks (PCN) in Europe and at 1900 MHz for Personal Communications Services (PCS) in the United States.

GSM Services

GSM telecommunication services are divided into teleservices, bearer services, and supplementary services.

Teleservices The most basic teleservice supported by GSM is telephony. There is an emergency service in which the nearest emergency service provider is notified by dialing three digits (similar to 911). Group 3 fax, an analog method described in ITU–T Recommendation T.30, is also supported by GSM through the use of an appropriate fax adapter.

Bearer Services A unique feature of GSM compared to older analog systems is the Short Message Service (SMS). SMS is a bidirectional service for sending short alphanumeric messages (up to 160 bytes) in a store-and-forward manner. For point-to-point SMS, a message can be sent to another subscriber to the service, and an acknowledgment of receipt is provided to the sender. SMS also can be used in cell broadcast mode for sending messages such as traffic updates or news updates. Messages can be stored in a smart card called the "Subscriber Identity Module" (SIM) for later retrieval.

Since GSM is based on digital technology, it allows synchronous and asynchronous data to be transported as a bearer service to or from an Integrated Services Digital Network (ISDN) terminal. The data rates supported by GSM are 300, 600, 1200, 2400, and 9600 bps. Data can use either the transparent service, which has a fixed delay but no guarantee of data integrity, or a nontransparent service, which guarantees data integrity through an automatic repeat request (ARQ) mechanism but with variable delay.

GSM has much more potential in terms of supporting data. The GSM standard for high-speed circuit-switched data (HSCSD) enables mobile phones to support data rates of up to 38.4 kbps, compared with 9.6 kbps for regular GSM networks. Transmission speeds of up to 171.2 kbps are available with mobile phones that support the GSM standard for General Packet Radio Service (GPRS). The high bandwidth is achieved by using eight timeslots, or voice channels, simultaneously. GPRS facilitates several new applications, such as Web browsing over the Internet.

Both HSCSD and GPRS are steps toward the third generation (3G) of mobile technology, called International Mobile Telecommunications (IMT), a framework for advanced mobile telephony that seeks to harmonize all national and regional standards for global interoperability, which is in various phases of implementation around the world. IMT includes standards that eventually will allow mobile phones to operate at up to 2 Mbps, enabling broadband applications such as videoconferencing.

Supplementary Services Supplementary services are provided on top of teleservices or bearer services and include such features as caller identification, call forwarding, call waiting, and multiparty conversations. There is also a lock-out feature that prevents the dialing of certain types of calls, such as international calls.

Network Architecture

A GSM network consists of the following elements: mobile station, base station subsystem, and mobile services switching center (MSC). Each GSM network also has an operations and maintenance center that oversees the proper operation and setup of the network. There are two air interfaces: the Um interface is a radio link over which the mobile station and the base station subsystem communicate; the A interface is a radio link over which the base station subsystem communicates with the MSC.

The Mobile Station The mobile station (MS) consists of the radio transceiver, display and digital signal processors, and the SIM. The SIM provides personal mobility so that the subscriber can have access to all services regardless of the terminal's location or the specific terminal used. By removing the SIM from one GSM cellular phone and inserting it into another GSM cellular phone, the user is able to receive calls at that phone, make calls from that phone, or receive other subscribed services. The SIM card may be protected against unauthorized use by a password or personal identification number (PIN).

An International Mobile Equipment Identity (IMEI) number uniquely identifies each mobile station. The SIM card contains an International Mobile Subscriber Identity (IMSI) number identifying the subscriber, a secret key for authentication, and other user information. Since the IMEI and IMSI are independent, this arrangement provides users with a high degree of security.

The SIM comes in two form factors: credit-card size (ISO format) or postage-stamp size (plug-in format). Both sizes are offered together to fit any kind of cell phone the user happens to have (Figure G-4). There is also a micro SIM adapter (MSA) that allows the user to change back from the plug-in format SIM card into an ISO format card.

The SIM cards also allow services to be individually tailored and updated over the air and activated without requiring the user to find a point-of-sale location in order to carry out the update. SIM cards' remote control and modification possibilities allow the carriers to offer their subscribers such new, interactive services as remote phonebook loading and remote recharging of prepaid SIMs. The cards also can contain company/private or parent/children subscriptions with separate PIN codes that can be changed over the air.

Figure G-4 SIM issued to subscribers of Vodafone, the largest cellular service provider in the United Kingdom. Within the larger card is a detachable postage-stamp-sized SIM. Both use the same gold contact points.

Base Station Subsystem The base station subsystem consists of two parts: the base transceiver station (BTS) and the base station controller (BSC). These communicate across the A–bis interface, enabling operation between components made by different suppliers.

The base transceiver station contains the radio transceivers that define a cell and handles the radio link protocols with the mobile stations. In a large urban area, there typically will be a number of BTSs to support a large subscriber base of mobile service users.

The base station controller provides the connection between the mobile stations and the mobile service switching center (MSC). It manages the radio resources for the BTSs, handling such functions as radio channel setup, frequency hopping, and handoffs. The BSC also translates the 13-kbps voice channel used over the radio link to the standard 64-kbps channel used by the land-based Public Switched Telephone Network (PSTN) or ISDN.

Mobile Services Switching Center The mobile services switching center (MSC) acts like an ordinary switching node on the PSTN or ISDN and provides all the functionality needed to handle a mobile subscriber, such as registration, authentication, location updating, handoffs, and call routing to a roaming subscriber. These services are provided in conjunction with several other components, which together form the network subsystem. The MSC provides the connection to the public network (PSTN or ISDN) and signaling between various network elements that use Signaling System 7 (SS7).

The MSC contains no information about particular mobile stations. This information is stored in two location registers that are essentially databases. The Home Location Register (HLR) and Visitor Location Register (VLR), together with the MSC, provide the call routing and roaming (national and international) capabilities of GSM.

The HLR contains administrative information for each subscriber registered in the corresponding GSM network, along with the current location of the mobile device. The current location of the mobile device is in the form of a Mobile Station Roaming Number (MSRN), which is a regular ISDN number used to route a call to the MSC where the mobile device is currently located. Only one HLR is needed per GSM network, although it may be implemented as a distributed database.

The Visitor Location Register (VLR) contains selected administrative information from the HLR that is necessary for call control and provision of the subscribed services for each mobile device currently located in the geographic area controlled by the VLR.

There are two other registers that are used for authentication and security purposes. The Equipment Identity Register (EIR) is a database that contains a list of all valid mobile equipment on the network, where each mobile station is identified by its IMEI. An IMEI is marked as invalid if it has been reported stolen or is not type approved. The authentication center is a protected database that stores a copy of the secret key stored in each subscriber's SIM card, which is used for authentication.

Channel Derivation and Types

Since radio spectrum is a limited resource shared by all users, a method must be devised to divide up the bandwidth among as many users as possible. The method used by GSM is a combination of Time and Frequency Division Multiple Access (TDMA/FDMA).

The FDMA part involves the division by frequency of the total 25-MHz bandwidth into 124 carrier frequencies of 200-kHz bandwidth. One or more carrier frequencies are then assigned to each base station. Each of these carrier frequencies is then divided in time, using a TDMA scheme, into eight time slots. One time slot is used for transmission by the mobile device and one for reception. They are separated in time so that the mobile unit does not receive and transmit at the same time.

Within the framework of TDMA, two types of channels are provided: traffic channels and control channels. Traffic channels carry voice and data between users, while the control channels carry information that is used by the network for supervision and management.

Among the control channels are the following:

- *Fast Associated Control Channel (FACCH).* Robs slots from traffic channels to transmit power control and call handoff messages.

- *Broadcast Control Channel (BCCH).* Continually broadcasts on the downlink, information including base station identity, frequency allocations, and frequency hopping sequences.

- *Stand-Alone Dedicated Control Channel (SDCCH).* Used for registration, authentication, call setup, and location updating.

- *Common Control Channel (CCCH)* Comprises three control channels used during call origination and call paging.

- *Random Access Channel (RACH).* Used to request access to the network.

- *Paging Channel (PCH).* Used to alert the mobile station of an incoming call.

Authentication and Security

Since radio signals can be accessed by virtually anyone, authentication of users to prove their identity is a very important feature of a mobile network. Authentication involves two functional entities, the SIM card in the mobile unit and the authentication center (AC). Each subscriber is given a secret key, one copy of which is stored in the SIM card and the other in the AC. During authentication, the AC generates a random number that it sends to the mobile unit. Both the mobile unit and the AC then use the random number, in conjunction with the subscriber's secret key and an encryption algorithm called

A3, to generate a number that is sent back to the AC. If the number sent by the mobile unit is the same as the one calculated by the AC, the subscriber is authenticated.

The calculated number is also used, together with a TDMA frame number and another encryption algorithm called A5, to encrypt the data sent over the radio link, preventing others from listening in. Encryption provides an added measure of security, since the signal is already coded, interleaved, and transmitted in a TDMA manner, thus providing protection from all but the most technically astute eavesdroppers.

Another level of security is performed on the mobile equipment, as opposed to the mobile subscriber. As noted, a unique IMEI number is used to identify each GSM terminal. A list of IMEIs in the network is stored in the EIR. The status returned in response to an IMEI query to the EIR is one of the following:

- *White listed.* Indicates that the terminal is allowed to connect to the network.

- *Gray listed.* Indicates that the terminal is under observation from the network for possible problems.

- *Black listed.* Indicates that the terminal either has been reported as stolen or is not type approved (i.e., not the correct type of terminal for a GSM network). Such terminals are not allowed to connect to the network.

Summary

By mid-2001, there were 404 GSM networks in operation in 171 countries, providing mobile telephone service to 538 million subscribers. GSM accounts for 70 percent of the world's digital market and 65 percent of the world's wireless market. One new subscriber signs up for service every second of the day and night. GSM in North America has some 11 million customers across the United States and Canada. GSM service is available in 6500 cities in 48 states, the District of Columbia, and six Canadian provinces. According to the North American GSM Alliance, GSM coverage reaches more than half the Canadian population and two-thirds of the U.S. population.

See also

Digital Enhanced Cordless Telecommunications
International Mobile Telecommunications
PCS 1900

HERTZ

The term *hertz* is a measure of frequency, or the speed of transmission. The frequency of electromagnetic waves generated by radio transmitters is measured in cycles per second (cps), but this designation was officially changed to hertz (Hz) in 1960.

An electromagnetic wave is composed of complete cycles. The number of cycles that occur each second gives radio waves their frequency, while the peak-to-peak distance of the waveform gives the amplitude of the signal (Figure H-1).

The frequency of standard speech is between 3000 cycles per second, or 3 kilohertz (kHz), and 4000 cycles per second, or 4 kHz. Some radio waves may have frequencies of many millions of hertz (megahertz, or MHz), and even billions of hertz (gigahertz, or GHz). Table H-1 provides the range of frequencies and their band classification.

The term *hertz* was adopted in 1960 by an international group of scientists and engineers at the General Conference of Weights and Measures in honor of Heinrich R. Hertz (1857–1894), a German physicist (Figure H-2). Hertz is best known for proving the existence of electromagnetic waves,

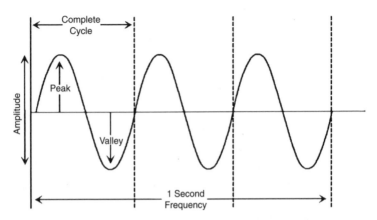

Figure H-1 Each cycle per second equates to 1 hertz (Hz). In this case, 3 cycles occur in 1 second, which equates to 3 Hz.

TABLE H-1 List of Frequency Ranges and Corresponding Band Classifications

Frequency	Band Classification
Less than 30 kHz	Very low frequency (VLF)
30 to 300 kHz	Low frequency (LF)
300 kHz to 3 MHz	Medium frequency (MF)
3 to 30 MHz	High frequency (HF)
30 to 300 MHz	Very high frequency (VHF)
300 MHz to 3 GHz	Ultrahigh frequency (UHF)
3 to 30 GHz	Superhigh frequency (SHF)
More than 30 GHz	Extremely high frequency (EHF)

which had been predicted by British scientist James Clerk Maxwell in 1864.

Hertz used a rapidly oscillating electric spark to produce UHF waves. These waves caused similar electrical oscillations in a distant wire loop. The discovery of electromagnetic waves and how they could be manipulated paved the way for the development of radio, microwave, radar, and other forms of wireless communication.

Figure H-2 German physicist Heinrich R. Hertz (1857–1894) proved the existence of electromagnetic waves, which led to the development of radio, microwave, radar, and other forms of wireless communication.

Summary

As interest in electromagnetic waves grew in the nineteenth century, a physical model to describe it was proposed. It was suggested that electromagnetic waves, including light, were like sound waves but that they propagated through some previously unknown medium called the "luminiferous ether" that filled all unoccupied space throughout the universe. The experiments of Albert A. Michelson and Edward W. Morley in 1887 proved that the ether did not exist. Albert Einstein's theory of relativity, proposed in 1905, eliminated the need for a light-transmitting medium, so today the term *ether* is used only in a historical context, as in the term *Ethernet*.

See also

Decibel

HOME RADIO-FREQUENCY (RF) NETWORKS

One method of implementing a wireless network in the home is to use products that adhere to the standards of the Home Radio Frequency Working Group. HomeRF is positioned as a global extension of Digitally Enhanced Cordless Telephony (DECT), the popular cordless phone standard that allows different brands to work together so that certified handsets from one vendor can communicate with base stations from another. DECT has been largely confined to Europe because its native 1.9-GHz frequency band requires a license elsewhere, but HomeRF extends DECT to other regions by using the license-free 2.4-GHz frequency band. It also adds functionality by blending several industry standards, including IEEE 802.11 frequency hopping for data and DECT for voice. This convergence makes HomeRF useful for broadband households.

As more PCs, peripherals, and intelligent devices are installed in the home, and as network connections proliferate, users are faced with new opportunities for accessing information as well as challenges for sharing resources. For example, users want to

- Access information delivered via the Internet from anywhere in the home
- Share files between PCs and share access to peripherals no matter where they are located within the home
- Control electrical systems and appliances whether in, around, or away from home

- Effectively manage communications channels for phone, fax, and Internet usage

Each of these capabilities requires a common connection between the various devices and networks found in the home. However, in order to truly be effective, any home network must meet certain criteria:

- It must not require additional home wiring. Most existing homes are not wired for networking, and retrofitting them would too labor-intensive and expensive. A wireless solution is a viable alternative.

- The wireless connections must be immune to interference, especially with the growing number of wireless devices and appliances emitting RF noise in the home.

- The range of the wireless connection must be adequate to allow devices to communicate from anywhere within and around a typical family home.

- The network must be safe and protected from unwanted security breaches.

- It must be easy to install, configure, and operate for non-technical users. Most home users do not have the expertise to handle complex network installation and configuration procedures.

- The entire system must be easily and spontaneously accessible—anytime and from anywhere in or even away from the home.

These issues have been addressed by a consortium of vendors called the Home Radio Frequency Working Group (HomeRF WG), which has developed a platform for a broad range of interoperable consumer devices. Its specification, called the Shared Wireless Access Protocol (SWAP), is an open standard that allows PCs, peripherals, cordless telephones, and other consumer electronic devices to communicate and interoperate with one another without the complexity and expense associated with installing new wires.

The SWAP is designed to carry both voice and data traffic and to interoperate with the Public Switched Telephone Network (PSTN) and the Internet. It operates in the 2.4-GHz ISM (industrial, scientific, medical) band and uses frequency-hopping spread-spectrum radio for security and reliability. The SWAP technology was derived from extensions of DECT and wireless local area network (LAN) technologies to enable a new class of home cordless services. It supports both a Time Division Multiple Access (TDMA) service to provide delivery of interactive voice and other time-critical services, and a Carrier Sense Multiple Access/Collision Avoidance (CSMA-CA) service for delivery of high-speed packet data. Table H-2 summarizes the main characteristics of HomeRF.

Applications

The SWAP specification provides the basis for a broad range of new home networking applications, including

TABLE H-2   HomeRF Characteristics

Frequency-hopping network	50 hops/second
Frequency range	2.4-GHz ISM band
Transmission power	100 mW
Data rate	1.6 Mbps with HomeRF 1.0; 10 Mbps with HomeRF 2.0; 25 Mbps with HomeRF 3.0 (future)
Range	Covers up to 150 feet for typical home and yard
Total network devices	Up to 127
Voice connections	Up to 4 active handsets
Data security	Blowfish encryption algorithm (over 1 trillion codes)
Data compression	LZRW3-A algorithm
48-bit network ID	Enables concurrent operation of multiple colocated networks

Source: HomeRF Working Group.

- Shared access to the Internet from anywhere in the home, allowing a user to browse the Web from a laptop on the deck or have stock quotes delivered to a PC in the den.

- Automatic intelligent routing of incoming telephone calls to one or more cordless handsets, fax machines, or voice-mail boxes of individual family members.

- Cordless handset access to an integrated message system to review stored voice mail, faxes, and electronic mail.

- Personal intelligent agents running on the PC for each family member, accessed by speaking into cordless handsets. This new voice interface would allow users to access and control their PCs and all the resources on the home wireless network spontaneously, from anywhere within the home, using natural language commands.

- Wireless LANs allowing users to share files and peripherals between one or more PCs, no matter where they are located within the home.

- Spontaneous control of security and electrical, heating, and air-conditioning systems from anywhere in or around the home.

- Multiuser computer games playable in the same room or in multiple rooms throughout the home.

Network Topology

The SWAP system can operate either as an ad-hoc network or as a managed network under the control of a connection point. In an ad-hoc network, where only data communication is supported, all stations are equal, and control of the network is distributed between the stations. For time-critical communications such as interactive voice, a connection point is required to coordinate the system. The connection point, which provides the gateway to the public switched telephone network (PSTN), can be connected to a PC via a standard interface such as the Universal Serial Bus (USB) that will enable enhanced voice and data services. The SWAP system also can use the connection point to support power management for prolonged battery life by scheduling device wakeup and polling.

The network can accommodate a maximum of 127 nodes. The nodes are of four basic types:

- Connection point that supports voice and data services
- Voice terminal that only uses the TDMA service to communicate with a base station
- Data node that uses the CSMA-CA service to communicate with a base station and other data nodes
- Integrated node thath can use both TDMA and CSMA/CA services

HomeRF uses intelligent hopping algorithms that detect wideband static interference from microwave ovens, cordless phones, baby monitors, and other wireless LAN systems. Once detected, the HomeRF hop set adapts so that no two consecutive hops occur within this interference range. This means that, with very high probability, a packet lost due to interference will get through when it retries on the next hop. While these algorithms benefit data applications, they are especially important for voice, which requires extremely low bit error rates and low latency.

Future Plans

Work has already begun on the future HomeRF 2.1 specification, which will add features designed to reinforce its advantages for voice. Planned enhancements also will allow HomeRF to complement other wireless standards, including IEEE 802.11, also known as Wi-Fi.

HomeRF 2.0 already supports up to eight phone lines, eight registered handsets, and four active handsets with voice quality and range comparable to leading 2.4-GHz phone systems. With this many lines, each family member can have a personal phone number. HomeRF 2.1 plans to increase the number of active handsets with the same or better voice quality, thus supporting the needs of small businesses.

The 150-foot range of HomeRF already covers most homes and into the yard. HomeRF 2.1 will extend this range for larger homes and businesses by using wireless repeaters that are similar to enterprise access points but without the need to connect each one to Ethernet. HomeRF frequency hopping technology also avoids the complexity of assigning RF channels to multiple access points (or repeaters) and offers easy and effective security and interference immunity. This is especially important since households and small businesses do not usually have network administrators.

To allow individuals to roam across very large homes and fairly large offices while talking on the phone and without loosing their voice connection, HomeRF 2.1 also will support voice roaming with soft handoff between repeaters.

HomeRF 2.0 supports Ethernet speeds up to 10 Mbps with fallback speeds and backward compatibility to earlier versions of HomeRF. Performance can be further enhanced to about 20 Mbps. The HomeRF WG is evaluating the need for such enhancements at 2.4 GHz in light of its planned support of 5 GHz.

A proposed change to Federal Communications Commission (FCC) Part 15 rules governing the 2.4-GHz ISM band will allow adaptive frequency hopping. While not legal today, these proposed techniques allow hoppers such as Bluetooth and HomeRF to recognize and avoid interference from static frequency technologies such as Wi-Fi. Since HomeRF already adjusts its hopping pattern based on interference to ensure that two consecutive hops do not land on interference, supporting this FCC proposal seems trivial.

The HomeRF WG believes in the peaceful coexistence of 2.4 and 5 GHz since each frequency band and technology has specific strengths that complement each other. Rather than draft a specification for 5 GHz, the group simply endorses IEEE

802.11a (also known as Wi-Fi5) for high-bandwidth applications such as high-definition video streaming and MPEG2 compression. It plans to write application briefs describing how to bridge between 2.4- and 5-GHz technologies, including how to handle differences in quality of service (QoS). This information, while written for IEEE 802.11a, also can apply to HiperLAN-2, IEEE 802.11h, and proprietary IEEE 802.11a extensions.

Some analysts expect IEEE 802.11a to eventually take over as the wireless standard for enterprise offices, gain needed QoS support from IEEE 802.11e, and start a slow migration into homes. It already supports 54 Mbps, and proprietary extensions increase performance to about 100 Mbps. But because of the higher frequencies used, IEEE 802.11a has disadvantages in cost, power consumption, range, and signal attenuation through materials. A combination of HomeRF and IEEE 802.11a brings together the strengths of both technologies.

Summary

Home users have a need for a wireless network that is easy to use, cost-effective, spontaneously accessible, and can carry voice and data communications. Certified HomeRF products are available today from consumer brands such as Compaq, Intel, Motorola, Proxim, and Siemens through retail, online, and service provider channels. They come in a variety of form factors such as USB and PC card adapters, residential gateways, and a growing variety of devices that embed HomeRF.

See also

Bluetooth
Digitally Enhanced Cordless Telephony
Spread Spectrum Radio
Wireless Fidelity

I-MODE

i-Mode means "information mode" and refers to a type of Internet-enabled mobile phone service that is currently available in Japan from NTT DoCoMo, the world's largest cellular provider. With the push of a button, i-mode connects users to a wide range of online services, many of which are interactive, including mobile banking, news and stock updates, telephone directory service, restaurant guide, and ticket reservations. The i-mode phones also feature the Secure Sockets Layer (SSL) protocol, which provides encryption for the safe transmission of personal information such as credit card and bank account numbers.

Services

All services are linked directly to the DoCoMo i-mode portal Web site. Content can be accessed virtually instantly simply by pushing the cell phone's dedicated i-mode button. Once connected, users also can access hundreds of other i-mode sites via standard Web addresses. Since i-mode is based on packet data transmission technology, users are charged only for how much information they retrieve, not by how long they are online.

Customers can access many different kinds of content, including news, travel, information, database services, and entertainment. In addition, i-mode can be used to exchange e-mail with computers, personal digital assistants (PDAs), and other i-mode cellular phones. In Japan, the e-mail address is simply the cellular phone number followed by @docomo.ne.jp. And since i-mode is always active, e-mails are displayed automatically when they arrive.

Initially, the transmission speed was only 9.6 kbps, but this increased to 28.8 kbps in mid-2002. The next phase in development is underway with the company's introduction of first third-generation (3G) wireless service, which delivers data between 64 and 384 kbps. At these rates, it is possible to deliver music or video over wireless networks. Restaurant location programs also are able to deliver three-dimensional maps of the restaurant that describe the ambiance.

Simplicity

The i-mode service was launched in February 1999, and by mid-2002, the number of subscribers exceeded 30 million. Over 800 companies provide information services through i-mode. In addition, there are over 38,000 i-mode Web sites that offer content to mobile phone users. This makes i-mode a worthy contender to the Wireless Application Protocol (WAP), which is used by almost 22 million users worldwide.

The primary reason for i-mode's growing success is its simplicity. Unlike the WAP, which provides access to Web content from cell phones in the United States, content providers catering to the i-mode market can use standard HyperText Markup Language (HTML) to develop their Web sites. The Web sites are linked to DoCoMo's i-mode portal, where users go automatically on hitting the cell phone's dedicated i-mode button. An i-mode cell phone typically weighs less than 4 ounces, has a comparatively large liquid-crystal display, and features a four-point navigation button that moves a pointer on the display.

The i-mode platform also supports Java technology. Java supports stand-alone applications that can be downloaded and stored, eliminating the need to continually connect to a Web site to play video games, for example. Java also supports agent-type applications for constantly changing information, such as stock quotes, weather forecasts, and sports scores, that can be updated automatically at set times by the agent.

Summary

i-Mode services are now available in the United States. AT&T Wireless' mMode service is based on the i-mode technology developed by Japan's NTT DoCoMo. mMode provides consumers with a variety of communication, information, and entertainment services. The services include e-mail, news, weather, sports, and games. Pricing is based on volume of data, with plans starting at $2.99 per month. The service works over the General Packet Radio Service (GPRS) network of AT&T Wireless using cell phones that are specifically designed to support i-mode. Other thin application environments include Binary Runtime Environment for Wireless BREW) from Qualcomm, Java 2 Micro Edition (J2ME) from Sun Microsystems, and the WAP from the WAP Forum.

See also

General Packet Radio Service

Wireless Application Protocol

INCUMBENT LOCAL EXCHANGE CARRIERS

Incumbent Local Exchange Carriers (ILECs) is a term that refers to the 22 former Bell Operating Companies (BOCs) divested from AT&T in 1984, as well as Cincinnati Bell, Southern New England Telephone (SNET), and the larger independent telephone companies of GTE and United

Telecommunications. In addition, some 1300 smaller telephone companies are also in operation, serving mostly rural areas. These, too, are considered incumbents, but the small markets they serve do not attract much competition.

After being spun off by AT&T in 1984, the BOCs were assigned to seven regional holding companies: Ameritech, Bell Atlantic, BellSouth, Nynex, Pacific Telesis, Southwestern Bell Communications (SBC), and US West. Over the years, some of these regional companies merged to the point that today only four are left. Bell Atlantic and Nynex were the first to merge in 1994. Bell Atlantic also completed a $53 billion merger with GTE in mid-1999 and changed its name to Verizon. SBC Communications merged with Pacific Telesis in 1997 and then Ameritech in 1999. It also acquired Southern New England Telephone (SNET).

Regulatory Approval

All the mergers passed regulatory approval at the state and national level. The Federal Communications Commission (FCC) approves mergers with input from the Department of Justice (DoJ). In the case of the SBC-Ameritech merger, the FCC imposed 28 conditions on SBC in exchange for approving the transaction.

The approval package contained a sweeping array of conditions designed to make SBC-Ameritech's markets the most open in the nation, boosting local competition by providing competitors with the nation's steepest discounts for resold local service and full access to operating support systems (OSS).

It also required SBC to accelerate by 6 months its entry into new markets, forcing the company to compete in 30 new markets within 30 months after completion of the merger. The FCC's rationale was that increased competition in out-of-region territories would help offset reduced competition in the SBC-Ameritech service areas.

The conditions also required stringent performance monitoring, reporting, and enforcement provisions that could trigger more than $2 billion in fines if these goals were not met. Fortunately for SBC, the agreement required it to serve only three customers in each out-of-region market. According to SBC, it will not begin to seriously market its out-of-region services until it has obtained approval to offer long-distance services in its 13 home states.

Summary

The monopoly status of the ILECs officially ended with passage of the Telecommunications Act of 1996. Not only can other types of carriers enter the market for local services in competition with them, but also their regional parent companies can compete in each other's territories. Through mergers, the reasoning went, the combined companies can enter out-of-region markets on a broad scale quickly and efficiently enough to become effective national competitors.

Unfortunately, this has not occurred on a significant scale. In fact, the lack of out-of-region competition among the Baby Bells means that consumers and businesses do not have as much choice in service providers, especially now that many Competitive Local Exchange Carriers (CLECs) are being hit hard by financial problems and the lack of venture capital. The ILECs are more concerned with being able to qualify for long-distance services in their own markets so that they can bundle local and long-distance services and Internet access—a package few, if any, competitors would be able to match.

See also

Competitive Local Exchange Carriers

Interexchange Carriers

Local Exchange Carriers

INFRARED NETWORKING

Infrared (IR) technology is used to implement wireless local area networks (LANs) as well as the wireless interface to connect laptops and other portable machines to the desktop computer equipped with an IR transceiver. IR LANs are proprietary in nature, so users must rely on a single vendor for all the equipment. However, the IR interface for connecting portable devices with the desktop computer is standardized by the Infrared Data Association (IrDA).

Infrared LANs

IR LANs typically use the wavelength band between 780 and 950 nanometers (nm). This is due primarily to the ready availability of inexpensive, reliable system components.

There are two categories of IR systems that are commonly used for wireless LANs. One is directed IR, which uses a very narrow laser beam to transmit data over one to three miles. This approach may be used for connecting LANs in different buildings. Although transmissions over laser beam are virtually immune to electromechanical interference and would be extremely difficult to intercept, such systems are not widely used because their performance can be impaired by atmospheric conditions, which can vary daily. Such effects as absorption, scattering, and shimmer can reduce the amount of light energy that is picked up by the receiver, causing the data to be lost or corrupted.

The other category is nondirected IR, which uses a less focused approach. Instead of a narrow beam to convey the signal, the light energy is spread out and bounced off narrowly defined target areas or larger surfaces such as office walls and ceilings.

Nondirected IR links may be further categorized as either line of sight or diffuse (Figure I-1). Line-of-sight links require a clear path between transmitter and receiver but generally offer higher performance.

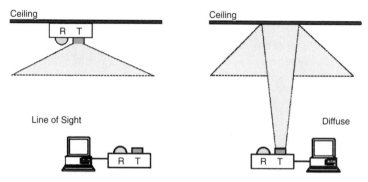

Figure I-1 Line-of-sight versus diffuse configurations for infrared links.

The line-of-sight limitation may be overcome by incorporating a recovery mechanism in the IR LAN that is managed and implemented by a separate device called a "multiple access unit" (MAU) to which the workstations are connected. When a line-of-sight signal between two stations is temporarily blocked, the MAU's internal optical link control circuitry automatically changes the link's path to get around the obstruction. When the original path is cleared, the MAU restores the link over that path. No data are lost during this recovery process.

Diffuse links rely on light bounced off reflective surfaces. Because it is difficult to block all the light reflected from large surface areas, diffuse links are generally more robust than line-of-sight links. The disadvantage of diffused IR is that a great deal of energy is lost, and consequently, the data rates and operating distances are much lower.

System Components

Light-emitting diodes (LEDs) or laser diodes (LDs) are used for transmitters. LEDs are less efficient than LDs, typically exhibiting only 10 to 20 percent electrooptical power conversion efficiency, while LDs offer an electrooptical conversion efficiency of 30 to 70 percent. However, LEDs are much less expensive than LDs, which is why most commercial systems use them.

Two types of low-capacitance silicon photodiodes are used for receivers: positive-intrinsic-negative (PIN) and avalanche. The simpler and less expensive PIN photodiode is typically used in receivers that operate in environments with bright illumination, whereas the more complex and more expensive avalanche photodiode is used in receivers that must operate in environments where background illumination is weak. The difference in the two types of photodiodes is their sensitivity.

The PIN photodiode produces an electric current in proportion to the amount of light energy projected onto it. Although the avalanche photodiode requires more complex receiver circuitry, it operates in much the same way as the PIN diode, except that when light is projected onto it, there is a slight amplification of the light energy. This makes it more appropriate for weakly illuminated environments. The avalanche photodiode also offers a faster response time than the PIN photodiode.

Operating Performance

Current applications of IR technology yield performance that matches or exceeds the data rate of wire-based LANs: 10 Mbps for Ethernet and 16 Mbps for Token Ring. However, IR technology has a much higher performance potential—transmission systems operating at 50 and 100 Mbps have already been demonstrated.

Because of its limited range and inability to penetrate walls, nondirected IR can be easily secured against eavesdropping. Even signals that go out windows are useless to eavesdroppers because they do not travel far and may be distorted by impurities in the glass as well as by the glass's placement angle.

IR offers high immunity from electromagnetic interference, which makes it suitable for operation in harsh environments like factory floors. Because of its limited range and inability to penetrate walls, several IR LANs may operate in different areas of the same building without interfering with each other. Since there is less chance of multipath fading (large fluctuations in received signal amplitude and phase), IR links are highly robust.

Many indoor environments have incandescent or fluorescent lighting that induces noise in IR receivers. This is overcome by using directional IR transceivers with special filters to reject background light.

Media Access Control

IR supports both contention-based and deterministic media access control techniques, making it suitable for Ethernet as well as Token Ring LANs.

To implement Ethernet's contention protocol, carrier-sense multiple access (CSMA), each computer's IR transceiver is typically aimed at the ceiling. Light bounces off the reflector in all directions to let each user receive data from other users (Figure I-2). CSMA ensures that only one station can transmit data at a time. Only the stations to which packets are addressed can actually receive them.

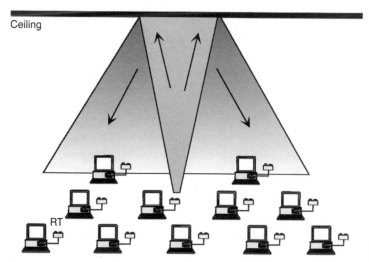

Figure I-2   Ethernet implementation using diffuse IR.

Deterministic media access control relies on token passing to ensure that all stations in turn get an equal chance to transmit data. This technique is used in Token Ring LANs, where each station uses a pair of highly directive (line-of-sight) IR transceivers. The outgoing transducer is pointed at the incoming transducer of a station down line, thus forming a closed ring with the wireless IR links among the computers (Figure I-3). With this configuration, much higher data rates can be achieved because of the gain associated with the directive IR signals. This approach improves overall throughput, since fewer bit errors will occur, which minimizes the need for retransmissions.

Infrared Computer Connectivity

Most notebook computers and personal digital assistants (PDAs) have IR ports. Every major mobile phone brand has

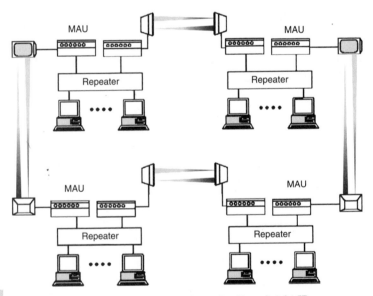

Figure I-3 Token Ring implementation using line-of-sight IR.

at least one IR-enabled handset, and even wristwatches are beginning to incorporate IR data ports. IR products for computer connectivity conform to the standards developed by the Infrared Data Association (IrDA). The standard protocols include Serial Infrared (SIR) at 115 kbps, Fast Infrared (FIR) at 4 Mbps, and Very Fast Infrared (VFIR) at 16 Mbps. The complete IrDA protocol suite contains the following mandatory protocols and optional protocols.

Mandatory protocols:

- *Infrared Physical Layer* Specifies IR transmitter and receiver optical link, modulation and demodulation schemes, and frame formats.

- *Infrared Link Access Protocol (IrLAP)* Has responsibility for link initiation, device address discovery, address conflict resolution, and connection startup. It also ensures reliable data delivery and provides disconnection services.

- *Infrared Link Management Protocol (IrLMP)* Allows multiple software applications to operate independently and concurrently, sharing a single IrLAP session between a portable PC and network access device.

- Information Access Service (IAS) Used along with IrLMP and IrLAP, this protocol resolves queries and responses between a client and server to determine the services each device supports.

Optional protocols:

- *Infrared Transport Protocol (IrTTP) or Tiny TP* Has responsibility for data flow control and packet segmentation and reassembly.

- *IrLAN* Defines how a network connection is established over an IrDA link.

- *IrCOM* Provides COM (serial and parallel) port emulation for legacy COM applications, printing, and modem devices.

- *IrOBEX* Provides object exchange services similar to the HyperText Transfer Protocol (HTTP) used to move information around the Web.
- *IrDA Lite* Provides methods of reducing the size of IrDA code while maintaining compatibility with full implementations.
- *IrTran-P* Provides image exchange for digital image capture devices/cameras.
- IrMC Specifies how mobile telephony and communication devices can exchange information. This includes phonebook, calendar, and message data.

In addition, there is a protocol called IrDA Control that allows cordless peripherals such as keyboards, mice, game pads, joysticks, and pointing devices to interact with many types of intelligent host devices. Host devices include PCs, home appliances, game machines, and television/Web set-top boxes.

An extension called Very Fast IR (VFIR) provides a maximum transfer rate of 16 Mbps, a fourfold increase in speed from the previous maximum data rate of 4 Mbps. The extension provides users with faster throughput without an increase in cost and is backward compatible with equipment employing the previous data rate. The higher speed is intended to address the new demands of transferring large image files between digital cameras, scanners, and PCs. Table I-1 summarizes the performance characteristics of the IrDA's IR standard.

The IrDA has developed a "point and pay" global wireless point-of-sale (POS) payment standard for hand-held devices, called Infrared Financial Messaging (IrFM). In an electronic wallet application, consumers use their IR-enabled PDAs to make purchases at the point of sale. Users "beam" their financial information to pay for a purchase and receive a digital receipt. The IrFM protocol defines payment usage models, profiles, architecture, and protocol layers to enable hardware, software, and systems designers to develop IrFM-compliant products and ensure interoperability and compat-

TABLE I-1   Performance Characteristics of the Infrared Data
Association's Infrared Standard

Feature/Function	Performance
Connection type	Infrared, narrow beam (30º angle or less)
Spectrum	Optical, 850 nanometers (nm)
Transmission power	100 milliwatts (mW)
Data rate	Up to 16 Mbps using Very Fast Infrared (VFIR)
Range	Up to 3 feet (1 meter)
Supported devices	Two
Data security	The short range and narrow angle of the IR beam provide a simple form of security; otherwise, there are no security capabilities at the link level.
Addressing	Each device has a 32-bit physical ID that is used to establish a connection with another device.

ibility globally. IrFM uses IrDA's Object Exchange (OBEX)
protocol to facilitate interoperability between devices.

The Infrared Data Association has formed a Special
Interest Group (SIG) to produce a standard for interappli-
ance MP3 data exchange using IR technology. The popularity
of MP3-capable appliances begs for a standard connection
between the MP3 players, computers, and the network,
allowing consumers to easily move music from device to
device without a cable or docking port. The hand-held player
should be able to transfer a song into a car stereo or home
entertainment system. The MP3 SIG is identifying concerns
specific to transferring MP3 data and building solutions into
the protocol. Among the issues that must be addressed is how
to identify copyrighted content and describe distribution
restrictions to handle the MP3 content appropriately.

Summary

IR's primary impact will take the form of benefits for mobile
professional users. It enables simple point-and-shoot

connectivity to standard networks, which streamlines users' workflow and allows them to reap more of the productivity gains promised by portable computing. IrDA technology is supported in over 100 million electronic devices, including desktop, notebook, and palm PCs; printers; digital cameras; public phones/kiosks; cellular phones; pagers; PDAs; electronic books; electronic wallets; and other mobile devices.

When used on a LAN, IR technology also confers substantial benefits to network administrators. IR is easy to install and configure, requires no maintenance, and imposes no remote-access tracking hassles. It does not disrupt other network operations, and it provides data security. And because it makes connectivity so easy, it encourages the use of high-productivity network and groupware applications on portables, thus helping administrators amortize the costs of these packages across a larger user base.

See also

Spread Spectrum Radio

Wireless LANs

INTEGRATED DIGITAL ENHANCED NETWORK

Introduced in 1994 by Motorola, Integrated Digital Enhanced Network (iDEN) is a wireless network technology designed for vertical market mobile business applications. iDEN operates in the 800-MHz, 900-MHz, and 1.5-GHz bands and is based on Time Division Multiple Access (TDMA) and Global System for Mobile (GSM) architectures. It uses Motorola's Vector Sum Excited Linear Predictors (VSELP) voice encoder for compression and quadrature amplitude modulation (QAM) to deliver 64 kbps over a 25-kHz channel.

iDEN is promoted as being "four-in-one," allowing users to take advantage of two-way digital radio, digital wireless phone, alphanumeric messaging, and data, fax, and Internet capabilities with one pocket-sized digital handset. This eliminates the need to carry around multiple communication devices and the time-consuming task of synchronizing them.

The iDEN phones are full-featured business phones that provide features such as call forwarding, call hold, automatic answer, and speakerphone. Using the group call feature, users can communicate with one or hundreds of people with the push of a single button without having to set up a conference call or waste time with costly individual calls. The data-ready phones also provide fax and e-mail capabilities and access to the Internet.

Motorola iDEN phones incorporate a number of valuable messaging services, including voice messaging, text messaging, and alphanumeric paging. This gives the user the ability to receive messages even when the phone is turned off. Some iDEN phones are voice-activated and allow the user to record voice memos for playback later.

The newest phones, such as the i90c, feature J2ME technology, enabling users to download interactive content and applications—from business tools to graphically rich games. Users select "Java Apps" from the phone's main menu, and the applications execute directly on the handset instead of on a server within the network, as is the case with applications based on the complementary Wireless Application Protocol (WAP).

Call Setup

When a mobile radio places a call on an iDEN system, it goes through a series of system handshakes to establish the call as follows:

- When the mobile radio is powered up, it scans and locks onto a control channel.
- The mobile radio registers on the system.
- When it initiates a call, the mobile radio places a service request on the control channel.

- The fixed end system assigns the mobile radio to the dedicated control channel.

- The mobile radio uses the dedicated control channel to transmit the information required by the fixed end system to complete the call.

- The fixed end system assigns the mobile radio to a traffic channel that is used for communication of voice or data.

Each time the mobile radio initiates or receives a call, it measures the strength of the received signal. Based on this measurement and the power control constant defined on the control channel, it adjusts its transmit power level to a level just high enough to ensure clear reception of its signal by the intended recipients.

Future Directions

In keeping with the industry trend of upgrading wireless networks to third-generation (3G) technologies, Nextel and Motorola are working to double the voice capacity of iDEN and implement data compression on the network.

The two most important factors driving carriers to upgrade to 3G technologies are the need to increase voice capacity and deliver packet data service at acceptable speeds. After considering its various technology and strategic options, Nextel decided to leverage its existing infrastructure by making enhancements to its iDEN network rather than overlay a standards-based technology onto the proprietary Motorola iDEN platform.

Nextel also expects to provide a two- to fivefold increase in the end-user experience of most Web-enabled and general office data applications. While 2.5G technologies like General Packet Radio Service (GPRS) are capable of achieving up to 115 or 144 kbps with extensions to TDMA or CDMA, respectively, Nextel expects to achieve data speeds up to 60 to 70 kbps with the addition of data compression.

Summary

iDEN is designed to give the mobile user quick access to information without having to carry around several devices. Currently, iDEN systems work in more than a dozen countries, but only two service providers in the United States use iDEN in their wireless networks: Nextel and Southern Company.

See also

> Code Division Multiple Access
>
> General Packet Radio Service
>
> Global System for Mobile Telecommunications
>
> Time Division Multiple Access

INTERACTIVE TELEVISION

Although the concept of interactive television (ITV) has been around for more than two decades, the market is still emerging in fits and starts, with no clear business model in sight. The term *ITV* refers to a set of real and potential capabilities that are designed to improve the television viewing experience. But platform makers, application developers, content providers, and the major networks differ in their opinions of what capabilities will accomplish this.

Numerous entrants have appeared on the scene with offerings that incorporate a hodge-podge of different functions. Despite the feverish pace of innovation, a consensus seems to be forming that ITV, at a minimum, should include Internet-on-the-TV-set, video-on-demand services, interactive program guides, and a consumer electronic device that permits viewers to interact with their television sets. What remains to be seen is whether any particular enhancement or combination of potential ITV features will succeed in the marketplace.

To date, many attempts at ITV have failed either because of lack of consumer interest or limits of the technology. Nevertheless, there seems to be industry-wide interest in pursuing interactive TV, due in large part to technical advancements that make all things possible. And then there is the persistent conviction that consumers now have a greater interest and appreciation for interactive services in general, which equates to "pent-up demand."

One of the largest players in the ITV market is Microsoft, which has garnered close to 1 million subscribers for its WebTV interactive service. WebTV provides consumers with an interactive electronic program guide, interactive content, Internet access, e-mail, chat, and Microsoft's own instant messenger. But after 5 years, WebTV's growth has flattened out, indicating that something is wrong. The service was expected to have 1 million subscribers by year-end 1998 and eventually become as ubiquitous as the VCR. As of mid-2001, WebTV still had about the same number of subscribers, despite several major upgrades.

Interactive Applications

There is already a level of "enhanced" television possible through the traditional set-top box. Viewers can order pay-per-view movies, read messages from their service provider, and search through the evening's programming guide for their favorite sitcom. With the addition of another device and a subscription fee, viewers can access the Internet from their televisions, as with WebTV. And with still another device and subscription fee, a VCR-like device searches and records television programs automatically for playback at a convenient time, as with TiVo's offering. Moving from enhanced television to interactive television promises to offer more interesting applications.

Among the possible ITV applications are home shopping, bank account access, e-mail, advertised product information, voting in viewer surveys, and playing along with game shows. Another application for ITV is the delivery of enhanced content. While watching a sports program, for example, viewers can call up more information about the game, buy merchandise, or click on pop-up advertisements of interest (Figure I-4).

Telephone calls can be made through the television as well. Subscribers can use a television set-top box equipped with a speakerphone and voice over Internet Protocol (VoIP) software to make telephone calls through a graphical dial pad that appears on the screen. Notification of incoming calls is displayed on the screen, and the viewer can decide to take the call or let it go to a voice mailbox.

Online gaming in Europe and the United States is expected to fuel growth of the ITV industry. The apparent appeal of ITV games is that they allow television viewers to participate in contests linked to popular TV shows or a series

Figure I-4 With interactive television, viewers can call up more information about a sports program, shop for merchandise, or click on advertisements of interest.

they watch on a regular basis. Revenues can be generated through advertisers or by charging users a nominal fee to enter these contests in the hope of winning prizes or participating in live shows.

Advertisers are intrigued with the information-gathering potential of ITV technology, gleaned from testing television programs, hosts or trailers, and advertising spots. A viewer panel would receive questions and provide answers directly onto their television screens. A special set-top box connects the television via the Internet to a research database. This method allows for better representation, faster response and processing of results, and lowers the costs of collecting this kind of information. With the results, advertisers could target their messages more effectively.

Market Complexity

The complicated nature of the ITV market, in which different cable operators use different set-top boxes with different operating systems, has hindered market growth so far. In addition, cable carriers, advertisers, and the TV networks seem to be having difficulty figuring out how to make ITV pay off. This is reminiscent of the dot-com market a few years ago, where companies could not quite figure out how to make money on the services they put up on the Web.

Even the top players in the ITV market remain uncertain about it. This was illustrated in mid-2001, when AT&T scaled back its plans for the service. Microsoft had pumped $5 billion into AT&T's cable operation in 1999 with the hope of creating an ITV service for about 10 million users. Things never really progressed beyond the planning stage, as Microsoft experienced delays in getting the operating system ready for its ITV systems based on Motorola's DCT5000 set-top box.

Meanwhile, AT&T did additional research and concluded that the average consumer did not consider surfing the Web as a priority for ITV, which is Microsoft's strong point.

Instead, consumers expressed the desire for video on demand (VoD) and personal video recorders that automatically save TV shows to a hard drive. AT&T then scaled back its service model and considered software from other vendors before launching its ITV service.

Potential Market Barriers

A possible clue about the difficulties that await ITV comes from the lackluster performance of WebTV. Microsoft acquired the Internet-on-TV service in 1997, and since then, the number of subscribers to the service reportedly has reached a peak of about 1 million.

Some of WebTV's lack of success has been blamed on poor corporate vision. But the cost of innovation may have played a part as well. Adding more functionality forces higher prices for equipment and service, which is not a good thing to do when market growth is on the decline and the economy is sputtering.

One source of trouble for ITV companies has been how to keep costs low enough to be attractive. Industry surveys reveal that less than 3 percent of consumers would pay more than $300 for a set-top box that supports time-shifting and personalized TV services, as well as several other leading ITV applications. Only 6 percent of all consumers would pay more than $9 per month to subscribe to a service that offers such capabilities. With this much consumer resistance, there is not much hope of ITV achieving widespread success anytime soon, and key players in the market will continue to struggle.

Among the over-40 population, ITV comes under more severe scrutiny from the content perspective. To this market segment, ITV does not appear to improve content—it is perceived merely as a way to throw more of the same low-quality stuff at viewers while asking them to pay more for it.

There will be a limited number of venues where ITV will succeed. Interactive-friendly areas include sports, game shows, and news, but interactivity in drama and sitcom venues will be more difficult to penetrate. If it turns out that ITV just offers more ways for advertisers to improve the targeting of their messages, however interactive, market penetration will be limited.

This also raises the privacy issue, even among young people, who are more likely to be receptive to ITV. The ITV service providers will count on subscriber revenues as well as advertising dollars. Privacy advocates are concerned that ITV combines the worst aspects of the Internet and mass media because the new systems are being designed to track not only every activity of users as they surf the Net but also the programs and commercials they watch. They fear that if ITV systems are to realize their promise as the "advertising nirvana" for marketers, privacy necessarily must collapse as ITV becomes a spy in the home, collecting information on age, discretionary income, and parental status, along with psychographic and demographic data, that will be analyzed and made available to marketers, advertisers, programmers, and others.

Even the more recent reports on the ITV market have not factored in the dismal state of the global economy and the consequent decline in discretionary income for new and novel forms of entertainment. The interactive TV industry is not immune to general economic conditions, and the computer and telecommunications markets have been hit the hardest throughout 2001 and 2002.

Summary

Consumer demand for ITV has proved to be too complex for any one product or service to appeal to everyone. The players in this market are realizing that certain consumer segments have very different preferences and attitudes toward ITV. The challenges for these companies is to understand how consumers differ, why these differences exist, and how they can determine the success of a given product or service.

Failure to do so will inhibit market growth, despite all the optimistic revenue projections and technologies that seemingly can fulfill any request for entertainment and information—passively or interactively.

See also

High-Definition Television

Interactive Video and Data Service

INTERACTIVE VIDEO AND DATA SERVICE

Interactive Video and Data Service (IVDS) is a point-to-multipoint, multipoint-to-point, short-distance communication service that operates in the 218- to 218.5-MHz and 218.5- to 219-MHz frequency bands. In September 1993, the FCC assigned the licenses for IVDS by lottery in the top 10 markets. The following year, two licenses per market were offered for auction at the same time, with the highest bidder given a choice between the two available licenses, and the second highest bidder winning the remaining license. More than 95 percent of all IVDS licenses were won by small businesses or businesses owned by members of minority groups or women. Additional auctions, ending in 1997, were held for the available spectrum in other markets.

Applications

As envisioned, an IVDS licensee would be able to use IVDS to transmit information and product and service offerings to its subscribers and receive interactive responses. Potential applications included ordering goods or services offered by television services, viewer polling, remote meter reading, vending inventory control, and cable television theft deterrence. An IVDS licensee was able to develop other applications without specific approval from the FCC. An IVDS channel, however, was insufficient for the transmission of conventional full-motion video.

Initially, mobile operation of IVDS was not permitted. In 1996, however, the FCC amended its rules to permit IVDS licensees to provide mobile service to subscribers. This action authorized mobile operation of response transmitter units (subscriber units) operated with an effective radiated power of 100 milliwatts or less. The FCC found that this change would increase the flexibility of IVDS licensees to meet the communications needs of the public without increasing the likelihood of interference with TV channel 13, which was of concern at the time.

According to the FCC's rules at the time IVDS spectrum was awarded, licenses cancel automatically if a licensee does not make its service available to at least 30 percent of the population or land area within the service area within 3 years of grant of the system license. Each IVDS system licensee had to file a progress report at the conclusion of 3- and 5-year benchmark periods to inform the FCC of the construction status of the system. This arrangement was intended to reduce the attractiveness of licenses to parties interested in them only as a speculative vehicle.

Summary

The spectrum for IVDS was initially allocated in 1992 to provide interactive capabilities to television viewers. This did not occur largely because the enabling technology was too expensive and market demand too low or nonexistent. IVDS is now known as 218- to 219-MHz service. In September 1999, the FCC revised its rules for the 218- to 219-MHz service to maximize the efficient and effective use of this frequency band. The FCC simply reclassified service from a strictly private radio service—one that is used to support the internal communications requirements of the licensee—to a service that can be used in both common carrier and private operations. Licensees are now free to design any service offering that meets market demand. In addition, licensees now have up to 10 years from the date of the license grant to build out their service without meeting the 3- and 5-year construction benchmarks.

See also

Interactive Television

Local Multipoint Distribution Service

Multichannel Multipoint Distribution Service

INTEREXCHANGE CARRIERS

Interexchange Carriers (IXCs), otherwise known as "long-distance carriers," include the big three—AT&T, Worldcom, and Sprint—all of which also operate wireless networks and are migrating them to 3G capabilities.

In addition to providing long-distance telephone service over wired and wireless networks, the IXCs offer business services like Integrated Services Digital Network (ISDN), Frame Relay, leased lines, and a variety of other digital services. Many IXCs are also Internet service providers (ISPs), which offer Internet access services, virtual private networks, electronic mail, Web hosting, and other Internet-related services.

Bypass

Traditionally limited to providing service between local access and transport areas (LATAs), the Telecommunications Act of 1996 allows IXCs to offer local exchange services in competition with the Incumbent Local Exchange Carriers (ILECs). But because the ILECs charge too much for local loop connections and services and do not deliver them in a consistently timely manner, the larger IXCs have implemented technologies that allow them to bypass the local exchange. Among the methods IXCs use to bypass the local exchange include CATV networks and broadband wireless technologies, such as Local Multipoint Distribution Service (LMDS) and Multichannel Multipoint Distribution Service (MMDS).

With regard to cable, AT&T, for example, has acquired the nation's two largest cable companies, TCI and MediaOne, to bring local telephone services to consumers, in addition to television programming and broadband Internet access. As these bundled services are introduced in each market, they are provided to consumers at an attractive price with the added convenience of a single monthly bill. Sprint uses MMDS to offer Internet access to consumers and businesses that are out of range for Digital Subscriber Line (DSL) services. XO Communications, a nationwide integrated communications provider (ICP) uses LMDS to reach beyond its metropolitan fiber loops to reach buildings that are out of the central business districts.

Long-Distance Market

In January 2001, the FCC released the results of a study on the long-distance telecommunications industry. Among the findings from the report:

- In 1999, the long-distance market had more than $108 billion in revenues, compared to $105 billion in 1998. In 1999, long-distance carriers accounted for over $99 billion and local telephone companies accounted for the remaining $9 billion.

- Interstate long-distance revenues increased by 12.8 percent in 1999 compared to 1.5 percent the year before.

- Since 1984, international revenues have grown more than fivefold from less than $4 billion in 1984 to over $20 billion in 1999. The number of calls has increased from about half a billion in 1984 to almost 8 billion in 1999.

- In 1984, AT&T's market share was about 90 percent of the toll revenues reported by long-distance carriers. By 1999, AT&T's market share had declined to about 40 percent, WorldCom's share was 25 percent, Sprint's was 10 percent, and more than 700 other long-distance carriers had the remaining quarter of the market.

- According to a sampling of residential telephone bills, in 1999 the average household spent $64 monthly on telecommunications. Of this amount, $21 was for services provided by long-distance carriers, $34 for services by local exchange carriers, and the remainder for services by wireless carriers.

- According to the same sampling of residential telephone bills, 38 percent of toll calls in 1999 were interstate and accounted for 50 percent of toll minutes. Also, 33 percent of residential long-distance minutes were on weekdays, 30 percent on weekday evenings, and 37 percent on weekends.

Summary

Growing competition in long-distance services has eroded AT&T's market share from its former monopoly level to about 40 percent. With this competition has come increasing availability of low-cost calling plans for a broad range of consumers. As a result, average revenue per minute earned by carriers has been declining steadily for several years, while long-distance usage has increased substantially to make up for that revenue shortfall. As more ILECs get permission from the FCC to enter the in-region long-distance market, IXCs will come under increasing competitive pressure because the ILECs will be able to bundle local and long-distance calling and Internet access into attractively priced service packages.

See also

Competitive Local Exchange Carriers
Incumbent Local Exchange Carriers

INTERNATIONAL MOBILE TELECOMMUNICATIONS

The International Telecommunication Union (ITU) has put together a framework for 3G mobile communications systems that are capable of bringing high-quality mobile multimedia services to a worldwide mass market based on a set of standardized interfaces. Known as International Mobile Telecommunications-2000 (IMT-2000), this framework encompasses a small number of frequency bands, available on a globally harmonized basis, that make use of existing national and regional mobile and mobile-satellite frequency allocations.

IMT-2000 is the largest telecommunications project ever attempted, involving regulators, operators, manufacturers, media, and information technology (IT) players from all regions of the world as they attempt to position themselves to serve the needs of an estimated 2 billion mobile users worldwide by 2010. Originally conceived in the early 1990s when mobile telecommunications provided only voice and low-speed circuit-switched data, the IMT-2000 concept has adapted to the changing telecommunication environment as its development progressed. In particular, the advent of Internet, intranet, e-mail, e-commerce, and video services has significantly raised user expectations of the responsiveness of the network and the terminals and, therefore, the bandwidth of the mobile channel.

Spanning the Generations

Over the years, mobile telecommunications systems have been implemented with great success all over the world. Many are still first-generation systems—analog cellular systems such as the Advanced Mobile Phone System (AMPS), Nordic Mobile Telephone (NMT), and the Total Access Communication System (TACS). Most systems are now in the second generation, which is digital in nature. Examples of digital cellular systems include Global System for Mobile

(GSM) communications, Digital AMPS (DAMPS), and Japanese Digital Cellular (JDC). Although both first- and second-generation systems were designed primarily for speech, they offer low-bit-rate data services as well. However, there is little or no compatibility between the different systems, even within the same generation.

The spectrum limitations and various technical deficiencies of second-generation systems and the potential fragmentation problems they could cause in the future led to research on the development and standardization of a global 3G platform. The ITU and regional standards bodies came up with a "family of systems" concept that would be capable of unifying the various technologies at a higher level to provide users with global roaming and voice-data convergence, leading to enhanced services and support for innovative multimedia applications.

The result of this activity is IMT-2000, a modular concept that takes full account of the trends toward convergence of fixed and mobile networks and voice and data services. The 3G platform represents an evolution and extension of current GSM systems and services available today, optimized for high-speed packet data-rate applications, including high-speed wireless Internet services, videoconferencing, and a host of other data-related applications.

Vendor compliance with IMT-2000 enables a number of sophisticated applications to be developed. For example, a mobile phone with color display screen and integrated 3G communications module becomes a general-purpose communications and computing device for broadband Internet access, voice, videotelephony, and videoconferencing (Figure I-5). These applications can be used by mobile professionals on the road, in the office, or at home. The number of Internet Protocol (IP) networks and applications is growing fast. Most obvious is the Internet, but private IP networks (i.e., intranets and extranets) show similar or even higher rates of growth and usage. With an estimated billion Internet users worldwide expected in 2010, there exists tremendous pent-up demand for 3G capabilities.

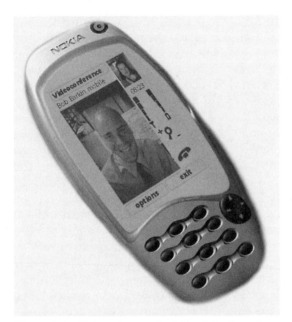

Figure I-5 This prototype of a 3G mobile phone from Nokia supports digital mobile multimedia communications, including videotelephony. Using the camera eye in the top right corner of the phone, along with the thumbnail screen below it, the local user can line up his or her image so that it can appear properly centered on the remote user's phone.

3G networks will become the most flexible means of broadband access because they allow for mobile, office, and residential use in a wide range of public and nonpublic networks. Such networks can support both IP and non-IP traffic in a variety of transmission modes, including packet (i.e., IP), circuit-switched (i.e., PSTN), and virtual circuit (i.e., ATM).

Goals of IMT-2000

Under the IMT-2000 model, mobile telephony will no longer be based on a range of market-specific products but will be founded on common standardized flexible platforms that will meet the basic needs of major public, private, fixed, and mobile markets around the world. This approach should result in a longer product life cycle for core network and transmission components and offer increased flexibility and cost-effectiveness for network operators, service providers, and manufacturers.

In developing the family of systems that would be capable of meeting the future communications demands of mobile users, the architects of IMT-2000 identified several key issues that would have to be addressed to ensure the success of the third-generation of mobile systems.

High Speed Any new system must be able to support high-speed broadband services, such as fast Internet access or multimedia-type applications. Users will expect to be able to access their favorite services just as easily from their mobile equipment as they can from their wire line equipment.

Flexibility The next generation of integrated systems must be as flexible as possible, supporting new kinds of services such as universal personal numbering and satellite telephony while providing for seamless roaming to and from IMT-2000-compatible terrestrial wireless networks. These and other features will greatly extend the reach of mobile systems, benefiting consumers and operators alike.

Affordability The system must be as affordable as today's mobile communications services, if not more so. Economies of scale achievable with a single global standard will drive down the price to users.

Compatibility Any new-generation system has to offer an effective evolutionary path for existing networks. While the advent of digital systems in the early 1990s often prompted the shutting down of first-generation analog networks, the enormous investments that have been made in developing the world's 2G cellular networks over the last decade make a similar scenario for adoption of 3G systems untenable.

Differentiation In coordinating the design of the IMT-2000 framework, the ITU was mindful of the need to preserve a competitive domain for manufacturers to foster incentive and stimulate innovation. Accordingly, the aim of IMT-2000 standards is not to stifle the evolution of better technologies or innovative approaches but to accommodate them.

Spectrum Allocations

The 2500- to 2690-MHz band was identified by the 2000 World Radio Conference (WRC-2000) as candidate spectrum for 3G systems, along with the 806- to 960-MHz and 1710- to 1885-MHz bands. The WRC-2000 results allow countries flexibility in deciding how to implement 3G systems. The conference recognized, however, that in many countries the frequency bands identified for 3G systems might already be in use by equally vital services.

In the United States, the 2500- to 2690-MHz band is currently used by the Instructional Television Fixed Service (ITFS) and the Multipoint Distribution Service (MDS), which are experiencing and are expected to see significant future growth, particularly in the provision of new broadband fixed access to the Internet. Given the ubiquitous nature of ITFS/MDS, the FCC found that sharing of this spectrum for 3G does not appear feasible. Further, the FCC found that reallocating a portion of the 2500- to 2690-MHz band from incumbent services for new 3G mobile wireless services would raise significant technical and economic difficulties.

The 1710- to 1755-MHz band is now used by federal government operations and is scheduled for transfer to the private sector on a mixed-use basis by 2004. The 2110- to 2150-MHz and 2160- to 2165-MHz bands are currently used by the private sector for fixed microwave services. The FCC identified these bands several years ago for reallocation to emerging technologies.

The 1710- to 1850-MHz band would be the preferred choice for 3G services. This would partially harmonize U.S. spectrum allocations with those in use or planned internationally. Harmonization would permit economies of scale and reduce costs in manufacturing equipment, as well as facilitate international roaming.

Parts of the 1710- to 1850-MHz band also could be used to harmonize with 2G GSM systems, which are currently used extensively throughout the world and are expected to transition eventually to 3G systems. Other parts of the 1710- to 1850-MHz band could be paired with the 2110- to 2150-MHz band to achieve partial harmonization with spectrum recently auctioned in Europe and elsewhere for 3G systems.

Although decisions have not yet been finalized on allocating these bands to 3G wireless communications at this writing, it looks as if there is general agreement that this is the direction that will be pursued. In addition, the FCC is committed to making spectrum available for new advanced wireless services in the United States, as is the World Radio Conference at the international level.

Radio Interface Technology

A key ingredient of the IMT-2000 framework is the air interface technology for 3G systems. For the radio interface technology, the ITU considered 15 submissions from organizations and regional bodies around the world. These proposals were examined by special independent evaluation groups, which submitted their final evaluation reports to the ITU in September 1998. The final selection of key characteristics for the IMT-2000 radio interfaces occurred in March 1999, which led to the development of more detailed ITU specifications for IMT-2000.

The decision of the ITU was to provide essentially a single flexible standard with a choice of multiple access methods, which include CDMA, TDMA, and combined TDMA/CDMA—all potentially in combination with Space Division Multiple Access—to meet the many different mobile operational environments around the world. Although 2G mobile systems involve both TDMA and CDMA technologies, very little use is currently being made of SDMA. However, the ITU expects the advent of adaptive antenna technology linked to systems designed to optimize performance in the space dimension to significantly enhance the performance of future systems.

The IMT-2000 key characteristics are organized, for both the terrestrial and satellite components, into the radio frequency (RF) part (front end), where impacts are primarily on the hardware part of the mobile terminal, and the baseband part, largely defined in software. In addition to RF and baseband, the satellite key characteristics also cover the architecture and the system aspects. According to the ITU, the use of common components for the RF part of the terminals, together with flexible capabilities that are primarily software defined in baseband processing, should provide the mobile terminal functionality to cover the various radio interfaces needed in the twenty-first century as well as provide economies of scale in their production.

U.S. proposals submitted to the ITU for consideration as the radio interface technology in the IMT-2000 framework included wideband versions of CDMA, of which there are three competing standards in North America: wideband cdmaOne, WIMS W-CDMA, and WCDMA/NA. All three have been developed from 2G digital wireless technologies and are evolving to 3G technologies. Early on, WIMS W-CDMA and WCDMA/NA, however, were merged into a single proposed standard and, along with wideband cdmaOne, were submitted to the ITU for inclusion into its IMT-2000 family-of-systems concept for globally interconnected and interoperable 3G networks. Also submitted to the ITU was a separate proposal for a TDMA-based radio interface. Eventually, all these proposals were accepted by the ITU and included in the IMT-2000 family of standards.

Summary

IMT-2000 addresses the key needs of the increasingly global economy—specifically, cross-national interoperability, global roaming, high-speed transmission for multimedia applications and Internet access, and customizable personal services. The markets for all of these exist now and will grow by leaps and bounds through the next millenium. IMT-2000 puts into place standards that permit orderly migration from current 2G networks to 3G networks while providing a growth path to accommodate more advanced mobile services.

See also

Code Division Multiple Access
Global System for Mobile (GSM) Telecommunications
Time Division Multiple Access

LASER TRANSMISSION

A relatively new category of wireless communication uses laser, sometimes called "free-space optics," operating in the near-infrared region of the light spectrum. Utilizing coherent laser light, these wireless line-of-sight links are used in campus environments and urban areas where the installation of cable is impractical and the performance of leased lines is too slow. Laser links between sites can be operated at the full local area network (LAN) channel speed. And unlike microwave transmission, laser transmission does not require a Federal Communications Commission (FCC) license, and data traveling by laser beam cannot be intercepted.

Performance Impairments

The lasers at each location are aligned with a simple bar graph and tone lock procedure. Fiberoptic repeaters are used to connect the LANs to the laser units. Alternatively, a bridge equipped with a fiberoptic to attachment unit interface (AUI) transceiver may be used. Connections to and from the laser are made using standard fiberoptic cable, protecting data from radio frequency interference (RFI) and electromagnetic interference (EMI). Monitors can be attached to the laser units to provide operational status, such as signal strength, and to implement local and remote loop-back diagnostics.

The reason that laser products are not used very often for business applications is that transmission is affected by atmospheric conditions that produce such effects as absorption, scattering, and shimmer. All three can reduce the amount of light energy that is picked up by the receiver and corrupt the data being sent.

Absorption refers to the ability of various frequencies to pass through the air. Absorption is determined largely by the water vapor and carbon dioxide content of the air along the transmission path, which, in turn, depends on humidity and altitude. The gases that form in the atmosphere have many resonance bands that allow specific frequencies of light to pass. These transmission windows occur at various wavelengths, such as the visible light range. Another window occurs at the near-infrared wavelength of approximately 820 nanometers (nm). Laser products tuned to this window are not greatly affected by absorption.

Scattering has a much greater effect on laser transmission than absorption. The atmospheric scattering of light is a function of its wavelength and the number and size of scattering particles in the air. The optical visibility along the transmission path is directly related to the number and size of these particles. Fog and smog are the main conditions that tend to limit visibility for optical-infrared transmission, followed by snow and rain.

Shimmer is caused by localized differences in the air's index of refraction. This is caused by a combination of factors, including time of day (daytime heat), terrain, cloud cover, wind, and the height of the optical path above the source of shimmer. These conditions cause fluctuations in the received signal level by directing some of the light out of its intended path. Beam fluctuations may degrade system performance by producing short-term signal amplitudes that approach threshold values. Signal fades below these threshold values result in error bursts.

Vendors have taken steps to mitigate the effects of absorption, scatter, and shimmer. For example, such techniques as frequency modulation (FM) in the transmitter and an automatic gain control (AGC) in the receiver can minimize the effects of shimmer. Also, selecting an optical path several meters above heat sources can greatly reduce the effects of shimmer. However, all of these distorting conditions can vary greatly within a short time span or persist for long periods, requiring onsite expertise to constantly fine-tune the system.

Many businesses simply cannot risk frequent or extended periods of downtime while the necessary compensating adjustments are being made. As if all this were not enough, there are other potential problems to contend with, such as thermal window coatings and the laser beam's angle of incidence, both of which can disrupt transmission. These problems are being overcome with newer lasers that operate in the 1550-nanometer (nm) wavelength. A 1550-nanometer delivery system is powerful enough to go through windows, can deliver signals under the fog blanket, and is safe enough that it does not blind the casual viewer who happens to look into the beam. Up to 1 Gbps of bandwidth is available with these systems—the equivalent bandwidth capacity of 660 T1 lines (Figure L-1).

There is also a distance limitation associated with laser. The link generally cannot exceed 1.5 kilometers (km), and 1 kilometer is preferred. With 1550-nanometer systems, the practical distance of the link is only 500 meters.

Summary

Despite its limitations, laser, or free-space optics, can provide a valuable last link between the fiber network and the end user—including as a backup to more conventional methods, such as fiber. Free-space optics, unlike other transmission technologies, are not tied to standards or standards development. Vendors simply attach their equipment into existing fiber-based networks and then use any laser transmission

Figure L-1   Terabeam Magna, a free-space optics system from TeraBeam Corp.

methods they like. This encourages innovation, differentiation, and speed of deployment.

See also

Infrared Networking

LOCAL MULTIPOINT DISTRIBUTION SERVICE

Local Multipoint Distribution Service (LMDS) is a two-way millimeter microwave technology that operates in the 27- to 31-GHz range. This broadband service allows communications providers to offer a variety of high-bandwidth services to homes and businesses, including broadband Internet access. LMDS offers greater bandwidth capabilities than a predecessor technology called "Multichannel Multipoint Distribution Service" (MMDS) but has a maximum range of only 7.5 miles from the carrier's hub to the customer premises. This range can be extended, however, through the use of optical fiber links.

Applications

LMDS provides enormous bandwidth—enough to support 16,000 voice conversations plus 200 channels of television programming. Figure L-2 contrasts LMDS with the bandwidth available over other wireless services.

Competitive Local Exchange Carriers (CLECs) can deploy LMDS to completely bypass the local loops of the Incumbent Local Exchange Carriers (ILECs), eliminating access charges and avoiding service-provisioning delays. Since the service entails setting up equipment between the provider's

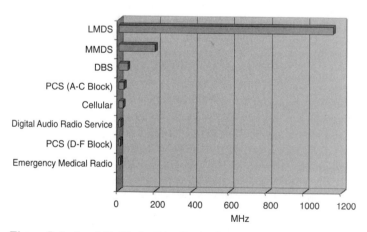

Figure L-2   Local Multipoint Distribution Service (LMDS) operates in the 27- to 31-GHz range and offers 1150 MHz of bandwidth capacity, which is over two times more than all other auctioned spectrum combined.

hub location and customer buildings for the microwave link, LMDS costs far less to deploy than installing new fiber. This allows CLECs to very economically bring customer traffic onto their existing metropolitan fiber networks and, from there, to a national backbone network.

The strategy among many CLECs is to offer LMDS to owners of multitenant office buildings and then install cable to each tenant who subscribes to the service. The cabling goes to an on-premises switch, which is run to the antenna on the building's roof. That antenna is aimed at the service provider's antenna at its hub location. The line-of-sight wireless link between the two antennas offers a broadband "pipe" for multiple voice, data, and video applications. Subscribers can use LMDS for a variety of high-bandwidth applications, including television broadcast, videoconferencing, LAN interconnection, broadband Internet access, and telemedicine.

Operation

LMDS operation requires a clear line of sight between the carrier's hub station antenna and the antenna at each customer location. The maximum range between the two is 7.5 miles. However, LMDS is also capable of operating without having a direct line-of-sight with the receiver. This feature, highly desirable in built-up urban areas, may be achieved by bouncing signals off buildings so that they get around obstructions. At the receiving location, the data packets arriving at different times are held in queue for resequencing before they are passed to the application. This scheme does not work well for voice, however, because the delay resulting from queuing and resequencing disrupts two-way conversation.

At the carrier's hub location there is a roof-mounted multisectored antenna (Figure L-3). Each sector of the antenna receives/transmits signals between itself and a specific customer location. This antenna is very small, some measuring only 12 inches in diameter. The hub antenna brings the multiplexed traffic down to an indoor switch (Figure L-4) that processes the data into 53-byte Asynchronous Transfer Mode (ATM) "cells" for transmission over the carrier's fiber network. These individually addressed cells are converted back to their native format before going off the carrier's network to their proper destinations—the Internet, Public Switched Telephone Network (PSTN), or customer's remote location.

At each customer's location, there is a rooftop antenna that sends/receives multiplexed traffic. This traffic passes through an indoor network interface unit (NIU) that provides the gateway between the RF (radio frequency) components and the in-building equipment, such as a LAN hub, Private Branch Exchange (PBX), or videoconferencing system. The NIU includes an up/down converter that changes the frequency of the microwave signals to a lower intermediate frequency (IF) that the electronics in the office equipment can manipulate more easily (and inexpensively).

Spectrum Auctions

In May 1999, the FCC held the last auction for LMDS spectrum. Over 100 companies qualified for the auctions, bidding against each other for licenses in select basic trading areas

Figure L-3 A multisectored antenna at the carrier's hub location transmits/receives traffic between the antennas at each customer location.

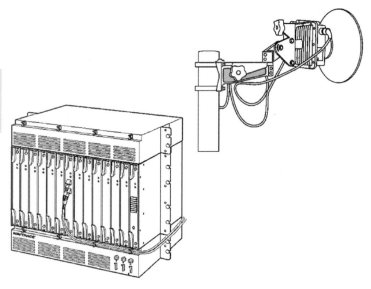

Figure L-4   A microwave transceiver (*top right*) handles multiple point-to-point downstream and upstream channels to customers. The transceiver is connected via coaxial cables to an indoor switch (*bottom left*) that provides the connectivity to the carrier's fiber network. The traffic is conveyed over the fiber network in the form of 53-byte ATM cells. (*Source: Wavtrace, Inc.*)

(BTAs). The FCC auctioned two types of licenses in each market: An "A-block" license permits the holder to provision 1150 MHz of spectrum for distribution among its customers, while a "B-block" license permits the holder to provision 150 MHz. Most of the A-block licenses in the largest BTAs were won by major CLECs, while the B-block licenses were taken by smaller companies, Internet service providers (ISPs), universities, and government agencies. The licenses are granted for a 10-year period, after which the FCC can take them back if the holder does not have service up and running.

Development History

Bernard Bossard is generally recognized as the inventor of LMDS. Bossard, who had worked with microwaves for the military, believed that he could make point-to-multipoint video work in the 28-GHz band. Not interested in sending high-powered, low-frequency signals over long distances, Bossard focused instead on sending low-powered, high-frequency signals over a short distance. The result was LMDS. In 1986, he received funding and formed CellularVision with his financial backers. CellularVision then spun off the technical rights to the technology into a separate subsidiary, CT&T, that licenses it to other companies.

CellularVision was awarded a pioneer's preference license by the FCC for its role in developing LMDS. CellularVision began operating a commercial LMDS in metropolitan New York, providing video programming to subscribers in the Brighton Beach area. In 1998, CellularVision changed its name to SPEEDUS.COM. The company has a network operations center and recently has been expanding the number of operating cells in the New York area and now claims more than 12,000 residential and business subscribers.

SPEED service is delivered via 14 fully functional Internet broadcast stations in operation under SPEEDUS.COM's FCC license covering metropolitan New York. SPEED subscribers are able to browse the Web using the company's SPEED modem capable of downstream speeds of up to 48 Mbps, which is 31 times faster than a full T1 line.

In the SPEEDUS.COM system, cable programming is downlinked from satellites to the company's head-end facility, where local broadcast transmissions are also received. At the company's master control room, the programming signals are then amplified, sequenced, scrambled, and up-converted to 28 GHz. The SPEED.COM transmitters and repeaters then broadcast a polarized FM signal in the 28-GHz band over a radius of up to 3 miles to subscribers and to adjacent cells for transmission. A 6-inch-square, highly directional, flat-plate, window- , roof- , or wall-mounted antenna receives the scrambled signal and delivers it to the addressable set-top converter, which decodes the signals. The subscriber receives 49 channels of high-quality video and audio programming, including pay-per-view and premium channels.

Over 100 companies own licenses for LMDS. XO Communications (formerly known as Nextlink) is one of the largest single holders of LMDS licenses in the United States, having invested over $800 million in such systems, largely through the acquisition of other companies that held LMDS licenses. XO is a CLEC and is using LMDS to feed traffic to its fiber networks. Its approach to building out a city is to install fiber. In areas where that will take too long or where permits are too hard to come by, XO will use, in this order, LMDS, Digital Subscriber Line (DSL), and ILEC facilities.

Potential Problems

A potential problem for LMDS users is that the signals can be disrupted by heavy rainfall and dense fog—even foliage can block a signal. In metropolitan areas where new construction is a fact of life, a line-of-sight transmission path can disappear virtually overnight. For these reasons, many information technology (IT) executives are leery of trusting mission-critical applications to this wireless technology. Service providers downplay this situation by claiming that LMDS is just one local access option and that fiber links are the way to go for mission-critical applications. In fact, some LMDS providers offer fiber as a backup in case the microwave links experience interference.

There is controversy in the industry about the economics of the point-to-multipoint architecture of LMDS, with some experts claiming that the business model of going after low-usage customers is fundamentally flawed and will never justify the service provider's cost of equipment, installation, and provisioning. With an overabundance of fiber in the ground and metropolitan area Gigabit Ethernet services coming online at a competitive price, the time for LMDS may have come and gone. In addition, newer wireless technologies like free-air laser hold a significant speed advantage over LMDS, as does submillimeter transmission in the 60- and 95-GHz bands.

Another problem that has beset LMDS is that the major license holders have gotten caught up in financial problems, some declaring Chapter 11 bankruptcy. These carriers built their networks quickly, incurring massive debt, without lining up customers fast enough. This strategy worked well as long as the capital markets were willing to continue funding these companies. But once the capital markets dried up in 2000, so did the wireless providers' coffers and their immediate prospects. The uncertain future of these financially strapped carriers has discouraged many companies from even trying LMDS.

Summary

Fiberoptics is the primary transmission medium for broadband connectivity today. However, of the estimated 4.6 million commercial buildings in the United States, 99 percent are not served by fiber. Businesses are at a competitive disadvantage in today's information-intensive world unless they have access to broadband access services, including high-speed Internet access. These businesses, including many data-intensive high-technology companies, can be served adequately with LMDS. Despite the financial problems of LMDS providers, the technology has the potential to become a significant portion of the global access market, which will include a mix of many technologies, including DSL, cable modems, broadband satellite, and fiberoptic systems.

See also

Microwave Communications

Multichannel Multipoint Distribution Service

LOW-POWER FM RADIO SERVICE

The FCC in January 2000 created two new classes of noncommercial radio stations, referred to as "low-power frequency-modulated (LPFM) radio services." LPFM radio services are designed to serve very localized communities or underrepresented groups within communities.

The LP100 service operates in the power range of 50 to 100 watts and has a service radius of about 3.5 miles. The LP10 service operates in the power range of 1 to 10 watts and has a service radius of about 1 to 2 miles. In conjunction with the new radio services, the FCC adopted interference protection requirements based on distance separation between stations. This is intended to preserve the integrity of existing FM radio stations and safeguard their ability to transition to digital transmission capabilities.

The FCC put into place minimum distance separations as the best practical means of preventing interference between low-power radio and full-power FM stations. It requires minimum distances between stations using the same or first adjacent channels. However, third adjacent channel and possibly second adjacent channel separations may not be necessary in view of the low power levels of LPFM radio.

License Requirements

Eligible LPFM licensees can be noncommercial government or private educational organizations, associations, or entities; nonprofit entities with educational purposes; or government or nonprofit entities providing local public safety or transportation services. However, LPFM licenses will be awarded throughout the FM radio band and will not be limited to the channels reserved for use by noncommercial educational radio stations.

To further its goals of diversity and creating opportunities for new voices, no existing broadcaster or other media entity can have an ownership interest or enter into any program or operating agreement with any LPFM station. In addition, to encourage locally originated programming, LPFM stations will be prohibited from operating as translators.

To foster local ownership and diversity, during the first 2 years of LPFM license eligibility, licensees will be limited to local entities certifying that they are physically headquartered, have a campus, or have 75 percent of their board members residing within 10 miles of the station they seek to operate. During this time, no entity may own more than one LPFM station in any given community. After 2 years from the date the first applications are accepted, in order to bring into use whatever low-power stations remain available but unapplied for, applications will be accepted from nonlocal entities.

For the first 2 years, no entity will be permitted to operate more than one LPFM station nationwide. After the second year, eligible entities will be able to own up to five stations nationwide, and after 2 more years, up to 10 nationwide.

LPFM stations are licensed for 8-year renewable terms. These licenses are not transferable. Licensees receive four-letter call signs with the letters *LP* appended.

In the event multiple applications are received for the same LPFM license, the FCC will implement a selection process that awards applicants one point each for

- Certifying an established community presence of at least 2 years prior to the application
- Pledging to operate at least 12 hours daily
- Pledging to air at least 8 hours of locally originated programming daily

If applicants have the same number of points, time-sharing proposals will be used as a tiebreaker. Where ties have not been resolved, a group of up to eight mutually exclusive applicants will be awarded successive license terms of at least 1 year for a total of 8 years. These 8-year licenses will not be renewable.

LPFM stations will be required to broadcast a minimum of 36 hours per week, the same requirement imposed on full-power noncommercial educational licensees. They will be subject to statutory rules, such as sponsorship identification, political programming, prohibitions of airing obscene or indecent programming, and requirements to provide periodic call sign announcements. They also will be required to participate in the national Emergency Alert System.

Summary

According to the FCC, the new LPFM service will enhance community-oriented radio broadcasting. During the proceedings leading up to the new classes of radio service, broad national interest in LPFM service was demonstrated by the thousands of comments received from state and local government entities, religious groups, students, labor unions, community organizations, musicians, and others supporting the introduction of a new LPFM service. The FCC expects that the local nature of the LPFM service, coupled with the eligibility and selection criteria, will ensure that LPFM licensees will meet the needs and interests of their communities.

See also

Federal Communications Commission

Spectrum Auctions

LOW-POWER RADIO SERVICE

Low-Power Radio Service (LPRS) is one of the Citizens' Band (CB) Radio Services. It is a one-way short-distance very-high-frequency (VHF) communication service providing auditory assistance to persons with disabilities, persons who require language translation, and persons in educational settings. It also provides health care assistance to the ill, law enforcement tracking services in cooperation with a law enforcement agency, and point-to-point network control communications for Automated Marine Telecommunications System (AMTS) coast stations. In all applications, two-way voice communications are prohibited.

A license from the FCC is not needed to use most LPRS transmitters. To operate an LPRS transmitter for AMTS purposes, however, the user must hold an AMTS license. Otherwise, provided the user is not a representative of a foreign government, anyone can operate an FCC type-accepted LPRS transmitter for voice, data, or tracking signals.

An LPRS transmitter may be operated within the territorial limits of the 50 United States, the District of Columbia, and the Caribbean and Pacific insular areas. It also may be operated on or over any other area of the world, except within the territorial limits of areas where radio communications are regulated by another agency of the United States or within the territorial limits of any foreign government. The transmitting antenna must not exceed 30.5 meters (100 feet) above ground level. This height limitation does not apply, however, to LPRS transmitter units located indoors or where the antenna is an integral part of the unit.

There are 260 channels available for LPRS. These channels are available on a shared basis only and are not assigned for the exclusive use of any entity. Certain channels (19, 20, 50, and 151 to 160) are reserved for law enforcement tracking purposes. Further, AMTS-related transmissions are limited to the upper portion of the band (216.750 to 217.000 MHz).

Users must cooperate in the selection and use of channels in order to reduce interference and make the most effective use of the authorized facilities. Channels must be selected in an effort to avoid interference with other LPRS transmissions. This means that if users are experiencing interference on a particular channel, they should change to another channel until a clear one is found.

Finally, operation is subject to the conditions that no harmful interference is caused to the U.S. Navy's SPASUR radar system (216.88 to 217.08 MHz) or to a Channel 13 television station.

Summary

LPRS can operate anywhere CB station operation is permitted. An LPRS station is not required to transmit a station identification announcement. The LPRS transmitting device may not interfere with TV reception or federal government radar and must accept any interference received, including interference that may cause undesired operation. On request, system equipment must be available for inspection by an authorized FCC representative.

See also

Citizens Band Radio Service

Family Radio Service

General Mobile Radio Service

M

MARITIME MOBILE SERVICE

The Maritime, or Marine, Radio Services have evolved from the earliest practical uses of radio. In 1900, just 6 years after Marconi demonstrated his "wireless" radio, devices were being installed aboard ships to enable them to receive storm warnings transmitted from stations on shore. Today, the same principle applies in using both shipboard and land stations in the marine services to safeguard life and property at sea. Both types of stations are also used to aid marine navigation, commerce, and personal business, but such uses are secondary to safety, which has international priority.

The Marine Radio Services include the Maritime Mobile Service, the Maritime Mobile-Satellite Service, the Port Operations Service, the Ship Movement Service, the Maritime Fixed Service, and the Maritime Radiodetermination Service.

- Maritime Mobile Service is an internationally allocated radio service providing for safety of life and property at sea and on inland waterways.

- Maritime Mobile-Satellite Service provides frequencies for public correspondence between ships and public coast stations as well as between aircraft and public coast stations and coast earth stations. The transmission of public correspondence from aircraft must not cause interference to maritime communications.

- Port Operations Service provides frequencies for inter-ship communications related to port operations in coastal harbors, allowing the vessel traffic to be managed more efficiently while protecting the marine environment from vessel collisions and groundings.

- Ship Movement Service provides frequencies for communications relating to the operational handling of the movement and the safety of ships and, in emergency, to the safety of persons.

- Maritime Fixed Service provides frequencies for communications equipment installed on oil drilling platforms, lighthouses, and maritime colleges.

- Maritime Radiodetermination Service provides frequencies for determining position, velocity, and other characteristics of vessels.

Summary

Together, shipboard and land stations in the Marine Services are meant to serve the needs of the entire maritime community. The Federal Communications Commission (FCC) regulates these services both for ships of U.S. registry that sail in international and foreign waters and for all marine activities in U.S. territory. For this and other reasons, the rules make a distinction between compulsory users of marine radio for safety at sea and noncompulsory uses for purposes other than safety. In addition, rules concerning domestic marine communications are matched to requirements of the U.S. Coast Guard, which monitors marine distress frequencies continuously to protect life and property in U.S. waters.

See also

Global Maritime Distress and Safety System

MICROWAVE COMMUNICATIONS

A microwave is a short radio wave that varies from 1 millimeter to 30 centimeters in length. Because microwaves can pass through the ionosphere, which blocks or reflects longer radio waves, microwaves are well suited for satellite communications. This reliability also makes microwave well suited to terrestrial communications as well, such as those delivered by Local Multipoint Distribution Service (LMDS) and Multichannel Multipoint Distribution Service (MMDS).

Much of the microwave technology in use today for point-to-point communications was derived from radar developed during World War II. Initially, microwave systems carried multiplexed speech signals over common carrier and military communications networks; but today they are used to handle all types of information—voice, data, facsimile, and video—in either an analog or digital format.

The first microwave transmission occurred in 1933, when European engineers succeeded in communicating reliably across the English Channel—a distance of about 12 miles (20 kilometers). In 1947, the first commercial microwave network in the United States came online. Built by Bell Laboratories, this was a New York to Boston system consisting of 10 relay stations carrying television signals and multiplexed voice conversations.

A year later, New York was linked to San Francisco via 109 microwave relay stations. By the 1950s, transcontinental microwave networks were routinely handling over 2000 voice channels on hops averaging 25 miles (41.5 kilometers) in length. By the 1970s, just about every single telephone call, television show, telegram, or data message that crossed the country spent some time on a microwave link.

Over the years, microwave systems have matured to the point that they have become major components of the nation's Public Switched Telephone Network (PSTN) and an essential technology with which private organizations can satisfy internal communications requirements.

Microwave systems can even exceed the 99.99 percent reliability standard set by the telephone companies for their phone lines.

Microwave Applications

Early technology limited the operations of microwave systems to radio spectrum in the 1-GHz range, but because of improvements in solid-state technology, today's government systems are transmitting in the 153-GHz region, while commercial systems are transmitting in the 40-GHz region with FCC approval. The 64- to 71-GHz band is reserved for intersatellite links. These frequency bands offer short-range wireless radio systems the means to provide communications capacities approaching those now achievable only with coaxial cable and optical fiber.

These spectrum allocations offer a variety of possibilities, such as use in short-range, high-capacity wireless systems that support educational and medical applications, and wireless access to libraries or other information databases. In addition to telecommunication service providers, short-haul microwave equipment is used routinely by hotel chains, CATV operators, and government agencies.

Corporations are making greater use of short-haul microwave, especially for extending the reach of local area networks (LANs) in places where the cost of local T1 lines is prohibitive. Common carriers use microwave systems for backup in the event of fiber cuts and in terrain where laying fiber is not economically feasible. Cellular service providers use microwave to interconnect cell sites with each other as well as to the regular telephone net-work. Some interexchange carriers (and corporations) even use short-haul microwave to bypass Incumbent Local Exchange Carriers (ILECs) to avoid lengthy service provisioning delays and to avoid paying hefty local access charges.

Network Configurations

There are more than 25,000 microwave networks in the United States alone. There are basically two microwave network configurations: point-to-point and point-to-multipoint. The first type meets a variety of low- and medium-density communications requirements, ranging from simple links to more complex extended networks, such as

- Sub-T1/E1 data links
- Ethernet/Token Ring LAN extensions
- Low-density digital backbone for wide area mobile radio and paging services
- PBX/OPX/FX voice, fax, and data extensions
- Facility-to-facility bulk data transfer

Point-to-multipoint microwave systems provide communications between a central command and control site and remote data units. A typical radio communications system provides connections between the master control point and remote data collection and control sites. Repeater configurations are also possible. The basic equipment requirements for a point-to-multipoint system include

- *Antennas* For the master, an omnidirectional antenna; for the remotes, a highly directional antenna aimed at the master station's location.
- *Tower (or other structure, such as a mast)* To support the antenna and transmission line.
- *Transmission line* Low-loss coaxial cable connecting the antenna and the radio.
- *Master station radio* Interfaces with the central computer; it transmits and receives data from the remote radio sites and can request diagnostic information from the remote transceivers. The master radio also can serve as a repeater.

- *Remote radio transceiver* Interfaces to the remote data unit; receives and transmits to the master radio.

- *Management station* A computer that can be connected to the master station's diagnostic system either directly or remotely for control and collection of diagnostic information from master and remote radios.

Wireless Cable

Traditionally, cable system operators have used microwave transmission systems to link cable networks. These Cable Antenna Relay Services (CARS) have experienced declining usage as cable operators have deployed more optical fiber in their transmission systems. However, improvements in microwave technology and the opening of new frequencies for commercial use have contributed to the resurgence in short-haul microwave. In the broadcast industry, short-haul microwave is often referred to as "wireless cable," which comes in the form of Local Multipoint Distribution Service (LMDS) and Multichannel Multipoint Distribution Service (MMDS).

These wireless cable technologies have two key advantages. One is availability—with an FCC license, they can be made available in areas of scattered population and other areas where it is too expensive to build a traditional cable station. The other is affordability—because of the lower costs of building a wireless cable station, savings can be passed on to subscribers.

Regulation

The radio spectrum is the part of the natural spectrum of electromagnetic radiation lying between the frequency limits of 9 kHz and 400 GHz. In the United States, regulatory responsibly for the radio spectrum is divided between the FCC and the National Telecommunications and Information Administration (NTIA).

The FCC, which is an independent regulatory agency, administers spectrum for non-federal government use, and the NTIA, which is an operating unit of the Department of Commerce, administers spectrum for federal government use. Within the FCC, the Office of Engineering and Technology (OET) provides advice on technical and policy issues pertaining to spectrum allocation and use. This office manages the spectrum and provides leadership to create new opportunities for competitive technologies and services for the American public.

Summary

Microwave is now almost exclusively a short-haul transmission medium, while optical fiber and satellite have become the long-haul transmission media of choice. Short-haul microwave is now one of the most agile and adaptable transmission media available, with the capability of supporting data, voice, and video. It is also used to back up fiberoptic facilities and to provide communications services in locations where it is not economically feasible to install fiber.

See also

Local Multipoint Distribution Service

Multichannel Multipoint Distribution Service

Satellite Communications

MOBILE TELEPHONE SWITCHING OFFICE

The Mobile Telephone Switching Office (MTSO) acts like an ordinary switching node on the Public Switched Telephone Network (PSTN) or Integrated Services Digital Network (ISDN) and provides all the functionality needed to handle a mobile subscriber, such as registration, authentication, location updating, handoffs, and call routing to a roaming subscriber.

All cell phones have special codes associated with them:

- *Electronic Serial Number (ESN)* A unique 32-bit number programmed into the phone when it is manufactured.
- *Mobile Identification Number (MIN)* A 10-digit number derived from the mobile phone number.
- *System Identification Code (SID)* A unique 5-digit number that is assigned to each carrier by the FCC.

While the ESN is considered a permanent part of the phone, both the MIN and SID codes are programmed into the phone on purchase and activation with a service plan.

When the mobile phone is powered up, it listens for the network operator's SID on the control channel. The control channel is a special frequency that the phone and base station use to talk to one another about such functions as call setup and channel changing. If the phone cannot find any control channels to listen to, it assumes that it is out of range and displays a "no service" message.

When it receives the SID, the phone compares it to the SID programmed into the phone. If the SIDs match, the phone knows that the cell it is communicating with is part of its home system.

Along with the SID, the phone also transmits a registration request. The MTSO then knows the location of the phone, which is recorded in a database so that it knows which cell to target when it wants to ring that phone for an incoming call.

When the MTSO gets the call, it looks up the location of the phone in its database. The MTSO picks a frequency pair the phone will use in that cell to take the call. The MTSO communicates with the phone over the control channel to tell it which frequencies to use, and once the phone and the tower switch on those frequencies, the call is connected.

As the mobile phone moves toward the edge of a cell, that cell's base station notices that its signal strength is diminishing. Meanwhile, the base station in the cell the mobile phone is moving toward is listening and measuring signal strength on all frequencies and sees that the approaching phone's signal strength increasing. The two base stations coordinate with each other through the MTSO, and at some point, the phone gets a signal on a control channel telling it to change frequencies. This handoff switches the phone to the new cell.

If the SID on the control channel does not match the SID programmed into the mobile phone, then the phone assumes that it is roaming. The MTSO of the cell that the phone is roaming in contacts the MTSO of its home system, which then checks its database to confirm that the SID is valid. The home system verifies the phone to the local MTSO, which then tracks it while it is moving through its cells.

Summary

The local wireless cellular network consists of a Mobile Telephone Switching Office (MTSO) with cell sites scattered throughout a geographic serving region. T-carrier or fiber lines are typically leased from the local carrier to interconnect the cell sites with the MTSO. These lines also provide the MTSO with connectivity to a local central office switch so that calls can be completed between the wireless network and the PSTN.

See also

Cell Sites
Cellular Voice Communication
Cellular Data Communication
Cellular Telephones

MULTICHANNEL MULTIPOINT DISTRIBUTION SERVICE

Multichannel Multipoint Distribution Service (MMDS) is a microwave technology that traces its origins to 1972 when it was introduced to provide an analog service called Multipoint Distribution Service (MDS). For many years, MMDS was used for one-way broadcast of television programming, but in early 1999, the FCC opened up this spectrum to allow for two-way transmissions, making it useful for delivering telecommunication services, including high-speed Internet access to homes and businesses.

This technology, which has now been updated to digital, operates in the 2- to 3-GHz range, enabling large amounts of data to be carried over the air from the operator's antenna towers to small receiving dishes installed at each customer location. The useful signal range of MMDS is about 30 miles, which beats Local Multipoint Distribution Service (LMDS) at 7.5 miles and Digital Subscriber Line (DSL) at 18,000 feet. Furthermore, MMDS is easier and less costly to install than cable service.

Operation

With MMDS, a complete package of TV programs can be transmitted to homes and businesses. Since MMDS operates within the frequency range of 2 to 3 GHz, which is much lower than LMDS at 28 to 31 GHz, it can support only up to 24 stations. However, operating at a lower frequency range means that the signals are not as susceptible to interference as those using LMDS technology.

Most of the time the operator receives TV programming via a satellite downlink. Large satellite antennas installed at the head end collect these signals and feed them into encoders that compress and encrypt the programming. The encoded video and audio signals are modulated, via amplitude modulation (AM) and frequency modulation (FM), respectively, to an intermediate frequency (IF) signal. These IF signals are up-converted to MMDS frequencies and then amplified and combined for delivery to a coaxial cable, which is connected to the transmitting antenna. The antenna can have an omnidirectional or sectional pattern.

The small antennas at each subscriber location receive the signals and pass them via a cable to a set-top box connected to the television. If the service also supports high-speed Internet access, a cable also goes to a special modem connected to the subscriber's PC. MMDS sends data as fast as 10 Mbps downstream (toward the computer). Typically, service providers offer downstream rates of 512 kbps to 2.0 Mbps, with burst rates up to 5 Mbps whenever spare bandwidth becomes available.

Originally, there was a line-of-sight limitation with MMDS technology. But this has been overcome with a complementary technology called Vector Orthogonal Frequency Division Multiplexing (VOFDM). Because MMDS does not require an unobstructed line of sight between antennas, signals bouncing off objects en route to their destination require a mechanism for being reassembled in their proper order at the receiving site. VOFDM handles this function by leveraging multipath signals, which normally degrade transmissions. It does this by combining multiple signals at the receiving end to enhance or recreate the transmitted signals. This increases the overall wireless system performance, link quality, and availability. It also increases service providers' market coverage through non-line-of-sight transmission.

Channel Derivation

MMDS equipment can be categorized into two types based on the duplexing technology used: Frequency Division Duplexing (FDD) or Time Division Duplexing (TDD). Systems based on FDD are a good solution for voice and bidirectional data because forward and reverse use separate and equally large frequency bands. However, the fixed nature of this scheme limits overall efficiency when used for Internet access. This is so because Internet traffic tends to be "bursty" and asymmetric. Instead of preassigning bandwidth with FDD, Internet traffic is best supported by a more flexible bandwidth allocation scheme.

This is where TDD comes in; it is more efficient because each radio channel is divided into multiple time slots through Time Division Multiple Access (TDMA) technology, which enables multiple channels to be supported. Because TDD has flexible timeslot allocations, it is better suited for data delivery—specifically Internet traffic. TDD enables service providers to vary uplink and downlink ratios as they add customers and services. Many more users can be supported by the allocation of bandwidth on a nonpredefined basis.

Summary

MMDS is being used to fill the gaps in market segments where cable modems and DSL cannot be deployed because of distance limitations and cost concerns. Like these technologies, MMDS provides data services and enhanced video services such as video on demand, as well as Internet access. MMDS will be another access method to complement a carrier's existing cable and DSL infrastructure, or it can be used alone for direct competition. With VOFDM technology, MMDS is becoming a workable option that can be deployed cost-effectively to reach urban businesses that do have line-of-sight access and in suburban and rural markets for small businesses and telecommuters.

See also

Cable Television Networks

Digital Subscriber Line Technologies

Local Multipoint Distribution Service

Microwave Communications

MULTICHANNEL VIDEO DISTRIBUTION AND DATA SERVICE

Multichannel Video Distribution and Data Service (MVDDS) supports broadband communication services that include local television programming and high-speed Internet access in the 12-GHz band. This is the same band used by Direct Broadcast Satellite (DBS).

The FCC adopted MVDDS service and technical rules that permit MVDDS operators to share the 12-GHz band with DBS on a coprimary basis subject to the condition that they do not cause impermissible interference to the DBS service. Specifically, the FCC adopted equivalent power flux density (EPFD) limits for MVDDS to protect DBS subscribers from interference. MVDDS operators are required to ensure that the adopted EPFD limits are not exceeded at any existing DBS customer location. If the EPFD limits are exceeded, the MVDDS operator will be required to discontinue service until the limits can be met. MVDDS power flux density limits, MVDDS spacing rules, and coordination requirements are intended to facilitate mutual sharing of the 12-GHz band. A "safety valve" allows individual DBS licensees or distributors to present evidence that the appropriate EPFD for a given service area should be different from the EPFD applicable in that zone.

With MVDDS, the FCC adopted a geographic licensing scheme and, in the event mutually exclusive applications are filed, will assign licenses via competitive bidding. A licensing system based on component economic areas (CEAs) will be used with one spectrum block of 500 MHz available per CEA.

In setting up the rules for MVDDS, the FCC restricted dominant cable operators from acquiring an interest in an MVDDS license for a service area where significant overlap is present. There is no restriction on DBS providers from acquiring MVDDS licenses.

Summary

MVDDS providers will share the 12-GHz band with incumbent Direct Broadcast Satellite (DBS) providers on a coprimary, non-harmful interference basis. The objective of this arrangement is to accommodate the introduction of innovative services and to facilitate the sharing and efficient use of spectrum. Furthermore, the FCC believes that this service will facilitate the delivery of communications services, such as video and broadband services, to various populations, including those deemed to be unserved or underserved.

See also

Direct Broadcast Satellite

Federal Communications Commission

Interactive Video and Data Service

O

OVER-THE-AIR SERVICE ACTIVATION

Over-the-air service activation, sometimes called "over-the-air service provisioning," allows a potential wireless, both cellular and personal communications services (PCS), service subscriber to activate new wireless service without the intervention of a third party (e.g., authorized dealer). The use of special software enables wireless service providers to offer over-the-air service provisioning capabilities—including initial activation—plus provisioning of other innovative wireless features, such as paging and voice mail. The process is made secure by restricting the phone's initial use to activation only. Once subscribers have the phone in hand, they can immediately dial a customer service representative who can activate the phone and accept account information.

Over-the-air activation also enables the service provider to activate a potential service subscriber's unit by downloading over the air the required parameters, such phone number and features, into the unit. The service subscriber does not have to bring the unit into a dealer or service agent. This allows service providers the capability to start marketing subscriber units through nontraditional mass-market

225

retailers who do not have the personnel to individually program subscriber units.

Another capability of over-the-air service activation is the
ability to load an authentication key into a subscriber unit
securely. Authentication is the process by which information
is exchanged between a subscriber unit and the network for
the purpose of confirming and validating the identity of the
subscriber unit. The over-the-air service activation feature
incorporates an authentication key exchange agreement
algorithm. This algorithm enhances security for the subscriber and reduces the potential for fraudulent use of cellular service.

New customers simply place a call to the cellular operator,
and the information is transferred automatically to the cellular phone over the cellular airwaves. This method of activation enables cellular operators to explore new distribution
channels for subscriber units and substantially reduce distribution and service provisioning costs.

These features operate over a digital control channel
(DCC). In addition to over-the-air programming, the DCC
supports such advanced services as calling line ID, message
waiting indication, and Short Message Service (SMS). It also
offers tiered services, allowing operators to tailor pricing
packages for residential and business customers based on
location and usage. Sleep mode, another DCC feature,
improves handset battery life by allowing mobile phones to
"sleep" while idle to conserve power. DCC also improves network performance and supports advanced voice coder technology for improved audio quality.

Summary

In 1995, Code Division Multiple Access (CDMA) became the
first digital cellular technology to offer instant activation to
customers based on specifications defined by the CDMA
Development Group (CDG) in 1994. In the future, cellular
operators will look to leverage this capability to offer more

value-added services, providing them with the latest applications software coming directly from the network.

See also

Cellular Telephones
Code Division Multiple Access
Mobile Telephone Switching Office
Time Division Multiple Access

P

PAGING

Paging is a wireless service that provides one- or two-way messaging to give mobile users continuous accessibility to family, friends, and business colleagues while they are away from their telephones. Typically, the mobile user carries a palm-sized device (the pager or some other portable device with a paging capability) that has a unique identification number. The calling party inputs this number, usually through the Public Switched Telephone Network (PSTN), to the paging system, which then signals the pager to alert the called party.

Alternatively, callback numbers and short-text messages can be sent to pagers via messaging software installed on a PC or input into forms accessed on the Web for delivery via an Internet gateway. Regardless of delivery method, the called party receives an audio or visual notification of the call, which includes a display of the phone number to call back. If the pager has an alphanumeric capability, messages may be displayed on the pager's screen.

Origin of the Pager

The pioneer of wireless telecommunications is Al Gross (Figure P-1). In 1938, the Canadian inventor developed the walkie-talkie. In 1948, he pioneered Citizens' Band (CB) radio. In 1949, he invented the pager from radio technology he used for blowing up bridges via remote control during World War II. His first attempt to sell pagers to doctors and nurses in 1960 failed because nurses did not want to disturb patients and doctors did not want to disturb their golf games.

Gross's ideas were so far advanced that most of his patents expired before the technology could catch up to make his inventions a reality. As a result, he did not make much money. Had he been born 35 years later, he could have capitalized on his ideas to become far wealthier than Bill Gates at Microsoft.

Figure P-1 Wireless pioneer Al Gross (1918–2000) invented many of the concepts used today in cordless and cellular telephony for which he held many patents.

Paging Applications

There are many applications for paging. Among the most popular are

- *Mobile messaging* Allows messages to be sent to mobile workers. They can respond with confirmation or a request for additional instructions.

- *Data dispatch* Allows managers to schedule work appointments for mobile workers. When they activate their pagers each morning, their itinerary will be waiting for them.

- *Single-key callback* Allows the user to read a message and respond instantly with a predefined stored message that is selected with a single key.

Some message paging services work with text messaging software programs, allowing users to send messages from their desktop or notebook computers to individuals or groups. This kind of software also keeps a log of all messaging activity. This method also offers privacy, since messages do not have to go through an operator before delivery to the recipient.

Types of Paging Services

Several types of paging services are available.

Selective Operator-Assisted Voice Paging Early paging systems were nonselective and operator-assisted. Operators at a central control facility received voice input messages, which were taped as they came in. After an interval of time—15 minutes or so—these messages were then broadcast and received by all the paging system subscribers. This meant that subscribers had to tune in at appointed times and listen to all messages broadcast to see if there were any messages for them. Not only did this method waste airtime, it also was inconvenient and labor-intensive and offered no privacy.

These disadvantages were overcome with the introduction of address encoders at the central control facility and associated decoders in the pagers. Each pager was given a unique address code. Messages intended for a particular called party were input to the system preceded by this address. In this way, only the party addressed was alerted to switch on his or her pager to retrieve messages.

With selective paging, tone-only alert paging became possible. The called party was alerted by a beep tone to call the operator or a prearranged home or office number to have the message read back.

Automatic Paging Traditionally, an operator was always needed either to send the paging signal or to play back or relay messages for the called party. With automatic paging, a telephone number is assigned to each pager, and the paging terminal can automatically signal for voice input, if any, from the calling party, after which it will automatically page the called party with the address code and relay the input voice message.

Tone and Numeric Paging Voice messages take up a lot of airtime, and as the paging market expands, frequency overcrowding becomes a potentially serious problem. Tone-only alert paging saves on airtime usage but has the disadvantage that the alerted subscriber knows only that he or she has to call certain prearranged numbers, depending on the kind of alert tone received.

With the introduction of numeric display pagers in the mid-1980s, the alert tone is followed by a display of a telephone number to call back or a coded message. This method resulted in great savings in airtime usage because it was no longer necessary to add a voice message after the alert tone. This is still the most popular form of paging.

Alphanumeric Paging Alphanumeric pagers display text or numeric messages entered by the calling party or operator using a modem-equipped computer or a custom page-entry device designed to enter short-text messages. Although alphanumeric pagers have captured a relatively small market in recent years, the introduction of value-added services that include news, stock quotes, sports scores, traffic bulletins, and other specialized information services has heated up the market for such devices.

Ideographic Paging Pagers capable of displaying different ideographic languages—Chinese, Japanese, and others—are also available. The particular language supported is determined by the firmware (computer program) installed in the pager and in the page-entry device. The pager is similar to that used in alphanumeric display paging.

Paging System Components

The key components of a paging system include an input source, the existing wireline telephone network, the paging encoding and transmitter control equipment, and the pager itself.

Input Source A page can be entered from a phone, a computer with modem, or other type of desktop page-entry device; a personal digital assistant (PDA); or an operator who takes a phone-in message and enters it on behalf of the caller. Various forms posted on the Web also can be used to input messages to pagers.

The Web form of WorldCom (Figure P-2), for example, allows users to send a text message consisting of a maximum of 240 characters to subscribers of its One-Way Alphanumeric service and 500 characters to subscribers of its Enhanced One-Way, Interactive (two-way), and QuickReply Interactive services. In addition, users can send a text message consisting of a maximum of 200 characters to subscribers of MobileComm. The form even provides a means to check the character count before the message is sent.

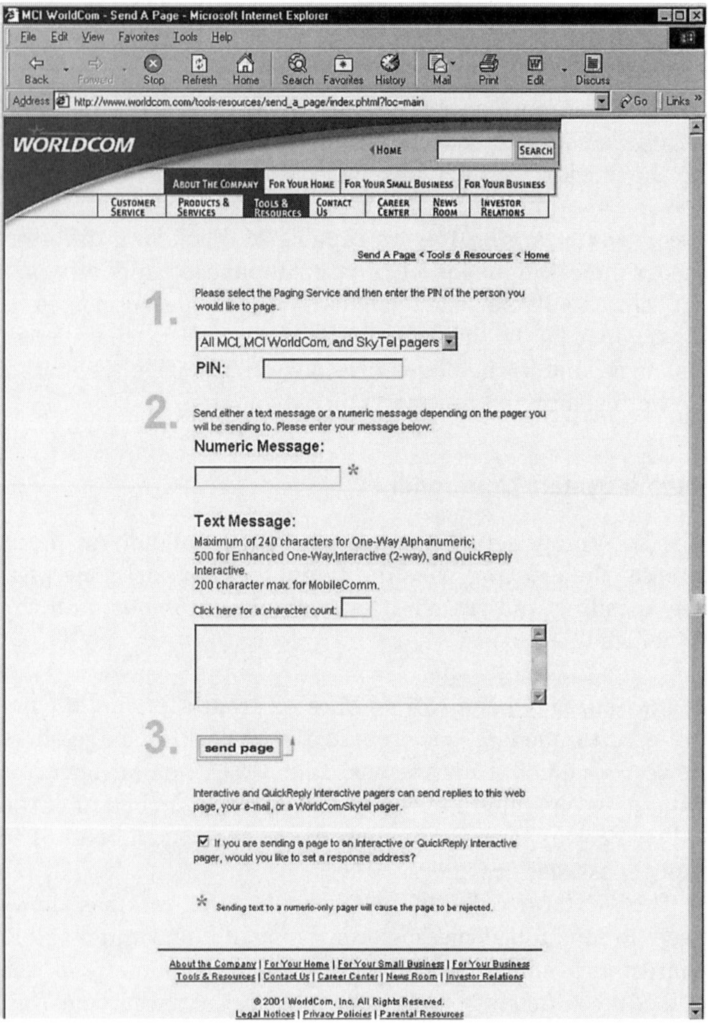

Figure P-2   WorldCom offers a Web pager that lets anyone send a message to anyone else who has a WorldCom, SkyTel, or MobileComm pager.

Telephone Network Regardless of exactly how the message is entered, it eventually passes through the PSTN to the paging terminal for encoding and transmission through the wireless paging system. Typically, the encoder accepts the incoming page, checks the validity of the pager number, looks up the directory or database for the subscriber's pager address, and converts the address and message into the appropriate paging signaling protocol. The encoded paging signal is then sent to the transmitters (base stations), through the paging transmission control systems, and broadcast across the coverage area on the specified frequency.

Encoder Encoding devices convert pager numbers into pager codes that can be transmitted. There are two ways in which encoding devices accept pager numbers: manually and automatically. In manual encoding, a paging system operator enters pager numbers and messages via a keypad connected to the encoder. In automatic encoding, a caller dials up an automatic paging terminal and uses the phone keypads to enter pager numbers. Regardless of the method used, the encoding device then generates the paging code for the numbers entered and sends the code to the paging base station for wireless transmission.

Base Station Transmitters The base station transmitters send page codes on an assigned radio frequency. Most base stations are specifically designed for paging, but those designed for two-way voice can be used as well.

Pagers Pagers are essentially FM receivers tuned to the same radio frequency as the paging base station. A decoder unit built into each pager recognizes the unique code assigned to the pager and rejects all other codes for selective alerting. However, pagers can be assigned the same code for group paging. There are also pagers that can be assigned multiple page codes, typically up to a maximum of four, allowing the same pager to be used for a mix of individual and group paging functions.

Despite all the enhancements built into pagers and paging services in recent years, the market is slowing down. Alphanumeric services—which provide word messages instead of just phone numbers—have failed to attract a wide audience largely because paging subscribers still need a phone to respond to messages. Some providers have tried to offer services that would allow callers to leave voice messages on pagers, but this too has failed to catch on. Consequently, about four out of five of today's paging customers still rely on cheap numeric services.

Two-way paging networks may be the industry's last hope for survival. They allow a pager—which comes equipped with a minikeyboard—to "talk" to another pager (Figure P-3) or with a telephone, e-mail address, or fax machine. Some two-

Figure P-3 With Motorola's PageWriter 2000, users can send text messages to other two-way pagers and alphanumeric pagers directly or through e-mail.

way services allow consumers to reply to messages with pre-determined responses.

Signaling Protocols

In a paging system, the paging terminal, after accepting an incoming page and validating it, will encode the pager address and message into the appropriate paging signaling protocol. The signaling protocol allows individual pagers to be uniquely identified/alerted and to be provided with the additional voice message or display message, if any.

Various signaling protocols are used for the different paging service types, such as tone-only or tone and voice. Most paging networks are able to support many different paging formats over a single frequency. Many paging formats are manufacturer-specific and often proprietary, but there are public-domain protocols, such as the Post Office Code Standardization Advisory Group (POCSAG), that allow different manufacturers to produce compatible pagers.

POCSAG is a public-domain digital format adopted by many pager manufacturers around the world. It can accommodate 2 million codes (pagers), each capable of supporting up to four addresses for such paging functions as tone-only, tone and voice, and numeric display. POCSAG operates at data rates of up to 2400 bps. At this rate, to send a single tone-only page requires only 13 milliseconds. This is about 100 times faster than two-tone paging.

With the explosion of wireless technology and dramatic growth in the paging industry in many markets, existing networks are becoming more and more overcrowded. In addition, RF spectrum is not readily available because of demands by other wireless applications. In response to this problem, Motorola has developed a one-way messaging protocol called Flex (feature-rich long-life environment for executing) messaging applications that is intended to transform and broaden paging from traditional low-end numeric services into a range of PCS/PCN and other wireless applications.

Relative to POCSAG, Flex can transmit messages at up to 6400 bps and permit up to 600,000 numeric pagers on a single frequency compared to POCSAG's 2400-bps transmission rate and 300,000 users per frequency. In addition, Flex provides enhanced bit error correction and much higher protection against the signal fades common in FM simulcast paging systems. The combination of increased bit error correction and improved fade protection increases the probability of receiving a message intact, especially longer alphanumeric messages and data files that will be sent over PCS/PCN.

Motorola also has developed ReFlex, a two-way protocol that will allow users to reply to messages, and InFlexion, a protocol that will enable high-speed voice messaging and data services at up to 112 kbps.

Summary

The hardware and software used in radio paging systems have evolved from simple operator-assisted systems to terminals that are fully computerized, with such features as message handling, scheduled delivery, user-friendly prompts to guide callers to a variety of functions, and automatic reception of messages. After tremendous growth in the last decade, from 10 million subscribers in 1990 to 60 million today, the paging industry has slowed down markedly—in large part because of cutthroat competition and increasing use of digital mobile phones.

See also

Electronic Mail

Personal Communication Services

PCS 1900

PCS 1900 is the American National Standards Institute (ANSI) radio standard for 1900-MHz Personal Communication Service

(PCS) in the United States. Also known as GSM 1900, it is compatible with the Global System for Mobile (GSM) communications, an international standard adopted by 404 networks supporting 538 million subscribers in 171 countries as of mid-2001. Network operators aligned to the GSM standard have 35 percent of the world's wireless market.

PCS 1900 can be implemented with either Time Division Multiple Access (TDMA) or Code Division Multiple Access (CDMA) technology. TDMA-based technology enjoys an initial cost advantage over rival technology CDMA equipment because suppliers making TDMA infrastructure equipment and handsets have already reached economies of scale. In contrast, CDMA equipment is still in its first generation and therefore is generally more expensive.

The CDMA (IS-95) standard has been chosen by about half of all the PCS licensees in the United States, giving it the lead in the total number of potential subscribers. However, the first operational PCS networks have been using PCS 1900 as their standard mainly because of the maturity of the GSM-based technology.

Although similar in appearance to analog cellular service, PCS 1900 is based on digital technology. As such, PCS 1900 provides better voice quality, broader coverage, and a richer feature set. In addition to improved voice quality, fax and data transmissions are more reliable. Laptop computer users can connect to the handset with a Personal Computer Memory Card International Association (PCMCIA) card and send fax and data transmissions at higher speeds with less chance of error.

Architecture

The PCS 1900 system architecture consists of the following major components:

- *Switching system* Controls call processing and subscriber-related functions.
- *Base station* Performs radio-related functions.

- *Mobile station* The end-user device that supports voice and data communications as well as short message services.
- *Operation and support system (OSS)* Supports the operation and maintenance activities of the network.

The switching system for PCS 1900 service contains the following functional elements (Figure P-4):

- *Mobile Switching Center (MSC)* Performs the telephony switching functions for the network. It controls calls to and from other telephone and data communications networks such as Public Switched Telephone Networks (PSTN), Integrated Services Digital Networks (ISDN), Public Land Mobile Radio Services (PLMRS) networks, Public Data Networks (PDN), and various private networks.
- *Visitor Location Register (VLR)* A database that contains all temporary subscriber information needed by the MSC to serve visiting subscribers.
- *Home Location Register (HLR)* A database for storing and managing subscriptions. It contains all permanent subscriber information, including the subscriber's service profile, location information, and activity status.
- *Authentication Center (AC)* Provides authentication and encryption parameters that verify the user's identity and ensure the confidentiality of each call. This functionality protects network operators from common types of fraud found in the cellular industry today.
- *Message Center (MC)* Supports numerous types of messaging services, for example, voice mail, facsimile, and e-mail.

Advanced Services and Features

Like GSM, PCS 1900's digital orientation makes possible several advanced services and features that are not effi-

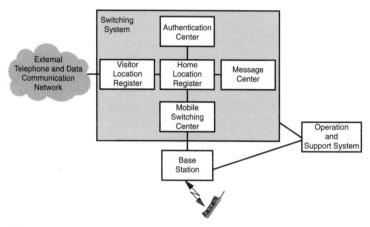

Figure P-4   PCS 1900 switching system architecture.

ciently and economically supported in analog cellular networks. Among them are

- *Short Message Service* Enables alphanumeric messages
  up to 160 characters to be sent to and from PCS
  1900–compatible handsets. Short Message Service applications include two-way point-to-point messaging, confirmed message delivery, cell-based messaging, and
  voice-mail alert. These messaging and paging capabilities
  create a broad array of potential new revenue-generating
  opportunities for carriers.

- *Voice Mail* The PCS 1900 network provides one central
  voice-mail box for both wired and wireless service. In
  addition, the voice-mail alert feature ensures that subscribers do not miss important messages.

- *Personal Call Management* Offers subscribers a single
  telephone number for all their physical telecommunication devices. For example, a single number can be
  assigned for home and mobile use or office and mobile use.
  This allows subscribers to receive all calls regardless of
  their physical location.

- *Data Applications* Wireless data applications that can be
  supported by PCS 1900 networks include Internet access,
  electronic commerce, and fax transmission.

Smart Cards

The PCS 1900 standard supports the smart card, which provides similar features as GSM's Subscriber Identity Module
(SIM). The size of a credit card, smart cards contain embedded computer chips with user-profile information. By removing the smart card from one PCS 1900 phone and inserting it
into another PCS 1900 phone, the user is able to receive calls
at that phone, make calls from that phone, or receive other
subscribed services such as wireless Internet access. The
handsets cannot be used to place calls (except 911 emergency
calls) until the subscriber inserts the smart card and enters
a personal identification number (PIN).

The profile information stored in the smart card also
enables international roaming. When traveling in the
United States, international GSM customers will be able to
rent handsets, insert their SIM, and access their services as
if they are back home. By the same token, when U.S. subscribers travel internationally to cities with compatible networks and mutual roaming agreements, they only need to
take their smart card with them to access the services they
subscribed to back home via the local GSM network.

Like SIMs, smart cards also provide storage for features
such as frequently called numbers and short messages.
Smart cards also include the AT command-set extensions,
which integrate computing applications with cellular data
communications. In the future, smart cards and PCS 1900
technology also will link subscribers to applications in electronic commerce, banking, and health care.

Summary

PCS 1900 is a frequency-adapted version of GSM that operates at 1800 MHz in Europe and elsewhere. While GSM

looks to be a perennial third in North America digital markets, this is mainly because the technology has not been adopted for use in cellular 800-MHz frequencies. Otherwise, PCS 1900 and GSM are similar in all other respects, including the network architecture and types of services supported. An advantage U.S. carriers have in supporting the PCS 1900 standard is that it is interoperable with the worldwide GSM standard, which means that users can roam globally. GSM phones are available in either dual-band (900/1900 MHz) or triband (900/1800/1900 MHz) models, enabling their use in countries with different frequency bands for GSM services.

See also

Global System for Mobile (GSM) Telecommunications

Personal Communication Services

PEER-TO-PEER NETWORKS

In a peer-to-peer network, computers are linked together for resource sharing. If there are only two computers to link together, networking can be done with a Category 5 crossover cable that plugs into the RJ45 jack of the network interface card (NIC) on each computer. If the computers do not have NICs, they can be connected with either a serial or parallel cable. Once connected, the two computers function as if they were on a local area network (LAN), and each computer can access the resources of the other. If three or more computers must be connected, a wiring hub is required.

The peer network can be extended to portable devices, including desktop and notebook computers and personal digital assistants (PDAs), through the use of wireless access points. The access point connects to the LAN through a hub or switch via Category 5 cabling, just like any other device on the wired network. The access point establishes the wireless link to one or more client devices, which are equipped with a wireless card. The wireless devices operate in either the 2.4-GHz band for 11 Mbps or the 5-GHz band for 54 Mbps. Newer access points have slots for wireless cards of both frequency bands, allowing users to protect investments in 2.4-GHz equipment while migrating to the higher speeds offered by 5-GHz equipment.

Regardless of exactly how the computers are interconnected, wired or wireless, each is an equal or "peer" and can share the files and peripherals of the others. For a small business doing routine word processing, spreadsheets, and accounting, this type of network is the low-cost solution to sharing resources like files, applications, and peripherals. Multiple computers can even share an external cable or Digital Subscriber Line (DSL) modem, allowing them to access the Internet at the same time.

Networking with Windows

Windows 95/98 and Windows NT/2000 are often used for peer-to-peer networking. In addition to peer-to-peer network access, both provide network administration features and memory management facilities, support the same networking protocols—including the Transmission Control Protocol/Internet Protocol (TCP/IP) for accessing intranets, virtual private networks (VPNs), and the public Internet—and provide such options as dial-up networking and fax routing.

One difference between the two operating systems is that in Windows 95/98 the networking configuration must be established manually, whereas in Windows NT/2000 the networking configuration is part of the initial program installation, on the assumption that NT/2000 will be used in a network. Although Windows 95/98 is good for peer-to-peer networking, Windows NT/2000 is more suited for larger client-server networks.

Windows supports Ethernet, Token Ring, Asynchronous Transfer Mode (ATM), and Fiber Distributed Data Interface (FDDI) data-frame types. Ethernet is typically the least expensive network to implement. The NICs can cost as little as $20 each, and a five-port hub can cost as little as $40. Category 5 cabling usually costs less than 50 cents per foot in 100-foot lengths with the RJ45 connectors already attached at each end. Snap-together wall plate kits cost about $6 each. If wireless connections are part of the peer network, the wireless NICs cost between $170 and $300, depending on whether the 2.4- or 5-GHz band is used. Access points can cost as little as $199 for 2.4-GHz units and $299 for 5-GHz units. Enterprise class versions cost quite a bit more.

Configuration Details

When setting up a peer-to-peer network with Windows 95/98, each computer must be configured individually. After installing an NIC and booting the computer, Windows will recognize the new hardware and automatically install the appropriate network-card drivers. If the drivers are not already available on the system, Windows will prompt the user to insert the manufacturer's disk containing the drivers, and they will be installed automatically (Figure P-5).

Next, the user must select the client type. If a Microsoft peer-to-peer network is being created, the user must add "Client for Microsoft Networks" as the primary network logon (see Figure P-5). Since the main advantage of networking computers is resource sharing, it is important to enable the sharing of both printers and files. The user does this by clicking on the "File and Print Sharing" button and choosing one or both of these capabilities (see Figure P-5). Through file and printer sharing, each workstation becomes a potential server.

Identification and security are the next steps in the configuration process. From the "Identification" tab of the dialog box, the user must select a unique name for the computer and the workgroup to which it belongs, as well as a brief description of the computer (Figure P-6). When others use Network Neighborhood to browse the network, they will see the menu trees of all active computers on the network.

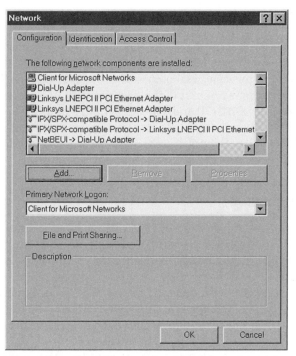

Figure P-5 To verify that the right drivers have been installed, the user opens the Network Control Panel to check the list of installed components. In this case, a Linksys LNEPCI II Ethernet Adapter has been installed.

From the "Access Control" tab of the dialog box, the user selects the security type. For a small peer-to-peer network, share-level access is adequate (Figure P-7). This allows printers, hard drives, directories, and other resources to be shared and enables the user to establish password access for each of these resources. In addition, read-only access allows users to view (not modify) a file or directory.

To allow a printer to be shared, for example, the user right-clicks on the printer icon in the Control Panel and selects "Sharing" from the drop-down list (Figure P-8). Next, the user clicks on the "Shared As" radio button and enters a unique name for the printer (Figure P-9). If desired, this

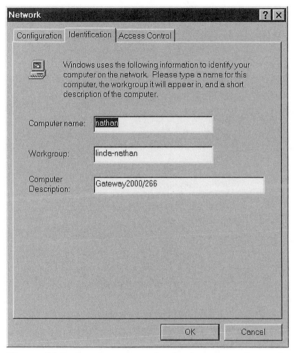

Figure P-6 A unique name for the computer, the work-group to which it belongs, and a brief description of the computer identify it to other users when they access Network Neighborhood to browse the network.

resource can be given a password as well. When another computer tries to access the printer, the user will be prompted to enter a password. If a password is not necessary, the password field is left blank.

Another security option in the "Access Control" tab is user-level access, which is used to limit resource access by user name. This function eliminates the need to remember passwords for each shared resource. Each user simply logs onto the network with a unique name and password; the network administrator governs who can do what on the network. However, this requires the computers to be part of a larger network with a central server—perhaps running

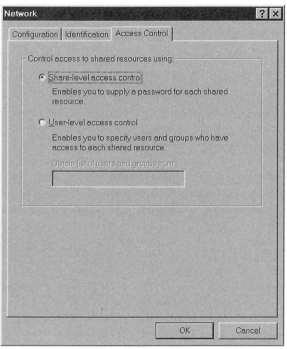

Figure P-7   Choosing share-level access allows the user to password-protect each shared resource.

Windows NT/2000 server—that maintains the access-control list for the whole network. Since Windows 95/98 and Windows NT/2000 workstations support the same protocols, Windows 95/98 computers can participate in a Windows NT/2000 server domain.

Peer services can be combined with standard client-server networking. For example, if a Windows 95/98 computer is a member of a Windows NT/2000 network and has a color printer to share, the resource "owner" can share that printer with other computers on the network. The server's access-control list determines who is eligible to share resources.

Once the networking infrastructure is in place, the NIC of each computer is individually connected to a hub with

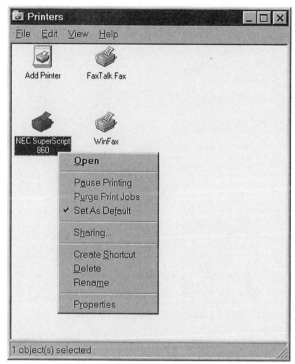

Figure P-8 A printer can be configured for sharing.

Category 5 cable or wirelessly via the access point. This cable has connectors on each end that insert into the RJ45 jacks of the hub and NICs. For small networks, the hub usually will be manageable with the Simple Network Management Protocol (SNMP), so no additional software is installed. Once the computers are properly configured and connected to the hub, the network is operational.

Summary

Peer-to-peer networking is an inexpensive way for small companies and households to share resources among a small

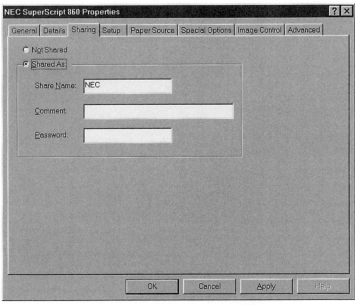

Figure P-9   A printer can be password-protected if necessary.

group of computers. This type of network provides most of the same functions as the traditional client-server network, including the ability to run network versions of popular software packages. Peer-to-peer networks also are easy to install. Under ideal conditions, installation of the cards, software, hub, and cabling for five users would take only a few hours. Wireless links provide the advantage of mobility within the home or office, allowing a notebook to be used in any room.

See also

HomeRF

Wireless Fidelity

Wireless LANs

PERSONAL ACCESS COMMUNICATIONS SYSTEMS

Personal Access Communications Systems (PACS) is a standard adopted by ANSI for Personal Communication Services (PCS). Adopted in June 1995, PACS provides an approach for implementing PCS in North America that is fully compatible with the local exchange telephone network and interoperable with existing cellular systems. Based on the Personal Handyphone System (PHS) developed in Japan and the Wireless Access Communications System (WACS) developed by Bellcore (now known as Telcordia Technologies), PACS is designed to support mobile and fixed applications in the 1900-MHz frequency range. It promises low installation and operating costs while providing very-high-quality voice and data services. In the United States, trials of PACS equipment began in 1995, and equipment rollout began in 1996.

Most of the standards—including up-banded versions of CDMA, TDMA, and GSM—look like cellular systems in that they have high transmit powers and receivers designed for the large delay spreads of the macrocellular environment and typically use low-bit-rate voice coders (vocoders). PACS fills the niche between these classes of systems, providing high-quality services, high data capability, and high user density in indoor and outdoor microcellular environments. PACS equipment is simpler and less costly than macrocellular systems yet more robust than indoor systems.

PACS capabilities include pedestrian- and vehicular-speed mobility, data services, and licensed and unlicensed spectrum systems, as well as simplified network provisioning, maintenance, and administration. The key features of PACS are summarized as follows:

- Voice and data services are comparable in quality, reliability, and security with wire-line alternatives.
- Systems are optimized to provide service to the in-building, pedestrian, and city traffic operating environments.

- It is most cost-effective to serve high-density traffic areas.

- Small, inexpensive line-powered radios provide for unobtrusive pole or wall mounting.

- There is low-complexity per-circuit signal processing.

- Low transmit power and efficient sleep mode require only small batteries to power portable subscriber units for hours of talk time and multiple days of standby time.

Like the Personal Handyphone System (PHS), PACS uses 32-kbps Adaptive Differential Pulse Code Modulation (ADPCM) waveform encoding, which provides near landline voice quality. ADPCM has demonstrated a high degree of tolerance to the cascading of vocoders, as experienced when a mobile subscriber calls a voice-mail system and the mailbox owner retrieves the message from a mobile phone. With other mobile technologies, the playback quality is noticeably diminished. With PACS, it is very clear. Similarly, the compounding of delays in mobile to PCS through satellite calls—a routine situation in Alaska and in many developing countries—can be troublesome. PACS provides extremely low delay.

The low complexity and transmit power of PACS yield limited cell sizes, which makes it well suited for urban and suburban applications where user density is high. Antennas can be installed inconspicuously, piggybacking on existing structures. This avoids the high costs and delays associated with obtaining permits for the construction of high towers.

Applications

Wireless local loop, pedestrian venues, commuting routes, and indoor wireless are typical PACS applications. Additionally, PACS is designed to offer high-capacity, superior voice quality and Integrated Services Digital Network (ISDN) data services. Interoperability with ISDN is provided by aggregating two 32-kbps time slots to form a single higher-speed 64-kbps channel. A 64-kbps channel also can support 28.8-kbps voice-band data using existing modems.

PACS also can be used for providing wireless access to the Internet. The packet data communications capabilities defined in the PACS standards, together with the ability to aggregate multiple 32-kbps channels, make it possible for users to access the Internet from their PCs equipped with suitable wireless modems at speeds of up to 200 kbps. When using the packet mode of PACS for Internet traffic, radio channels are not dedicated to users while they are on active Internet sessions, which can be very long. Rather, radio resources are used only when data are actually being sent or received, resulting in very efficient operation and minimally impacting the capacity of the PACS network to support voice communications.

PACS was designed to support the full range of advanced intelligent network (AIN) services, including custom calling features and personal mobility. As new AIN features are developed, the PACS-compliant technology will evolve to facilitate incorporation of the new services.

Summary

The market for PCS is very competitive. Already PCS is exerting downward price pressure on traditional analog cellular services where the two compete side by side. PACS enables PCS operators to differentiate their offerings through digital voice clarity, high-bit-rate data communications, and advanced intelligent network services—all in a lightweight handset. Moreover, the cost savings and ease of use associated with PACS make it very economical for residential and business environments compared to competitive high-powered wide area systems.

See also

Personal Handyphone System

Personal Communication Services

Voice Compression

Wireless Communications Services

PERSONAL AIR COMMUNICATIONS TECHNOLOGY

Personal Air Communications Technology (pACT) is a wireless two-way messaging and paging technology that is offered as an alternative to wireless Internet Protocol (IP), also known as Cellular Digital Packet Data (CDPD).

The pACT specification, released in 1995, was developed to enable compact, inexpensive devices to access low-cost, high-capacity network infrastructures. The protocol enhances one-way paging, response paging, two-way paging, voice paging, telemetry, and two-way messaging applications. The pACT protocol thus addresses the demands of a growing market for narrowband PCS. Despite huge investments in spectrum for narrowband PCS, some U.S. paging carriers are already running short of bandwidth. This situation has prompted them to look for ways to add capacity to their networks. In major U.S. cities, the solution has been to make better use of existing spectrum.

pACT supports two-way messaging and paging applications while still retaining all the strengths of one-way paging services, including long battery life, good in-building reception, and ubiquitous coverage. pACT also provides carriers with the ability to substantially increase system capacity by more efficiently utilizing spectrum, thereby allowing for more cost-effective paging and messaging services. Its cellular-like network design enables carriers to take advantage of capacity gains through frequency reuse, a fundamental difference from other two-way systems.

The difference between wireless IP and pACT is that the former is TCP/IP-centric, whereas the latter is User Datagram Protocol/Internet Protocol (UDP/IP)–centric. While wireless IP compresses the standard 40-byte TCP/IP header to an average of 3 bytes to conserve bandwidth, pACT compresses the header to only 1 byte. The greater compression is important for providing a short alphanumeric messaging service, especially when the message body is roughly the same size as the header. Consequently, the maximum number of subscribers that can be supported by the system is constrained not by protocol efficiency but by service traffic.

Applications

Two-way wireless messaging is working its way into a myriad of user applications where one-way messaging is no longer adequate. These are applications where time is of the essence, and guaranteed message delivery is critical. pACT uses the same upper layers of the protocol stack as wireless IP, making it suitable for a broad range of messaging applications, including two-way-paging, e-mail, fleet dispatch, telemetry, transaction processing (e.g., point-of-sale credit card authorizations), and voice messaging. Since pACT is based on the IP, it provides wireless network users with access to other IP-based networks such as corporate local area networks (LANs) and the Internet from remote locations.

Network Capacity Narrowband PCS and two-way protocols such as pACT give the paging industry new ways to increase the capacity of one-way paging systems, which introduce opportunities for providing new and enhanced services. Like traditional mobile telephony systems, pACT increases capacity by reusing frequencies—theoretically providing unlimited capacity. Given higher airlink speeds (8 kbps) and knowing a subscriber's exact location, operators can increase capacity in a large zone by 100 to 200 times and even more in networks that are made up of several zones.

While one-way paging over a two-way network offers no real benefits to subscribers, the benefits to operators are substantial. First, two-way networks enable paging devices to acknowledge that a message has been received. With this capability, operators can offer subscribers guaranteed delivery, which can result in competitive advantage.

The second benefit to operators is that they do not have to broadcast a message via every transmitter in the network to reach an intended recipient. With a two-way network, the exact location of each subscriber device is known because the devices register automatically as they move through the network. Since messages are sent solely via the transmitter that is closest to the subscriber, a great deal of network capacity is freed. This means that all other transmitters in the network can be used simultaneously to serve other subscribers.

Service Enhancement The two-way pACT architecture affects more than capacity: It also enables providers to enhance services, as well as to provide completely new services and applications. For example, paging devices that contain a transmitter are able to send information back to the network, to someone else in the network, or to any other network. The two-way paging and messaging paradigm provides five levels of acknowledgment:

- *System acknowledgment* The paging device acknowledges receipt of an error-free message. A transparent Link Layer acknowledgment between the device and the network enables guaranteed delivery service. The network stores and retransmits messages at periodic intervals that pagers have not acknowledged.

- *Message read* When a recipient reads a message, the paging device transmits a "message read" acknowledgment back to the host system or to the originator of the message.

- *Canned messages* The paging device contains several ready-to-use responses such as "Yes" or "No" that recipients can use when they reply to an inquiry.

- *Multiple choice* The originator of a message defines several possible responses to accompany his or her message. To reply, the recipient selects the most appropriate response.

- *Editing capabilities* Some devices may be used to create messages. Editing capabilities vary from device to device. Some devices are managed by a few simple keys and provide only minor editing capabilities, while others such as portable computers may contain full-feature keyboards.

Like wireless IP, pACT provides secure service—including encryption and authentication—to ensure that messages are delivered solely to intended subscribers.

System Overview

The pACT system is built from several flexible modules that can be combined and configured in different ways to meet specific operator demands (Figure P-10). Because the pACT network is based on the IP, operators and application providers can take full advantage of existing applications, application programming interfaces (APIs), and other development tools. With pACT, a single 50-kHz channel may accommodate up to

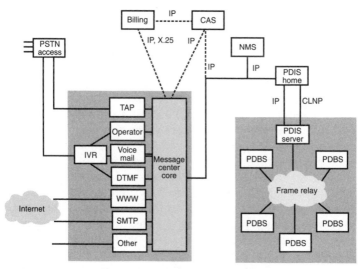

Figure P-10   The message center, which is the gateway or interface to the pACT network, permits several interfaces and applications, such as Internet applications, to be implemented. A pACT network may be configured in several ways using message centers and subnetworks of various sizes.

three individual radiofrequency (RF) carriers. Each base station is assigned a particular 12.5-kHz channel.

pACT Data Base Stations The pACT data base stations (PDBS) are located at the cell site and relay data between subscriber devices and the serving pACT data intermediate system (PDIS). Typically, the PDBSs are connected to the serving PDIS switch via a Frame Relay network. Cellular radio system design and roaming techniques enable pACT to determine which base station is closest to a subscriber device each time communication takes place between the device and the network. The mobile terminals determine cell handoff based on signal-strength measurements implemented by the base stations.

pACT Data Intermediate System The PDIS acts as the central switching site, routing data to and from the appropriate base stations. It also maintains routing information for each subscriber device in the network.

There are two versions of the PDIS: the home PDIS and the serving PDIS. Besides switching data packets, the home PDIS maintains a location directory and provides a forwarding service and subscriber authentication. Every subscriber is registered in a home PDIS database. The serving PDIS provides message forwarding, a registration directory, and readdress services. Other services or functions are multicast, broadcast, unicast, airlink encryption, header compression (to minimize airlink use), data segmentation, frame sequencing, and network management.

The serving PDIS is connected to the home PDIS switch. If necessary, the two switches may be located on the same hardware platform. Various configurations of computer processor power and memory are available for the PDIS, depending on requirements for computing capacity and on how the requirements relate to traffic load and the number of subscribers in the network. More computing capacity may be added easily if necessary.

Message Center The fixed entry point into the pACT network is provided by one or more message centers (MCs) that initiate, provision, and connect pACT services to private and public networks, including corporate intranets and the PSTN, respectively.

Every message passes through the message center, whose functionality and applications vary according to network operator requirements. The core of the message center is the message store, which handles virtually any data type and makes APIs accessible for building various applications, such as interactive voice response (IVR) and voice or fax mail.

The message store also provides functions for operation and maintenance, system monitoring, and event/alarm handling. Any of the message center's databases can be queried via the Structured Query Language (SQL). The message center also supports virtually any protocol. Typical protocols are the IP, Telocator Alphanumeric Protocol (TAP), Telocator Network Paging Protocol (TNPP), and X.400 (the ITU-T standard for message handling services).

Network Management System

The pACT network management system (NMS) gives operators full control of every component in a pACT network. Through the NMS, each base station is provided with a set of radio resource management (RRM) parameters that are used to control traffic and maintain links to the network, as well as to give instructions to mobile terminals that access the channel.

The NMS supports the Common Management Information Protocol (CMIP) and the Simple Network Management Protocol (SNMP). The NMS contains a database component that permanently stores parameters, configuration data, and historical records of traps and performance data. A pACT network may contain more than one NMS, allowing responsibility to be passed across time zones to other operators, ensuring 24-hour monitoring and control.

Customer Activation System The pACT Customer Activation System (CAS) enables customer service representatives to manage customer accounts and to dynamically activate pACT-related services for customers. Customer accounts are of two types: individual and business, where individual accounts are for single subscribers and business accounts are for multiple subscribers.

pACT End System (Mobile Terminals) Mobile terminals range from simple pagers to sophisticated two-way messaging devices such as PDAs or palmtop computers with wireless modems. When not being used to send messages, mobile terminals periodically check the designated forward channel for incoming messages; otherwise, they are usually in sleep mode to conserve battery life.

pACT System Protocol Stack The pACT protocol stack is based on the concepts and principles of the ITU-X.200 and ITU-X.210 reference models, as well as service conventions for Open Systems Interconnection (OSI). The Limited Size Messaging (LSM) protocol provides the functionality of a simple e-mail application protocol, such as the Simple Mail Transfer Protocol (SMTP), but is optimized for low-bandwidth channels so that unnecessary overhead over the airlink is minimized. In addition, the LSM protocol provides a platform for providing true two-way messaging and data communication services. Examples of services are embedded response messaging for simple pager devices, true two-way e-mail connectivity, and multicast and broadcast messaging.

pACT's subnetwork convergence layer provides a new approach to encryption. To ensure that airlink bandwidth is not spent on resynchronizing the encryption engines, pACT devices may employ a technique that automatically resynchronizes the engines, even when the underlying layers fail to deliver a packet. This technique is important because it provides a mechanism by which multicast and broadcast services may be encrypted.

The CDPD Link Layer is optimized for duplex, whereas pACT uses two-way simplex. The pACT Mobile Data Link Layer Protocol (MDLP) is an enhanced version of the link access procedure on the D-channel (LAPD) (i.e., ITU Q.920) that allows subscriber devices to adopt strategies for automatically resetting the link and saving power.

pACT Airlink Interface The pACT backbone network is similar to a CDPD network. The main differences between the two involve functionality—mostly for extending battery life in subscriber devices—and features such as group messaging, broadcast, and unicast. The pACT protocol shortens and reduces the number of transmissions and contains an efficient sleep mode for conserving battery power.

Summary

As an alternative to wireless IP, also known as CDPD, Personal Air Communications Technology (pACT) supports various paging and messaging applications while providing more efficient use of limited spectrum. This allows wireless carriers to stay competitive, even when they cannot easily add more capacity to their networks. Since pACT is based on the IP, wireless network users can access other IP-based networks such as corporate LANs and the Internet from remote locations. With integral authentication and encryption, data are protected as they traverse the pACT network.

See also

Cellular Data Communications

Paging

PERSONAL COMMUNICATIONS SERVICES

Personal Communications Services (PCS) is a set of wireless communication services personalized to the individual. Subscribers can tailor their service package to include only the services they want, which may include stock quotes, sports scores, headline news, voice mail, e-mail and fax notification, and caller ID. The service offers full roaming capability, allowing anywhere-to-anywhere communication.

Unlike many existing cellular networks, PCS is a completely digital service. The digital nature of PCS allows antennas, receivers, and transmitters to be smaller. It also allows for the simultaneous transmission/reception of data and voice with no performance penalty. Eventually, PCS will overtake analog cellular technology as the preferred method of wireless communication.

Several technologies are being used to implement PCS. In Europe, the underlying digital technology for Personal Communication Networks (PCNs) is Global System for Mobile (GSM) Telecommunications, where it has been assigned the 1800-MHz frequency band. GSM has been adapted for operation at 1900 MHz for PCS in the United States (i.e., PCS 1900). Other versions of GSM are employed to provide PCS services in other countries, such as the Personal Handyphone System (PHS) in Japan. Many service providers in the United States have standardized their PCS networks on Code Division Multiple Access (CDMA) or Time Division Multiple Access (TDMA) digital technology.

The PCS Network

A typical PCS network operates around a system of microcells—smaller versions of a cellular network's cell sites—each equipped with a base station transceiver. The microcell transceivers require less power to operate but cover a more limited range. The base stations used in the microcells can even be placed indoors, allowing seamless coverage as the subscriber walks into and out of buildings.

Similar to packet radio networks today, terminal devices stay connected to the network even when not in use, allowing the network to locate an individual within the network via the nearest microcell and routing calls and messages directly to the subscriber's location. For a "follow me" service, which incorporates more than one device, a subscriber may be required to turn on a pager, for example, to receive messages on that device. If the subscriber receives a phone call while the pager is on, the network may store the call, take a message, or send the call to a personal voice-mail system and simultaneously page the subscriber. In this way, PCS allows the concept of universal messaging to be fully realized.

Currently, cellular switching systems operate separately from the Public Switched Telephone Network (PSTN). When cellular subscribers call a landline phone (and vice versa), the two systems are interconnected to complete the communications circuit. Many PCS services are supported on the same switches that handle calls over the wire-line network, with the only distinction between a wireless and wire-line call being the medium at each end of the circuit. In some places, wireless PCS and cable television (CATV) are already integrated in a unified CDMA-based architecture called "PCS over cable." The use of CATV allows PCS to reach more potential subscribers with a lower startup costs for service providers.

Broadband and Narrowband

There are two technically distinct types of PCS: narrowband and broadband, each of which operates on a specific part of the radio spectrum and has unique characteristics.

Narrowband PCS is intended for two-way paging and other types of communications that handle small bursts of data. These services have been assigned to the 900-MHz frequency range, specifically, 901 to 912, 930 to 931, and 940 to

941 MHz. Broadband PCS is intended for more sophisticated data services. These types of services have been assigned a frequency range of 1850 to 1990 MHz.

Narrowband PCS and broadband PCS license ownerships have been determined by public auctions conducted by the Federal Communications Commission (FCC). PCS service areas are divided into 51 regional service areas, which are subdivided into a total of 492 metropolitan areas. There is competition in each service area by at least two service providers. There are 10 national service providers plus 6 regional providers in each of 5 multistate regions called "major trading areas."

An unlicensed portion of the PCS spectrum has been allocated from 1890 to 1930 MHz. This service is designed to allow unlicensed operation of short-distance—typically indoor or campus-oriented environments—voice and data services provided by wireless LANs and wireless Public Branch Exchanges (PBXs).

One of the largest PCS networks is operated by Sprint PCS. At year end 2001, the company's CDMA-based wireless network served close to 13 million customers, making it the fourth largest wireless carrier. In addition to offering voice services from 300 major metropolitan markets, including more than 4000 cities and communities in the United States, the company leads the wireless industry in the number of wireless Web users on its Internet-ready phones and devices (Figure P-11).

In addition, users may shop online at Amazon.com from their Internet-ready Sprint PCS phones. The service supports two-way transactional electronic commerce services to provide users with easy and convenient shopping on the Internet. Users also easily access the Sprint PCS Wireless Web to check e-mail, news, stock portfolios, or flight schedules. The service offers the ability to customize and receive important news, receive e-mail and information updates from Yahoo, as well as dial into a corporate intranet or the Internet using a Sprint PCS phone in place of a modem connected to a laptop, PDA, or other handheld computing device.

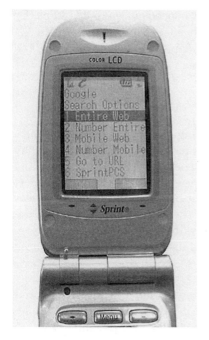

Figure P-11 Sprint PCS Wireless Web subscribers can access more than 1.3 billion pages of Web content via integrated search technology from Google, which automatically converts HyperText Markup Language (HTML) pages into a format optimized for Wireless Application Protocol (WAP) phones. This is in addition to 2.2 million Web pages that are already WAP-formatted.

Migration to 3G

PCS service providers are in the process of migrating their wireless networks to the global third generation (3G) framework. In the case of PCS based on CDMA, this entails a multiphase rollout of new technology that will increase the network's capacity for both voice and data.

The first phase of deployment will be to migrate to a Code Division Multiple Access 2000 (CDMA 2000) network, which will double the PCS network's capacity for voice communications, increase data transmission speeds from 14.4 to 144 kbps, and lower handset battery consumption. In early 2003, PCS service providers will move to the second stage of their transition to 3G and offer data speeds of up to 384 kbps. By late 2003, data transmission speeds will reach up to 2.4 Mbps, and in early 2004, transmission speeds for voice and data are expected to hit between 3 and 5 Mbps.

Summary

The popularity of PCS is bringing about a variety of new mobile and portable devices such as small, lightweight telephone handsets that work at home, in the office, or on the street; advanced "smart" paging devices; and wireless electronic mail and other Internet-based services. PCS services are available in all regions of the United States. Most of the smaller PCS networks are now interconnected to the nationwide PCS networks, providing users with extensive roaming coverage without having to switch to analog cellular on a dual-mode handset. At this writing, momentum toward 3G networks has stalled. Questions about how much demand there is for 3G services and delays in 3G network implementations are causing many PCS service providers to rethink their transition timetable.

See also

Cellular Voice Communications

Code Division Multiple Access

Global System for Mobile (GSM) Telecommunications

PCS 1900

Personal Handyphone System

PERSONAL DIGITAL ASSISTANTS

Personal digital assistants (PDAs) are hand-held computers equipped with operating system and applications software. PDAs can be equipped with communications capabilities for short-text messaging, e-mail, news updates, Web surfing, voice mail, and Internet telephony. Today's PDAs also can act as MP3 players, voice recorders, and digital cameras with the addition of multimedia modules. Some PDAs can even accommodate a module that provides location information via the Global Positioning System (GPS). PDAs are intended for mobile users who require instant access to information regardless of their location at any given time (Figure P-12).

The Newton MessagePad, introduced by Apple Computer in 1993, was the first true PDA. Trumpeted as a major milestone of the information age, the MessagePad was soon joined by similar products from such companies as Hewlett-Packard, Motorola, Sharp, and Sony.

These early hand-held devices were hampered by poor performance, excessive weight, and unstable software. Without a wireless communications infrastructure, there was no com-

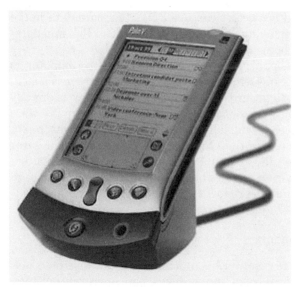

Figure P-12   Palm Computing offers one of the most popular lines of PDAs. This Palm V, shown with cradle charger and HotSync serial cable, weighs in at only 4 ounces.

pelling advantage of owning a PDA. With the performance limitations largely corrected and the emergence of new wireless PCS—plus continuing advances in operating systems, connectivity options, and battery technology—PDAs are now well on the way toward fulfilling their potential.

Applications

Real estate agents, medical professionals, field service technicians, and delivery people are just a few of the people using PDAs. Real estate agents can use PDAs to conveniently browse through property listings at client locations. Health care professionals can use PDAs to improve their ability to access, collect, and record patient information at the point of care. Numerous retailers and distributors can collect inventory data on the store and warehouse floor and later export tose data into a spreadsheet on a PC. Insurance agents, auditors, and inspectors can use PDAs to record data in the field and then instantly transfer those data to PCs and databases at the home office.

For professionals who tote around a laptop computer to give presentations with Microsoft PowerPoint, there is a module for the Springboard Visor that connects the device directly to digital projectors (or other VGA displays) with an interface cable. The user downloads the presentation material from a PC to the Visor and then taps an icon displayed on the Visor screen to start the 1024×768 resolution color presentation. The user can even control the presentation from anywhere in the room using the product's infrared remote control.

PDA Components

Aside from the case, PDA components include a screen, keypad, or other type of input device; an operating system; memory; and battery. Many PDAs can be outfitted with fax/modem cards and a docking station to facilitate direct connection to a PC or LAN for data transfers and file synchronization. Some PDAs, such as Palm Computing's Palm VII, have a wireless capability that allows information retrieval from the Internet. Of course, PDAs run numerous applications to help users stay organized and productive. Some PDAs have integral 56-kbps modems and serial ports that allow them to be attached via cable to other devices.

A unique PDA is Handspring's Visor, which uses the Palm OS operating system. What makes the Visor unique is that it is expandable via an external expansion slot called a Springboard. In addition to backup storage and flash storage modules, the slot lets users add software and hardware modules that completely change the function of the Visor. Springboard modules allow the Visor to become an MP3 player, pager, modem, GPS receiver, e-book, or video game device.

Display The biggest limitation of PDAs is the size of their screens. Visibility is greatly improved through the use of nonglare screens and backlighting, which aid viewing and entering information in any lighting condition. In a dim indoor environment, backlighting is a virtual necessity, but it drains the battery faster. Some PDAs offer user-controllable backlighting, while others let the user set a timer that shuts off the screen automatically after the unit has been idle for a specified period of time. Both features greatly extend battery life. Other PDAs, such as the Visor Prism, feature an active-matrix backlit display capable of displaying over 65,000 colors.

Keyboard Some PDAs have on-screen keyboards, but they are too small to permit touch typing. The use of a stylus speeds up text input and makes task selection easier. Of course, the instrument can be used for handwritten notes. The PDA's handwriting-recognition capability enables the notes to be stored as text for use by various applications, such as a date book, address book, and to-do list. As an option, foldout full-size keyboards are available that attach to the PDA. They weigh only 8 ounces, making it much easier to respond to e-mail, compose memos, and take notes without having to lug around a laptop.

Operating Systems A PDA's operating system provides the foundation on which applications run. The operating system may offer handwriting recognition, for example, and include solutions for organizing and communicating information via fax or electronic mail, as well as the ability to integrate with Windows and Mac OS-based computers in enterprise environments. The operating system also may include built-in support for a range of modems and third-party paging and cellular communication solutions. Because memory is limited in a PDA, usually between 2 and 64 MB, the operating system and the applications that run on it must be compact.

Some operating systems come with useful utilities. There are utilities that set up direct connections between the PDA and desktop applications to transfer files between them via a cable or infrared connection. A synchronization utility ensures that the user is working from the latest version of a file. Some operating systems offer tools called "intelligent agents" that automate routine tasks. An intelligent agent can be programmed to set up a connection to the Internet, for example, and check for e-mail. To activate this process, the user might only have to touch an icon on the PDA's screen with a pen.

There are two major operating systems in use today—Microsoft's Pocket PC platform and the Palm OS. Pocket PC's predecessor, Windows CE, was too difficult to use and not powerful enough to draw users away from the popular Palm OS. But the Pocket PC's redesigned interface overcomes most of Windows CE's previous problems.

The Pocket PC platform is a version of Windows that preserves the familiarity of the Windows-based desktop and integrates seamlessly with Outlook, Word, and Excel. The platform includes a version of Internet Explorer for browsing the Web over a wireless connection or ordinary phone line and Windows Media Player for listening to digital music and watching digital videos. It also includes Microsoft Reader for reading e-books downloaded from the Internet. Palm OS is a more efficient operating system than Pocket PC. Consequently, it requires less processing power and less memory than equivalent products using the Pocket PC operating system.

Memory Although PDAs come with a base of applications built into ROM—usually a file manager, word processor, and scheduler—users can install other applications as well. New applications and data are stored in RAM. At a minimum, PDAs come with only 2 MB, while others offer up to 64 MB. When equipped with 2 MB of memory, the PDA can store approximately 6000 addresses, 5 years of appointments, 1500 to-do items, and 1500 memos. Some PDAs have a PC card (formerly PCMCIA) slot that can accommodate storage cards that are purchased separately.

Even though many Pocket PC products come in higher memory configurations, an 8-MB Palm OS product can store as much or even more information than a 16-MB Pocket PC product with little performance degradation. The better performance is due to the efficiencies of the Palm OS, which uses less memory and processing power than equivalent products based on Pocket PC. Since greater memory capacity increases the overall price of the product, vendors like Handspring believe that 8 MB offers the most utility at the most competitive price.

Power Many PDAs use ordinary AAA alkaline batteries. Manufacturers claim a battery life of 45 hours when users search for data 5 minutes out of every hour the unit is turned on. Of course, using the backlight display will drain the batteries much faster. Using the backlight will reduce battery life by about 22 percent. Other power sources commonly used with PDAs include an ac adapter and rechargeable lithium-ion battery.

The rechargeable battery offers more flexible power management in a smaller space. The components that operate the color screen increase the power draw from the battery. With AAA batteries, the user would have to replace them rather frequently. The rechargeable battery solution enhances the user's experience by providing full power in a pocket-sized package. Lithium-ion rechargeable batteries offer 2 hours of continuous use.

Fax/Modems Some PDAs come with an external fax/modem to support basic messaging needs when hooked up to a telephone line. Others offer a PC card slot (formerly PCMCIA) that can accept not only fax/modems but also storage cards. With fax/modems, PDA users can receive a fax from their office, annotate it, and fax it back with comments written on it in "electronic ink."

There are wireless Ethernet modules available that allow the user to roam about the workplace or campus with secure connections, peer-to-peer links between devices, and high-speed access to the Internet, e-mail, and network resources. Transmissions of up to 11 Mbps are possible, but actual throughput is determined by the speed of the PDA's processor. The modules adhere to the IEEE 802.11b high-rate standard for wireless LANs and support 40- or 128-bit Wired-Equivalent Privacy (WEP) encryption. Transmission ranges of up to 1000 feet (300 meters) in open environments and 300 feet (90 meters) in office environments are supported.

Cradle A cradle allows the PDA to connect to a desktop PC via a standard serial cable or USB cable. The user simply drops the PDA into the cradle and presses a button to automatically synchronize desktop files with those held in the PDA. An alternative to cable is an infrared (IR) connection. With an IR-enabled PDA, users not only can swap and synchronize files with a PC but also can beam business cards, phone lists, memos, and add-on applications to other IR-enabled PDAs. IR-enabled PDAs also can use third-party beaming applications with IR-enabled phones, printers, and other devices.

Summary

Improvements in technology and the availability of wireless communications services, including PCS, overcome many of the limitations of early PDA products, making today's hand-held devices very attractive to mobile professionals. In the process, PDAs are finding acceptance beyond vertical markets and finally becoming popular among consumers, particularly those looking for an alternative to notebook computers and younger people who want a versatile device from which they also can play MP3 music files and games as well as read e-books.

See also

Cellular Telephones

Global System for Mobile (GSM) Telecommunications

Personal Communication Services

Personal Handyphone System

PERSONAL HANDYPHONE SYSTEM

The Personal Handyphone System (PHS) is a wireless technology that offers high-quality, low-cost mobile telephone services using a fully digital system operating in the 1.9-GHz range. Originally developed by NTT, the Japanese telecommunications giant, PHS is based on the Global System for Mobile (GSM) Telecommunications standard. PHS made its debut in Japan in July 1995, where service was initially offered in metropolitan Tokyo and Sapporo. Although PHS was originally developed in Japan, it is now considered a pan-Asian standard.

Advantages over Cellular

PHS phones (Handyphones) operate at 1.9 GHz, whereas cellular phones operate at 800 MHz. To achieve high voice quality, PHS uses a portion of its capacity to support a high-performance voice-encoding algorithm called Adaptive Differential Pulse Code Modulation (ADPCM). With this algorithm, PHS can support a much higher data throughput (32 kbps) than a cellular-based system, enabling PHS to support fax and voice-mail services and emerging multimedia applications such as high-speed Internet access and photo and video transmission.

PHS supports the handoff of calls from one microcell to the next during roaming. However, PHS goes a step farther than cellular by giving users the flexibility to make calls at home (just like conventional cordless phones), at school, in the office, while riding the subway, or while roaming through the streets. PHS also gives subscribers more security and complete privacy. And unlike cellular phones, PHS phones cannot be cloned for fraudulent use.

Another key advantage of PHS over cellular is cost—PHS can provide mobile communications more economically that cellular. Through its efficient microcell architecture and use of the public network, startup and expansion costs for operators are minimized. As a result, total per-subscriber costs tend to be much lower than with traditional cellular networks. Because PHS is a "low tier" microcellular wireless network, it offers far greater capacity per dollar of infrastructure than existing cellular networks, which results in lower calling rates. In Japan, the cost of a 3-minute call using a PHS handset is comparable to the cost of making that same call on a public phone.

Handsets

Not only are PHS handsets extremely small and light-weight—almost half the size and weight of cellular handsets—but the battery life of PHS handsets is superior to that offered by cellular handsets. PHS phones output 10 to 20 milliwatts, whereas most cellular phones output between 1 and 5 watts. Whereas the typical cellular handset has a battery life of 3 hours talk time, the typical PHS handset has a battery life of 6 hours talk time. Whereas the typical cellular handset has a battery life of 50 hours in standby mode, the typical PHS handset has a battery life of 200 hours in standby mode (more than a week). The low-power operation of PHS handsets is achieved through strict built-in power management and "sleep" functions in individual circuits.

Applications

There are a number of applications of PHS technology. In the area of mobile telecommunications, users can establish communications through public cell stations, which are installed throughout a serving area. PHS phones also can be used with a home base station as a residential cordless telephone at the Public Switched Telephone Network (PSTN) tariff.

When used in the local loop, PHS provides the means to access the PSTN in areas where conventional local loops—consisting of copper wire, optical fiber, or coaxial cable—are impractical or not available.

PHS also can be adopted as a digital cordless PBX for office use, providing readily expandable, seamless communications throughout a large office building or campus. Users carry PHS handsets with them and are no longer chained to their desks by their communications systems. As a digital system, PHS provides a level of voice quality not normally associated with a cordless telephone. In addition, the digital signal employed by PHS provides security for corporate communications, and the system's microcell architecture can be reconfigured easily to accommodate increases or decreases in the number of users. Another benefit of PHS is the ease and minimal expense with which the entire network can be dismantled and set up again in another facility if, for example, a business decides to relocate its offices.

For personal use as a cordless telephone at home, PHS is a low-cost mobile solution that allows the customer to use a single handset at home and out of doors with a digital signal that provides improved voice quality for a cordless phone and enough capacity for data and fax transmissions that are increasingly a part of users' home communications.

Network Architecture

The PHS radio interface offers four-channel Time Division Multiple Access with Time Division Duplexing (TDMA/TDD), which provides one control channel and three traffic channels for each cell station.

The base station allocates channels dynamically and is not constrained by a frequency reuse scheme, thus deriving the maximum advantage of carrier-switched TDMA. This means that PHS handsets communicating to a base station may all be on different carrier frequencies.

The PHS system uses a microcell configuration that creates a radio zone with a 100- to 300-meter diameter. The base stations themselves are spaced a maximum of 500 meters apart. In urban areas, the microcell configuration is capable of supporting several million subscribers. This configuration also makes possible smaller and lighter handsets and the more efficient reuse of radio spectrum to conserve frequency bands. In turn, this permits very low transmitter power consumption and, as a result, much longer handset talk times and standby times than are possible with cellular handsets.

A drawback of the lower operating power level is the smaller radius that a PHS base station can cover: only 100 to 300 meters, versus at least 1500 meters for cellular base stations. The extra power of cellular systems improves penetration of the signals into buildings, whereas PHS may require an extra base station inside some buildings. Another drawback of PHS is that the quality of reception can diminish significantly when mobile users are traveling at a rate greater than 15 miles per hour.

Since PHS uses the public network rather than dedicated facilities between microcells, the only service startup requirements are handsets, cell stations, a PHS server, and a database of services to support PHS network operation. With no separate transmission network needed for connecting cell stations and for call routing, carriers can introduce PHS service with little initial capital investment.

Service Features

PHS is a feature-rich service, giving Handyphone users access to a variety of call-handling features, including

- *Call forwarding* To a fixed line, to another PHS phone, or to a voice mail box.
- *Call waiting* Alerts the subscriber of an incoming call.
- *Call hold* Enables the subscriber to alternate between two calls.
- *Call barring* Restricts any incoming local or international call.
- *Calling line identification (CLI)* Displays the number of the incoming call, informing the subscriber of the caller's identity.
- *Voice mail* The subscriber receives recorded messages, even when the phone is busy or turned off.
- *Text messaging* Enables the subscriber to send and receive text messages through the PHS phone.
- *International roaming* Enables subscribers to use their PHS phones in another country but be billed by the service provider in their home country.

Depending on the implementation progress of the service provider, the following value-added data services also may be available to Handyphone users:

- *Virtual fax* Enables subscribers to retrieve fax messages anywhere, have fax messages sent to a Handyphone, or have them redirected to any fax machine.

- *Fax* By attaching the Handyphone to a laptop or desktop computer, subscribers can send and receive faxes anywhere.

- *E-mail/Internet access* Allows subscribers to retrieve e-mail from the Internet through the Handyphone.

- *Conference calls* Enables subscribers to talk to as many as four other parties at the same time.

- *News, sports scores, and stock quotes* Enables subscribers to obtain a variety of information on a real-time basis.

Many other types of services can be implemented over PHS. There is already the world's first consumer-oriented videophone service in Japan. Kyocera Corp. offers a mobile phone able to transmit a caller's image and voice simultaneously. Two color images are transmitted per second through a camera mounted on the top of the handset. The recipient can view the caller via a 2-inch active matrix liquid-crystal display (LCD). Since the transmission technology sends data at only 32 kbps, however, this makes for jerky video images.

Summary

While 32-kbps channels are now available in Japan, research is now under way to achieve a transmission rate of 64 kbps through the combined use of two channels. With this much capacity, PHS can be extended to a variety of other services in the future, including better-quality video. In combination with a small, lightweight portable data terminal, PHS also may be used to realize the concept of "mobile network computing," whereby users would access application software stored on the Internet. With the limited memory and disk storage capacity of the PHS terminals, the applications and associated programs would stay on the Internet, preventing the PHS devices from becoming overwhelmed.

See also

Cellular Data Communications

Cellular Voice Communications

Global System for Mobile (GSM) Telecommunications

Personal Access Communications Systems

Personal Communication Services

Personal Digital Assistants

PRIVATE LAND MOBILE RADIO SERVICES

Since the 1920s, Private Land Mobile Radio Services (PLMRS) have been meeting the internal communication needs of private companies, state and local governments, and other organizations. These services provide voice and data communications that allow users to control their business operations and production processes, protect worker and public safety, and respond quickly in times of natural disaster or other emergencies.

In 1934, shortly after its establishment, the Federal Communications Commission (FCC) identified four private land mobile services—Emergency Service, Geophysical Service; Mobile Press Service, and Temporary Service, which applied to frequencies used by the motion picture industry. Over the years, the FCC refined these categories. Until 1997, PLMRS consisted of 20 services spread among six service categories: Public Safety, Special Emergency, Industrial, Land Transportation, Radiolocation, and Transportation Infrastructure. In that year, the FCC did away with 20 discrete radio services and the six service categories and replaced them with two frequency pools: the Public Safety Pool and the Industrial/Business Pool. Table P-1

TABLE P-1 Radio Services in the Two Frequency Pools

Public Safety Pool	Industrial/Business Pool
Local Government Radio Service	Power Radio Service
Police Radio Service	Petroleum Radio Service
Fire Radio Service	Forest Products Radio Service
Highway Maintenance Radio Service	Film and Video Production Radio Service
Forestry-Conservation Radio Service	Relay Press Radio Service
Emergency Medical Radio Service	Special Industrial Radio Service
Special Emergency Radio Service	Business Radio Service
	Manufacturers Radio Service
	Telephone Maintenance Radio Service
	Motor Carrier Radio Service
	Railroad Radio Service
	Taxicab Radio Service
	Automobile Emergency Radio Service

summarizes the reorganization of the 20 radio services into the two frequency pools.

Public Safety Radio Pool

The Public Safety Radio Pool was created in 1997. It covers the licensing of the radio communications of state and local governmental entities and the following categories of activities: medical services, rescue organizations, veterinarians, persons with disabilities, disaster-relief organizations, school buses, beach patrols, establishments in isolated places, communications standby facilities, and emergency repair of public communications facilities.

The FCC has established an 800-MHz National Plan that specifies special policies and procedures governing the Public Safety Pool. The principal spectrum resource for the National Plan is the 821- to 824-MHz and the 866- to 869-MHz bands. The National Plan establishes planning regions covering all parts of the United States, Puerto Rico, and the U.S. Virgin Islands. The license application to provide service must be approved by the appropriate regional planning committee before frequency assignments will be made in these bands.

Industrial/Business Pool

The Industrial/Business Pool consists of a number of frequencies that were previously allotted to the Industrial or Land Transportation Radio Services, including the Business Radio Service. Anyone eligible in one of these radio services is eligible in the new Industrial/Business Pool for any frequency in that pool. In this regard, the FCC has adopted the eligibility criteria from the old Business Radio Service. The Industrial/Business Radio Pool covers the licensing of the radio communications of entities engaged in commercial activities; engaged in clergy activities; operating educational, philanthropic, or ecclesiastical institutions; or operating hospitals, clinics, or medical associations.

General Access Pool

Prior to 1977, channels in the 470- to 512-MHz band were allocated to seven frequency pools based on category of eligibility. The FCC eliminated the separate allocation to these pools and created a General Access Pool to permit greater flexibility and foster more effective and efficient use of the 470- to 512-MHz band. Frequencies in the 470- to 512-MHz band are shared with UHF TV Channels 14 to 20 and are available in only 11 cities, listed in Table P-2.

All unassigned channels, including those which subsequently become unassigned, are considered to be in the General Access Pool and available to all eligible parties on a first-come, first-served basis. If a channel is assigned in an area, however, subsequent authorizations on that channel will only be granted to users from the same category.

TABLE P-2   Cities with General Access Pool Frequencies

Urban Areas	Channels	Frequencies (MHz)
Boston, MA	14	470–476
	16	482–488
Chicago, IL	14	470–476
	15	476–482
Dallas/Fort Worth, TX	16	482–488
Houston, TX	17	488–494
Los Angeles, CA	14	470–476
	20	505–512
Miami, FL	14	470–476
New York/Northeast NJ	14	470–476
	15	476–482
Philadelphia, PA	19	500–506
	20	506–512
Pittsburgh, PA	14	470–476
	18	494–500
San Francisco/Oakland, CA	16	482–488
	17	488–494
Washington, DC/MD/VA	17	488–494
	18	494–500

Applications for PLMRS

PLMRS are used by organizations that are engaged in a wide variety of activities. Police, fire, ambulance, and emergency relief organizations such as the Red Cross use private wireless systems to dispatch help when emergency calls come in or disaster strikes. Utility companies, railroad and other transportation providers, and other infrastructure-related companies use their systems to provide vital day-to-day control of their systems (including monitoring and control and routine maintenance and repair), as well as to respond to emergencies and disasters—often working with public safety agencies. A wide variety of businesses, including package delivery companies, plumbers, airlines, taxis, manufacturers, and even the American Automobile Association

(AAA), rely on private wireless systems to monitor, control, and coordinate their production processes, personnel, and vehicles.

Although commercial services can serve some of the needs of these organizations, private users generally believe that their own systems provide them with capabilities, features, and efficiencies that commercial services cannot. Some of the requirements and features that PLMRS users believe make their systems unique include

- Immediate access to a radio channel (no dialing required)
- Coverage in areas where commercial systems cannot provide service
- Peak usage patterns that could overwhelm commercial systems
- High reliability
- Priority access, especially in emergencies
- Specialized equipment required by the job

Radio Trunking

In a conventional radio system, a radio can access only one channel at a time. If that channel is in use, the user must either wait for the channel to become idle or manually search for a free channel. A trunked radio system differs from a conventional system by having the ability to automatically search all available channels for one that is clear.

The FCC has recognized two types of trunking: centralized and decentralized. A centralized trunked system uses one or more control channels to transmit channel-assignment information to the mobile radios. In a decentralized trunked system, the mobile radios scan the available channels and find one that is clear.

The rules require that licensees take reasonable precautions to avoid causing harmful interference, including monitoring the transmitting frequency for communications in progress. This requirement is met in decentralized trunked systems because each mobile unit monitors each channel and finds a clear one on which to transmit. In a centralized trunked radio system, radios typically monitor the control channel(s), not the specific transmit frequencies. Therefore, this form of trunking generally has not been allowed in the shared bands below 800 MHz.

Under certain conditions, however, the FCC allows some licensees to implement centralized trunked radio systems in the shared bands below 800 MHz. Centralized trunking may be authorized if the licensee has an exclusive service area and uses the 470- to 512-MHz band only. If the licensee does not have an exclusive service area, it must obtain consent from all licensees who have cochannel and/or adjacent channel stations.

Summary

Private radio systems serve a great variety of communication needs that common carriers and other commercial service providers traditionally have not been able or willing to fulfill. Companies large and small use their private systems to support their business operations, safety, and emergency needs. The one characteristic that all these uses share—and that differentiates private wireless use from commercial use—is that private wireless licensees use radio as a tool to accomplish their missions in the most effective and efficient ways possible.

Private radio users employ wireless communications as they would any other tool or machine—radio contributes to their production of some other good or service. For commercial wireless service providers, by contrast, the services offered over the radio system are the end products. Cellular, PCS, and Specialized Mobile Radio (SMR) providers sell service or capacity on wireless systems, permitting a wide range of mobile and portable communications that extend the national communications infrastructure. This difference in purpose is significant because it determines how the services are regulated.

See also

Cellular Data Communications
Cellular Voice Communications
Federal Communications Commission
Personal Communication Services
Specialized Mobile Radio
Telemetry

R

RADIO COMMUNICATION INTERCEPTION

As the term implies, *radio communication interception* is the capture of radio signals by a scanning device for the purpose of eavesdropping on a voice call or learning the contents of data messages. When it comes to the interception of radio communications, the Federal Communications Commission (FCC) has the authority to interpret Section 705 of the Communications Act, 47 U.S.C. Section 605, which deals with "Unauthorized Publication of Communications."

Although the act of intercepting radio communications may violate other federal or state statutes, this provision generally does not prohibit the mere interception of radio communications. For example, if someone happens to over-hear a conversation on a neighbor's cordless telephone, this is not a violation of the Communications Act. Similarly, if someone listens to radio transmissions on a scanner, such as emergency service reports, this is not a violation of Section 705.

A violation of Section 705 would occur, however, if a person were to divulge or publish what he or she hears or use it for his or her own or someone else's benefit. An example of using an intercepted call for a beneficial use in violation of Section

705 would be someone listening to accident reports on a police channel and then driving or sending one of his or her own tow trucks to the reported accident scene in order to obtain business.

The Communications Act does allow for the divulgence of certain types of radio transmissions. The statute specifies that there are no restrictions on the divulgence or use of radio communications that have been transmitted for the use of the general public, such as transmissions of a local radio or television broadcast station. Likewise, there are no restrictions on divulging or using radio transmissions originating from ships, aircraft, vehicles, or persons in distress. Transmissions by amateur radio or citizens' band radio operators are also exempt from interception restrictions.

In addition, courts have held that the act of viewing a transmission (such as a pay television signal) that the viewer was not authorized to receive is a "publication" violating Section 705. This section also has special provisions governing the interception of satellite television programming transmitted to cable operators. The section prohibits the interception of satellite cable programming for private home viewing whether the programming is scrambled or not scrambled but is sold through a marketing system. In these circumstances, authorization must be obtained from the programming provider to legally intercept the transmission.

The act also contains provisions that affect the manufacture of equipment used for listening or receiving radio transmissions, such as scanners. Section 302(d) of the Communications Act, 47 U.S.C. Section 302(d), prohibits the FCC from authorizing scanning equipment that is capable of receiving transmissions in the frequencies allocated to domestic cellular services, that is capable of readily being altered by the user to intercept cellular communications, or that may be equipped with decoders that convert digital transmissions to analog voice audio. And since April 26, 1994 (47 CFR 15.121), such receivers may not be manufactured in the United States or imported for use in the United

States. FCC regulations also prohibit the sale or lease of such scanning equipment (47 CFR 2.803).

Government Interception

While intercepting radio communications for beneficial purposes is illegal in the United States, the federal government systematically engages in such monitoring for the purpose of learning industrial secrets. Under a program known as "Echelon," a satellite interception system, private and commercial communications are monitored around the world. The program is run by five nations—the United States, the United Kingdom, Canada, Australia, and New Zealand. France and Russia are also known to have systems of their own.

Business is often subject to surveillance involving economic data, such as details of developments in individual sectors of the economy, developments in commodity markets, or compliance with economic embargoes. Although this is ostensibly the purpose behind Echelon, some countries claim that it is also used for industrial espionage; specifically for spying on foreign businesses with the aim of securing a competitive advantage for firms in the home country. While it is often maintained that Echelon has been used in this way, no such case has been substantiated.

Summary

The FCC receives many inquiries regarding the interception and recording of telephone conversations. To the extent that these conversations are radio transmissions, there would be no violation of Section 705 if no divulgence or beneficial use of the conversation takes place. Again, however, the mere interception of some telephone-related radio transmissions—whether cellular, cordless, or landline conversations—may constitute a criminal violation of other federal or state statutes.

See also

Federal Communications Commission

Fraud Management

Wired Equivalent Privacy

Wireless Security

REMOTE MONITORING

The common platform from which to monitor multivendor wireless and wired networks is the Simple Network Management Protocol's (SNMP's) Remote Monitoring (RMON) Management Information Base (MIB). Although a variety of SNMP MIBs collect performance statistics to provide a snapshot of events, RMON enhances this monitoring capability by keeping a past record of events that can be used for fault diagnosis, performance tuning, and network planning. RMON works on wired, wireless and hybrid networks.

Hardware- and/or software-based RMON-compliant devices (i.e., probes) placed on each network segment monitor all data packets sent and received. The probes view every packet and produce summary information on various types of packets, such as undersized packets, and events, such as packet collisions. The probes also can capture packets according to predefined criteria set by the network manager or test technician. At any time, the RMON probe can be queried for this information by a network management application or an SNMP-based management console so that detailed analysis can be performed in an effort to pinpoint where and why an error occurred.

The original Remote Network Monitoring MIB defined a framework for the remote monitoring of Ethernet. Subsequent RMON MIBs have extended this framework to Token Ring and other types of networks. A map of the RMON MIB for Ethernet and Token Ring is shown in Figure R-1.

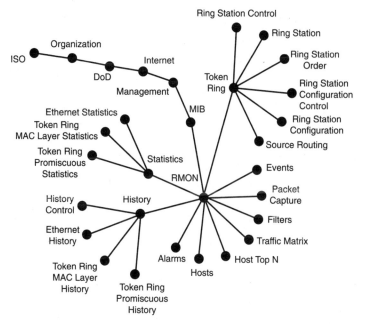

Figure R-1 A map of SNMP's remote monitoring management information base—RMON MIB.

RMON Applications

A management application that views the internetwork, for example, gathers data from RMON agents running on each segment in the network. The data are integrated and correlated to provide various internetwork views that provide end-to-end visibility of network traffic, both local area network (LAN) and wide area network (WAN). The operator can switch between a variety of views.

For example, the operator can switch between a Media Access Control (MAC) view (which shows traffic going through routers and gateways) and a network view (which shows end-to-end traffic) or can apply filters to see only traffic of a given protocol or suite of protocols. These traffic matrices provide the information necessary to configure or partition the internetwork to optimize LAN and WAN utilization.

In selecting the MAC level view, for example, the network map shows each node of a segment separately, indicating intrasegment node-to-node data traffic. It also shows total intersegment data traffic from routers and gateways. This combination allows the operator to see consolidated internetwork traffic and how each end node contributes to it.

In selecting the network level view, the network map shows end-to-end data traffic between nodes across segments. By connecting source and ultimate destination without clouding the view with routers and gateways, the operator can immediately identify specific areas contributing to an unbalanced traffic load.

Another type of application allows the network manager to consolidate and present multiple segment information, configure RMON alarms, and provide complete Token Ring RMON information, as well as perform baseline measurements and long-term reporting. Alarms can be set on any RMON variable. Notification via traps can be sent to multiple management stations. Baseline statistics allow long-term trend analysis of network traffic patterns that can be used to plan for network growth.

Ethernet Object Groups

The RMON specification consists of nine Ethernet/Token Ring groups and ten specific Token Ring RMON extensions (see Figure R-1).

Ethernet Statistics Group The Statistics Group provides segment-level statistics (Figure R-2). These statistics show packets, octets (or bytes), broadcasts, multicasts, and collisions on the local segment, as well as the number of occurrences of packets dropped by the agent. Each statistic is maintained in its own 32-bit cumulative counter. Real-time packet size distribution is also provided.

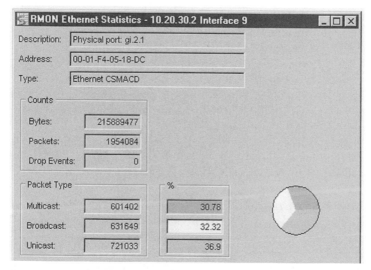

Figure R-2 The Ethernet Statistics window accessed from Enterasys Networks' NetSight Element Manager. This window would be used to view a detailed statistical breakdown of traffic on the monitored Ethernet network segment. The data provided apply only to the interface or network segment.

Ethernet History Group With the exception of packet size distribution, which is provided only on a real-time basis, the History Group provides historical views of the statistics provided in the Statistics Group. The History Group can respond to user-defined sampling intervals and bucket counters, allowing for some customization in trend analysis.

The RMON MIB comes with two defaults for trend analysis. The first provides for 50 buckets (or samples) of 30-second sampling intervals over a period of 25 minutes. The second provides for 50 buckets of 30-minute sampling intervals over a period of 25 hours. Users can modify either of these or add additional intervals to meet specific requirements for historical analysis. The sampling interval can range from 1 second to 1 hour.

Host Table Group The RMON MIB specifies a host table that includes node traffic statistics: packets sent and received, octets sent and received, as well as broadcasts, multicasts, and errored packets sent. In the host table, the classification "errors sent" is the combination of packet undersizes, fragments, cyclic redundancy check (CRC)/alignment errors, collisions, and oversizes sent by each node.

The RMON MIB also includes a host timetable that shows the relative order in which the agent discovered each host. This feature is not only useful for network management purposes but also assists in uploading those nodes to the management station of which it is not yet aware. This reduces unnecessary SNMP traffic on the network.

Host Top N Group The Host Top N Group extends the host table by providing sorted host statistics, such as the top 10 nodes sending packets or an ordered list of all nodes according to the errors sent over the last 24 hours. The data selected and the duration of the study are both defined at the network management station. The number of studies that can be run depends on the resources of the monitoring device.

When a set of statistics is selected for study, only the selected statistics are maintained in the Host Top N counters; other statistics over the same time intervals are not available for later study. This processing—performed remotely in the RMON MIB agent—reduces SNMP traffic on the network and the processing load on the management station, which would otherwise need to use SNMP to retrieve the entire host table for local processing.

Alarms Group The Alarms Group provides a general mechanism for setting thresholds and sampling intervals to generate events on any counter or integer maintained by the agent, such as segment statistics, node traffic statistics defined in the host table, or any user-defined packet match counter defined in the Filters Group. Both rising and falling thresholds can be set, each of which can indicate network faults. Thresholds can be established for both the absolute value of a statistic and its delta value, enabling the manager to be notified of rapid spikes or drops in a monitored value.

Filters Group The Filters Group provides a generic filtering engine that implements all packet capture functions and events. The packet capture buffer is filled with only those packets that match the user-specified filtering criteria. Filtering conditions can be combined using the Boolean parameters "and" or "not." Multiple filters are combined with the Boolean "or" parameter.

Packet Capture Group The types of packets collected depend on the Filter Group. The Packet Capture Group allows the user to create multiple capture buffers and to control whether the trace buffers will wrap (overwrite) when full or stop capturing. The user may expand or contract the size of the buffer to fit immediate needs for packet capturing rather than permanently commit memory that will not always be needed.

Notifications (Events) Group In a distributed management environment, the RMON MIB agent can deliver traps to multiple management stations that share a single community name destination specified for the trap. In addition to the three traps already mentioned—rising threshold and falling threshold (see Alarms Group) and packet match (see Packet Capture Group)—seven additional traps can be specified:

- *coldStart* This trap indicates that the sending protocol entity is reinitializing itself such that the agent's configuration or the protocol entity implementation may be altered.
- *warmStart* This trap indicates that the sending protocol entity is reinitializing itself such that neither the agent configuration nor the protocol entity implementation is altered.
- *linkDown* This trap indicates that the sending protocol entity recognizes a failure in one of the communication links represented in the agent's configuration.

- *linkUp* This trap indicates that the sending protocol entity recognizes that one of the communication links represented in the agent's configuration has come up.

- *authenticationFailure* This trap indicates that the sending protocol entity is the addressee of a protocol message that is not properly authenticated. While implementations of the SNMP must be capable of generating this trap, they also must be capable of suppressing the emission of such traps via an implementation-specific mechanism.

- *egpNeighborLoss* This trap indicates that an External Gateway Protocol (EGP) neighbor for whom the sending protocol entity was an EGP peer has been marked down and the peer relationship is no longer valid.

- *enterpriseSpecific* This trap indicates that the sending protocol entity recognizes that some enterprise-specific event has occurred.

The Notifications (Events) Group allows users to specify the number of events that can be sent to the monitor log. From the log, any specified event can be sent to the management station. The log includes the time of day for each event and a description of the event written by the vendor of the monitor. The log overwrites when full, so events may be lost if not uploaded to the management station periodically.

Traffic Matrix Group The RMON MIB includes a traffic matrix at the MAC layer. A traffic matrix shows the amount of traffic and number of errors between pairs of nodes—one source and one destination address per pair. For each pair, the RMON MIB maintains counters for the number of packets, number of octets, and error packets between the nodes. Users can sort this information by source or destination address.

Applying remote monitoring and statistics-gathering capabilities to the Ethernet environment offers a number of benefits. The availability of critical networks is maximized, since remote capabilities allow for a more timely resolution of the problem. With the capability to resolve problems remotely, operations staff can avoid costly travel to troubleshoot problems on site. With the capability to analyze data collected at specific intervals over a long period of time, intermittent problems can be tracked down that would normally go undetected and unresolved.

Token Ring Extensions

As noted, the first version of RMON defined media-specific objects for Ethernet only. Later, media-specific objects for Token Ring were added.

Token Ring MAC-Layer Statistics This extension provides statistics, diagnostics, and event notification associated with MAC traffic on the local ring. Statistics include the number of beacons, purges, and IEEE 803.5 MAC management packets and events; MAC packets; MAC octets; and ring soft error totals.

Token Ring Promiscuous Statistics This extension collects utilization statistics of all user data traffic (non-MAC) on the local ring. Statistics include the number of data packets and octets, broadcast and multicast packets, and data frame size distribution.

Token Ring MAC-Layer History This extension offers historical views of MAC-layer statistics based on user-defined sample intervals, which can be set from 1 second to 1 hour to allow short- or long-term historical analysis.

Token Ring Promiscuous History This extension offers historical views of promiscuous (i.e., unfiltered) statistics based on user-defined sample intervals, which can be set from 1 second to 1 hour to allow short-term or long-term historical analysis.

Ring Station Control Table This extension lists status information for each ring being monitored. Statistics include ring state, active monitor, hard error beacon fault domain, and number of active stations.

Ring Station Table This extension provides diagnostics and status information for each station on the ring. The type of information collected includes station MAC address, status, and isolating and nonisolating soft error diagnostics.

Source Routing Statistics The extension for source routing statistics is used for monitoring the efficiency of source-routing processes by keeping track of the number of data packets routed into, out of, and through each ring segment. Traffic distribution by hop count provides an indication of how much bandwidth is being consumed by traffic-routing functions.

Ring Station Configuration Control The extension for station configuration control provides a description of the network's physical configuration. A media fault is reported as a "fault domain," an area that isolates the problem to two adjacent nodes and the wiring between them. The network administrator can discover the exact location of the problem—the fault domain—by referring to the network map. Some faults result from changes to the physical ring—including each time a station inserts or removes itself from the network. This type of fault is discovered through a comparison of the start of symptoms and the timing of the physical changes.

The RMON MIB not only keeps track of the status of each station but also reports the condition of each ring being monitored by a RMON agent. On large Token Ring networks with several rings, the health of each ring segment and the number of active and inactive stations on each ring can be monitored simultaneously. Network administrators can be alerted to the location of the fault domain should any ring go into a beaconing (fault) condition. Network managers also can be alerted to any changes in backbone ring configuration that could indicate loss of connectivity to an interconnect device such as a bridge or to a shared resource such as a server.

Ring Station Configuration The ring station group collects Token Ring–specific errors. Statistics are kept on all significant MAC-level events to assist in fault isolation, including ring purges, beacons, claim tokens, and such error conditions as burst errors, lost frames, congestion errors, frame copied errors, and soft errors.

Ring Station Order Each station can be placed on the network map in a specified order relative to the other stations on the ring. This extension provides a list of stations attached to the ring in logical ring order. It lists only stations that comply with IEEE 802.5 active monitoring ring poll or IBM trace tool present advertisement conventions.

RMON II

The RMON MIB is basically a MAC-level standard. Its visibility does not extend beyond the router port, meaning that it cannot see beyond individual LAN segments. As such, it does not provide visibility into conversations across the network or connectivity between the various network segments. Given the trends toward remote access and distributed workgroups, which generate a lot of intersegment traffic, visibility across the enterprise is an important capability to have.

RMON II extends the packet capture and decoding capabilities of the original RMON MIB to Layers 3 through 7 of the Open Systems Interconnection (OSI) reference model. This allows traffic to be monitored via network-layer addresses—which lets RMON "see" beyond the router to the internetwork—and distinguish between applications.

Analysis tools that support the network layer can sort traffic by protocol rather than just report on aggregate traffic. This means that network managers will be able to determine, for example, the percent of Internet Protocol (IP) versus Internet Packet Exchange (IPX) traffic traversing the network. In addition, these higher-level monitoring tools can map end-to-end traffic, giving network managers the ability to trace communications between two hosts—or nodes—even if the two are located on different LAN segments. RMON II functions that will allow this level of visibility include

- *Protocol directory table* Provides a list of all the different protocols a RMON II probe can interpret.

- *Protocol distribution table* Permits tracking of the number of bytes and packets on any given segment that have been sent from each of the protocols supported. This information is useful for displaying traffic types by percentage in graphical form.

- *Address mapping* Permits identification of traffic-generating nodes, or hosts, by Ethernet or Token Ring address in addition to MAC address. It also discovers switch or hub ports to which the hosts are attached. This is helpful in node discovery and network topology applications for pinpointing the specific paths of network traffic.

- *Network-layer host table* Permits tracking of bytes, packets, and errors by host according to individual network-layer protocol.

- *Network-layer matrix table* Permits tracking, by network-layer address, of the number of packets sent between pairs of hosts.

- *Application-layer host table* Permits tracking of bytes, packets, and errors by host and according to application.

- *Application-layer matrix table* Permits tracking of conversations between pairs of hosts by application.

- *History group* Permits filtering and storing of statistics according to user-defined parameters and time intervals.

- *Configuration group* Defines standard configuration parameters for probes that includes such parameters as network address, serial line information, and SNMP trap destination information.

RMON II is focused more on helping network managers understand traffic flow for the purpose of capacity planning rather than for the purpose of physical troubleshooting. The capability to identify traffic levels and statistics by application has the potential to greatly reduce the time it takes to troubleshoot certain problems. Without tools that can pinpoint which software application is responsible for gobbling up a disproportionate share of the available bandwidth, network managers can only guess. Often it is easier just to upgrade a server or a buy more bandwidth, which inflates operating costs and shrinks budgets.

Summary

Applying remote monitoring and statistics-gathering capabilities to Ethernet and Token Ring environments via the RMON MIB offers a number of benefits. The availability of critical wireless and wired networks is maximized, since remote capabilities allow for timely problem resolution. With the capability to resolve problems remotely, operations staff can avoid costly travel to troubleshoot problems on site. With the capability to analyze data collected at specific intervals over a long period of time, intermittent problems can be tracked down that normally would go undetected and unresolved. And with RMON II, these capabilities are enhanced and extended up to the applications level across the enterprise.

See also

Wireless Management Tools

REPEATERS

A repeater is a device that extends the inherent distance limitations of various transmission media, including wireless links, by boosting signal power so that it stays at the same level regardless of the distance it must travel. As such, the repeater operates at the lowest level of the Open Systems Interconnection (OSI) reference model—the Physical Layer (Figure R-3).

Repeaters are necessary because signal strength weakens with distance: The longer the path a signal must travel, the weaker it gets. This condition is known as "signal attenuation." On a telephone call, a weak signal will cause low volume, interfering with the parties' ability to hear each other. In cellular networks, when a mobile user moves beyond the range of a cell site, the signal fades to the point of disconnecting the call. In the LAN environment, a weak signal can result in corrupt data, which can substantially reduce throughput by forcing retransmissions when errors are detected. When the signal level drops low enough, the chances of interference from external noise increase, rendering the signal unusable.

Repeaters also can be used to link different types of network media—fiber to coaxial cable, for example. Often LANs are interconnected in a campus environment by means of repeaters that form the LANs into connected network seg-

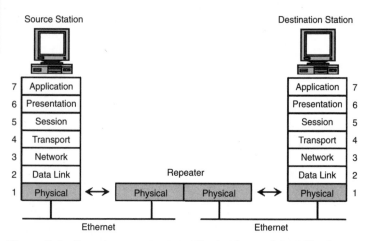

Figure R-3 Repeaters operate at the Physical Layer of the OSI reference model.

ments. The segments may employ different transmission media—thick or thin coaxial cable, twisted-pair wiring, or optical fiber. The cost of media converters is significantly less than full repeaters and can be used whenever media distance limitations will not be exceeded in the network.

Hubs or switches usually are equipped with appropriate modules that perform the repeater and media conversion functions on sprawling LANs. But the use of hubs or switches also can eliminate the need for repeaters, since most cable segments in office buildings will not run more than 100 feet (about 30 meters), which is well within the distance limitation of most LAN standards, including 1000BaseT Gigabit Ethernet running over Category 5 cable.

Regenerators

Often the terms *repeater* and *regenerator* are used interchangeably, but there is a subtle difference between the two. In an analog system, a repeater boosts the desired signal strength but also boosts the noise level as well. Consequently, the signal-to-noise ratio on the output side of the repeater remains the same as on the input side. This means that once noise is introduced into the desired signal, it is impossible to get the signal back into its original form again on the output side of the repeater.

In a digital system, regenerators are used instead of repeaters. The regenerator determines whether the information-carrying bits are 1s or 0s on the basis of the received signal on the input side. Once the decision of 1 or 0 is made, a fresh signal representing that bit is transmitted on the output side of the regenerator. Because the quality of the output signal is a perfect replication of the input signal, it is possible to maintain a very high level of performance over a range of transmission impairments. Noise, for instance, is filtered out because it is not represented as a 1 or 0.

Summary

Stand-alone repeaters have transceiver interface modules that provide connections to various media. There are fiberoptic transceivers, coaxial transceivers, and twisted-pair transceivers. Some repeaters contain the intelligence to detect packet collisions and will not repeat collision fragments to other cable segments. Some repeaters also can "deinsert" themselves from a hub or switch when there are excessive errors on the cable segment, and they can submit performance information to a central management station.

See also

Access Points

Wireless LANs

ROUTERS

A router operates at Layer 3 of the Open Systems Interconnection (OSI) reference model, the Network Layer. The device distinguishes among network layer protocols—such as IP, IPX, and AppleTalk—and makes intelligent packet delivery decisions using an appropriate routing protocol. Used on wired and wireless networks, routers can be used to segment a network with the goals of limiting broadcast traffic and providing security, control, and redundant paths.

A router also can provide multiple types of interfaces, including those for T1, Frame Relay, Integrated Services Digital Network (ISDN), Asynchronous Transfer Mode (ATM), cable networks, and Digital Subscriber Line (DSL) services, among others. Some routers can perform simple packet filtering to control the kind of traffic that is allowed to pass through them, providing a rudimentary firewall service. Larger routers can perform advanced firewall functions.

A router is similar to a bridge in that both provide filtering and bridging functions across the network. But while bridges operate at the Physical and Data Link Layers of the OSI reference model, routers join LANs at the Network Layer (Figure R-4). Routers convert LAN protocols into WAN protocols and perform the process in reverse at the remote location. They may be deployed in mesh as well as point-to-point networks and, in certain situations, can be used in combination with bridges.

Although routers include the functionality of bridges, they differ from bridges in the following ways: They generally offer more embedded intelligence and, consequently, more sophisticated network management and traffic control capabilities than bridges. Another distinction—perhaps the most significant one—between a router and a bridge is that a bridge delivers packets of data on a "best effort" basis, specifically, by discarding packets it does not recognize onto an adjacent network. Through a continual process of discarding unfamiliar packets, data get to theirs proper destination—on a network

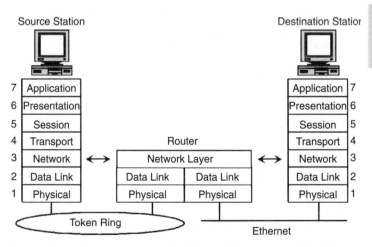

Figure- R-4 Routers operate at the Network Layer of the OSI reference model.

where the bridge recognizes the packets as belonging to a device attached to its network. By contrast, a router takes a more intelligent approach to getting packets to their destination—by selecting the most economical path (i.e., least number of hops) on the basis of its knowledge of the overall network topology, as defined by its internal routing table. Routers also have flow-control and error-protection capabilities.

Types of Routing

There are two types of routing: static and dynamic. In static routing, the network manager configures the routing table to set fixed paths between two routers. Unless reconfigured, the paths on the network never change. Although a static router will recognize that a link has gone down and issue an alarm, it will not automatically reroute traffic. A dynamic router, on the other hand, reconfigures the routing table automatically and recalculates the most efficient path in terms of load, line delay, or bandwidth.

In wired networks, some routers balance the traffic load across multiple access links, providing an $N \times$ T1 inverse multiplexer function. This allows multiple T1 access lines operating at 1.544 Mbps each to be used as a single higher-bandwidth facility. If one of the links fails, the other links remain in place to handle the offered traffic. As soon as the failed link is restored to service, traffic is spread across the entire group of lines as in the original configuration.

Routing Protocols

Each router on the network keeps a routing table and moves data along the network from one router to the next using such protocols as the Open Shortest Path First (OSPF) protocol and the Routing Information Protocol (RIP).

Although still supported by many vendors, RIP does not perform well in today's increasingly complex networks. As the network expands, routing updates grow larger under

RIP and consume more bandwidth to route the information. When a link fails, the RIP update procedure slows route discovery, increases network traffic and bandwidth usage, and may cause temporary looping of data traffic. Also, RIP cannot base route selection on such factors as delay and bandwidth, and its line-selection facility is capable of choosing only one path to each destination.

The newer routing standard, OSPF, overcomes the limitations of RIP and even provides capabilities not found in RIP. The update procedure of OSPF requires that each router on the network transmit a packet with a description of its local links to all other routers. On receiving each packet, the other routers acknowledge it, and in the process, distributed routing tables are built from the collected descriptions. Since these description packets are relatively small, they produce a minimum of overhead. When a link fails, updated information floods the network, allowing all the routers to simultaneously calculate new tables.

Types of Routers

Multiprotocol nodal, or hub, routers are used for building highly meshed internetworks. In addition to allowing several protocols to share the same logical network, these devices pick the shortest path to the end node, balance the load across multiple physical links, reroute traffic around points of failure or congestion, and implement flow control in conjunction with the end nodes. They also provide the means to tie remote branch offices into the corporate backbone, which might use such WAN services as Transmission Control Protocol/Internet Protocol (TCP/IP), T1, ISDN, and ATM.

Access routers are typically used at branch offices. These are usually fixed-configuration devices available in Ethernet and Token Ring versions that support a limited number of protocols and physical interfaces. They provide connectivity to high-end multiprotocol routers, allowing large and small nodes to be managed as a single logical enterprise network. Although low-cost, plug-and-play bridges can meet the need for branch office connectivity, low-end routers can offer more intelligence and configuration flexibility at comparable cost.

The newest access routers are multiservice devices that are designed to handle a mix of data, voice, and video traffic. They support a variety of WAN connections through built-in interfaces that include dual ISDN Basic Rate Interface (BRI) interfaces, dual analog ports, T1/Frame Relay ports, and an ISDN interface for videoconferencing. Such routers can run software that provides Internet Protocol Secure (IPSec) virtual private network (VPN), firewall, and encryption services.

Midrange routers provide network connectivity between corporate locations in support of workgroups or the corporate intranet, for example. These routers can be stand-alone devices or packaged as modules that occupy slots in an intelligent wiring hub or LAN switch. In fact, this type of router is often used to provide connectivity between multiple wiring hubs or LAN switches over high-speed LAN backbones such as ATM, Fiber Distributed Data Interface (FDDI), and Fast Ethernet.

There is a consumer class of routers for 2.4-GHz Wireless Fidelity (Wi-Fi) networks that are capable of providing shared access to the Internet over such broadband technologies as cable and DSL. The EtherFast Wireless AP + Cable/DSL Router from Linksys, for example, connects a wireless network to a high-speed broadband Internet connection and a 10/100 Fast Ethernet backbone (Figure R-5). Configurable through any networked PC's Web browser, the router can be set up for Network Address Translation (NAT), allowing it to act as an externally recognized Internet device with its own IP address for the home LAN. The Linksys device is also equipped with a four-port Ethernet switch. The combination of wireless router and switch technology eliminates the need to buy an additional hub or switch and extends the range of the wireless network.

Figure R-5 The EtherFast Wireless AP + Cable/DSL Router from Linksys

Summary

Whether used on a wired or wireless network or a hybrid network, routers fulfill a vital role in implementing complex mesh networks such as the Internet and private intranets using Layer 3 protocols, usually IP. They also have become an economical means of tying branch offices into the enterprise network and providing PCs tied together on a home network with shared access to broadband Internet services such as cable and DSL. Like other interconnection devices, enterprise-class routers are manageable via SNMP, as well as the proprietary management systems of vendors. Just as bridging and routing functions made their way into a single device, routing and switching functions are being combined in the same way.

See also

Access Points

Bridges

Repeaters

RURAL RADIOTELEPHONE SERVICE

Rural Radiotelephone Service is a fixed wireless service that allows common carriers to provide telephone service to the homes of subscribers living in extremely remote rural areas, where it is not feasible to provide telephone service by wire or other means. Rural Radiotelephone Stations, operating in the paired 152- to 158-MHz and 454- to 459-MHz bands, employ standard duplex analog technology to provide telephone service to subscribers' homes.

The quality of conventional rural radiotelephone service is similar to that of precellular mobile telephone service. Several subscribers may have to share a radio-channel pair (similar to party-line service), each waiting until the channel pair is not in use by the others before making or receiving a call.

Rural Radiotelephone Service is generally considered by state regulators to be a separate service that is interconnected to the Public Switched Telephone Network (PSTN). This service has been available to rural subscribers for more than 25 years.

Summary

Carriers must apply to the FCC for permission to offer Rural Radiotelephone Service. Among other things, each application for a central office station must contain an exhibit showing that it is impractical to provide the required communication service by means of landline facilities.

See also

    Air-Ground Radiotelephone Service

    Basic Exchange Telephone Radio Service

    Fixed Wireless Access

S

SATELLITE COMMUNICATIONS

The idea of using satellites as relay stations for an international microwave radiotelephone system goes back to 1945 when Arthur C. Clarke (Figure S-1) proposed the scheme in a British technical journal. Clarke, then a young scientist and officer of the Royal Air Force, later became a leading science fiction writer and coauthor of the motion picture *2001: A Space Odyssey*. However, it was not until 1957 that the first satellite was put into orbit. Although just a beacon whose primary purpose was to announce its presence in the sky, the successful orbital deployment of the 185-pound Russian Sputnik sparked a technological revolution in communications that continues to this day. There are now over 2560 satellites in orbit, along with over 6000 pieces of debris tracked by the National Aeronautics and Space Administration (NASA).

The United States launched the first communication satellites in the early 1960s. Echo 1 and Echo 2 were little more than metallic balloons that simply reflected microwave signals from point *A* to point *B*. These passive satellites could not amplify the signals. Reception was often poor and the range of transmission limited. Ground stations had to track them across the sky, and communication between two

Figure S-1 Arthur C. Clarke articulated his vision of satellites in a British technical journal in 1945.

ground stations was only possible for a few hours a day when both had visibility with the satellite at the same time. Later, geostationary satellites overcame this problem. Such systems were high enough in orbit to move with the earth's rotation, in effect giving them fixed positions so that they could provide communications coverage to specific areas. Satellites are now categorized by type of orbit and area of coverage as follows:

- Geostationary-earth-orbit (GEO) satellites orbit the equator in a fixed position about 23,000 miles above the earth. Three GEO satellites can cover most of the planet, with each unit capable of handling 20,000 voice calls simultaneously. Because of their large coverage "footprint," these satellites are ideal for radio and television broadcasting and long distance domestic and international communications.

- Middle-earth-orbit (MEO) satellites circle the earth at about 6000 miles up. It takes about 12 satellites to provide global coverage. The lower orbit reduces power requirements and transmission delays that can affect signal quality and service interaction.

- Low-earth-orbit (LEO) satellites circle the earth only 600 miles up (Figure S-2). As many as 200 satellites may be required to provide global coverage. Since their low altitude means that they have nonstationary orbits and they pass over a stationary caller rather quickly, calls must be handed off from one satellite to the next to keep the session alive. The omnidirectional antennas of these devices do not have to be pointed at a specific satellite. There is also very little propagation delay. And the low altitude of these satellites means that earthbound transceivers can be packaged as low-powered, inexpensive hand-held devices.

The International Telecommunication Union (ITU) is responsible for all frequency/orbit assignments. Through its International Bureau, the Federal Communications Commission (FCC) regulates all satellite service rates, competition among carriers, and international telecommunications traffic

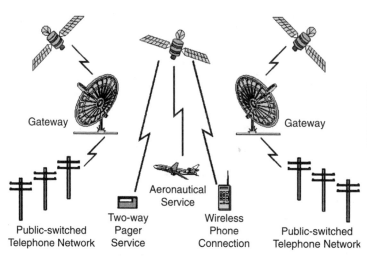

Figure S-2   Low-earth-orbit satellites hold out the promise of ubiquitous personal communications services, including telephone, pager, and two-way messaging services worldwide.

in the United States, ensuring that U.S. satellite operators conform to ITU frequency and orbit assignments. The FCC also issues licenses to domestic satellite service providers.

Satellite Technology

Each satellite carries transponders, which are devices that receive radio signals at one frequency and convert them to another for transmission. The uplink and downlink frequencies are separated to minimize interference between transmitted and received signals.

Satellite channels allow one sending station to broadcast transmissions to one or more receiving stations simultaneously. In a typical scenario, the communications channel starts at a host computer, which is connected through a traditional telephone company medium to the central office (i.e., master earth station or hub) of the satellite communications vendor. The data from this and other local loops are multiplexed into a fiberoptic or microwave signal and sent to the satellite vendor's earth station. This signal becomes part of a composite transmission that is sent by the earth station to the satellite (uplink) and then transmitted by the satellite to the receiving earth stations (downlink). At the receiving earth station, the data are transferred by a fiberoptic or microwave link to the satellite carrier's central office. The composite signal is then separated into individual communications channels that are distributed over the Public Switched Telephone Network (PSTN) to their destinations.

Satellite communication is very reliable for data transmission. The bit error rate (BER) for a typical satellite channel is in the range of 1 error in 1 billion bits transmitted. However, a potential problem with satellite communication is delay. Round-trip satellite transmission takes approximately 500 milliseconds, which can hamper voice communications and create significant problems for real-time interactive data transmissions. For voice communications, digital echo cancelers can correct voice echo problems caused by the transmission delay.

A number of techniques are employed to nullify the effects of delay during data transmissions via satellite. One technique employed by Mentat, Inc., increases the performance of Internet and intranet access over satellites by transparently replacing the Transmission Control Protocol (TCP) over the satellite link with a protocol optimized for satellite conditions. The company's SkyX Gateway intercepts the TCP connection from the client and converts the data to a proprietary protocol for transmission over the satellite. The gateway on the opposite side of the satellite link translates the data back to TCP for communication with the server (Figure S-3). The result is vastly improved performance while the process remains transparent to the end user and fully compatible with the Internet infrastructure. No changes are required to the client or server, and all applications continue to function without modification.

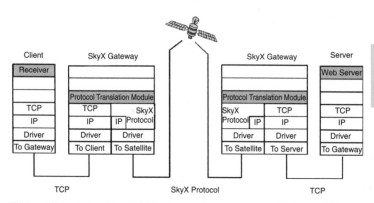

Figure S-3 Mentat's SkyX Gateway overcomes the effects of delay on Internet/intranet access by replacing TCP on the client side of the link with a proprietary protocol optimized for the satellite environment and then converting it back to TCP on the server side of the satellite link.

Very Small Aperture Terminals (VSATs)

VSAT networks have evolved to become mainstream communication networking solutions that are affordable to both large and small companies. Today's VSAT is a flexible, software-intensive system built around standard communications protocols. With a satellite as the serving office and using radiofrequency (RF) electronics instead of copper or fiber cables, these systems can be truly considered packet-switching systems in the sky.

VSATs can be configured for broadcast (one-way) or interactive (two-way) communications. The typical star topology provides a flexible and economical means of communications with multiple remote or mobile sites. Applications include broadcasting database information, insurance agent support, reservations systems, retail point-of-sale credit card checking, and interactive inventory data sharing.

Today's VSATs are used for supporting high-speed message broadcasting, image delivery, integrated data and voice, and mobile communications. VSATs are used increasingly for supporting local area network (LAN)–to-LAN connectivity and LAN-to–wide area network (WAN) bridging, as well as for providing route and media diversity for disaster recovery.

To make VSAT technology more affordable, VSAT providers offer compact hubs and submeter antennas that provide additional functionality at approximately 33 percent less than the cost of full-size systems. Newer submeter antennas are even supporting direct digital TV broadcasts to the home.

Network Management

The performance of the VSAT network is monitored increasingly at the hub location by the network control system. A failure anywhere on the network automatically alerts the network control operator, who can reconfigure capacity among individual VSAT systems. In the case of signal fade due to adverse weather conditions, for example, the hub detects the weak signal—or the absence of a signal—and alerts the network operations staff so that corrective action can be taken.

Today's network management systems indicate whether power failures are local or remote. They also increasingly locate the source of communications problems and determine whether the trouble is with the software or hardware. Such capabilities often eliminate the expense of dispatching technicians to remote locations. And when technicians must be dispatched, the diagnostic capabilities of the network management system can ensure that service personnel have the appropriate replacement parts, test gear, software patch, and documentation with them to solve the problem in a single service call.

Overall link performance is determined by the BER, network availability, and response time. Because of the huge amount of information transmitted by the hub station, uplink performance requirements are more stringent. A combination of uplink and downlink availability, coupled with BER and response time, provides the network control operator with overall network performance information on a continuous basis.

The VSAT's management system provides an interface to the major enterprise management platforms for single-point monitoring and control and offers a full range of accounting, maintenance, and data flow statistics, including those for inbound versus outbound data flow, peak periods, and total traffic volume by node. Also provided are capabilities for identifying fault conditions, performing diagnostics, and initiating service restoration procedures.

Communications Protocols

Within the VSAT network there are three categories of protocols—those associated with the backbone network, those of the host computer, and those concerned with transponder access. The scope and functionality of protocol handling differ markedly among VSAT network providers.

The backbone network protocol is responsible for flow control, retransmissions of bad packets, and running concurrent multiple sessions. The backbone network protocol could be associated with either the host or the communications link. The host protocol is related to the user application interface, which provides a compatible translation between the backbone protocol and the host communications protocol. Several host protocols are used in VSAT networks, including SNA/SDLC, 3270 Base Station Controller (BSC), Poll Select, and HASP. Multiple protocols can be used at the same VSAT location.

Transponder access protocols are used to assign transponder resources to various VSATs on the network. The three key transponder access protocols that are used on VSAT networks include Frequency Division Multiple Access (FDMA), Time Division Multiple Access (TDMA), and Code Division Multiple Access (CDMA).

Frequency Division Multiple Access With FDMA, the radiofrequency is partitioned so that bandwidth can be allocated to each VSAT on the network. This permits multiple VSATs to simultaneously use their portion of the frequency spectrum.

Time Division Multiple Access With TDMA, each VSAT accesses the hub via the satellite by the bursting of digital information onto its assigned radiofrequency carrier. Each VSAT bursts at its assigned time relative to the other VSATs on the network. Dividing access in this way—by time slots—is inherently wasteful because bandwidth is available to the VSAT in fixed increments whether or not it is needed. To improve the efficiency of TDMA, other techniques are applied to ensure that all the available bandwidth is used, regardless of whether the application contains bursty or streaming data. A reservation technique can even be applied to ensure that bandwidth is available for priority applications.

Code Division Multiple Access With CDMA, all VSATs share the assigned frequency spectrum and also can transmit simultaneously. This is possible through the use of spread-spectrum technology, which employs a wideband channel as opposed to the narrowband channels employed by other multiple access techniques such as FDMA and TDMA. Over the wideband channel, each transmission is assigned a unique code—a long row of numbers resembling a combination to a lock. The outbound data streams are coded so that they can be identified and received only by the station(s) having that code. This technique is also used in mobile communications as a means of cutting down interference and increasing available channel capacity by as much as 20 times.

Mobile Satellite Communications

Mobile satellite communications are used by the airline, maritime, and shipping industries. Among the key providers of mobile satellite communications services are Inmarsat, Intelsat, and Comsat. All began as government entities but, through global privatization efforts, have become private corporations that compete for market share.

Inmarsat The International Maritime Satellite Organization (Inmarsat) was formed in the late 1970s as a maritime-focused intergovernmental organization. Inmarsat completed its transition to a limited company in 1999 and now serves a broad range of markets. Today, Inmarsat delivers its solutions—including telex, voice, data, and video transmission—through a global distribution network of approximately 200 distributors and other service providers operating in over 150 countries to end users in the maritime, land, and aeronautical sectors. At year end 2000, approximately 212,000 terminals were registered to access Inmarsat services.

Inmarsat's primary satellite constellation consists of four Inmarsat-3 satellites in geostationary orbit. Between them, the global beams of the satellites provide overlapping coverage of the whole surface of the earth, apart from the poles. The Inmarsat-3 satellites are backed up by a fifth Inmarsat-3 and four previous-generation Inmarsat-2s, also in geostationary orbit. A key advantage of the Inmarsat-3s over their predecessors is their ability to generate a number of spot beams as well as single large global beams. Spot beams concentrate extra power in areas of high demand and make it possible to supply standard services to smaller, simpler terminals.

Among Imarsat's newest services is Swift64, which uses its Global Area Network platform to offer airlines and users of corporate jets the ability to operate an Integrated Services Digital Network (ISDN) connection of up to 64 kbps or a Mobile Packet Data connection using the existing Inmarsat antenna already installed on almost 80 percent of modern long-haul jets, as well as over 1000 corporate jets. With Swift64, airlines will be able to provide passengers with an ISDN 64-kbps bearer channel that is billed according to connection time. Mobile Packet Data, on the other hand, offers an "always on" connection that is billed according to the number of packets sent. These services enable users to surf the Web and send and receive e-mails and documents across all the continents of the world outside the north and south poles.

Intelsat Another provider on international mobile satellite communications services is Intelsat, which has the widest distribution network of any satellite communications company. Operating since 1964, Intelsat has a global fleet of 21 satellites from which it offers wholesale Internet, broadcast, telephony, and corporate network solutions to leading service providers in more than 200 countries and territories worldwide. Seven more satellites will be put into operation in the next 2 years to broaden coverage and add capacity. In mid-2001, Intelsat completed its transformation from a treaty-based organization to a privately held company with over 200 shareholders composed of companies from more than 145 countries.

Comsat In the United States, the government-sponsored satellite company was Comsat, which had been the only authorized U.S. organization that could directly access the Intelsat system. Lockheed Martin's acquisition of Comsat in August 2000 ended the 38-year history of quasi-government backing for Comsat. The federal government created the company in 1962 to prevent AT&T, then a telephone monopoly, from extending its monopoly to the satellite communications sector. Comsat became a publicly traded company the next year, but Congress ordered that no single investor could own a majority stake in the company because it was the only American firm with access to Intelsat. Congress eliminated the exclusive right to access Intelsat as part of the agreement that allowed Lockheed Martin to purchase Comsat.

Summary

Satellites provide a reliable, economical way of providing communications to remote locations and supporting mobile telecommunications. Taking into account the large number of satellites that can be employed, along with their corresponding radiofrequency assignments, it is clear that satellite communications systems offer ample room for expansion. Conversion of satellite transmissions from analog to digital and use of more sophisticated multiplexing techniques will further increase satellite transmission capacity. Other technological advances are focusing on the higher frequency bands—applying them in ways that decrease signal degradation.

See also

  Direct Broadcast Satellite
  Global Positioning System
  Microwave Communications

SHORT MESSAGING SERVICE

Short Messaging Service (SMS) is the ability to send and receive text messages between mobile telephones over cellular networks. Once used exclusively by carriers to push notifications of new voice messages down to smart cell phones, SMS is now on a fast track to universal adoption by the cellular subscribers around the world, who have adopted it as a two-way personal messaging medium.

Created as part of the Global System for Mobile (GSM) Phase 1 standard, the first short message was sent in December 1992 from a PC to a mobile phone on the Vodafone GSM network in the United Kingdom. With SMS, subscribers can send up to 160 characters of text when Latin alphabets are used and 70 characters when non-Latin alphabets such as Arabic and Chinese are used. The text can consist of words or numbers or an alphanumeric combination.

The success of SMS can be attributed to certain unique features, such as message storage if the recipient is not available, confirmation of short message delivery, and simultaneous transmission with GSM voice, data, and fax services. The person receiving the message also will know the phone number of the sender and the time at which the message was sent. While its principal use is still for personal messaging, SMS is also being used for receiving weather reports and traffic information, mobile shopping, banking, and stock trading. SMS is being enhanced to support the delivery of long-text messages, images, and video as well.

Although the prospects for SMS in the United States look promising, its rollout has been hampered by interoperability problems between different carriers. While all European carriers have standardized on GSM, carriers in the United States use several competing technologies. As a result, AT&T's message service will only work between other AT&T wireless customers or other carriers using its TDMA transmission technology. Likewise, Sprint PCS message-service customers would be able to communicate only with other Sprint PCS customers or other carriers using its CDMA transmission technology. Fortunately, gateway systems are being deployed that permit messages to be transmitted between customers on different carrier networks.

Most premises-based mobile-access gateways and servers already provide interfaces to SMS. A growing number of service providers offer SMS backbone-routing services to bridge otherwise-incompatible SMS carrier services. For example, TeleCommunication Systems introduced technology that allows intercarrier text messaging through use of a customer's phone number. InfoSpace added wireless network SMS interoperability to its technology platform that already enables Internet content and mobile-commerce services via several wireless carriers. Software developed by InphoMatch enables AT&T Wireless customers to send text messages to other wireless carriers' customers.

The growth prospects for SMS look so good that traditional paging services may one day disappear. Motorola has already announced that it is pulling out of the market for traditional paging infrastructure and handsets in favor of developing two-way technologies for use in wireless phones. The company's Wireless Messaging Division will now focus on providing SMS products for use on networks for GSM, General Packet Radio Services (GPRS), and CDMA protocols. At the same time, Motorola announced that it is discontinuing products that support the ReFlex protocol, which Motorola developed in the 1990s for traditional two-way paging carriers such as Skytel and Arch Wireless.

AT&T Wireless was the first national carrier to bring SMS to the United States and is the current market leader. AT&T claims to handle 35 million text messages a month. Two plans are available for the two-way text-messaging service. The first has no monthly charge, and the user pays $0.10 per message sent. The second has a monthly charge of $4.99 per month with 100 included outgoing messages and $0.10 per message thereafter. Both plans offer unlimited incoming messages and allow messages to be sent to subscribers of other carriers at no additional charge.

The service also offers features that let users reply, forward, store, and retrieve messages right from their phone. Messages of up to 150 characters can be sent between compatible phones and to Internet e-mail addresses—including e-mail addresses for pagers, handheld devices, or Web-enabled phones. Among the mobile phones offered by AT&T Wireless that are compatible with its 2-Way Text Messaging service are the Nokia 5165, 3360, and 8260 (Figure S-4).

Since November 2001, AT&T Wireless customers have had the ability to send text messages to virtually any wireless phone regardless of the carrier simply by knowing the

Figure S-4   Among the mobile phones offered by AT&T Wireless that are compatible with its 2-Way Text Messaging service is the Nokia 3360.

recipient's phone number. The intercarrier text messaging capability is available to postpaid and prepaid subscribers in all the company's TDMA and GSM markets. InphoMatch, a wireless messaging solution provider, supplies the software for the intercarrier messaging services. InphoMatch's routing system sends messages between and across U.S. carriers simply by using a wireless phone number while remaining transparent to the customer.

Other service providers in the United States now offer SMS. The two-way text messaging service of Cingular is Interactive Messaging, which allows up to 150 characters. Users send text messages via the 10-digit mobile number of the recipient, regardless of which wireless service the recipient happens to have. Sprint PCS offers an SMS-based service called Short Mail, which allows users to send messages of up to 100 characters from one Sprint PCS phone to another. Verizon Wireless offers its SMS-based service, called Mobile Messenger, which allows users to send short text messages of up to 160 characters between handsets or any Internet e-mail system. Voicestream's two-way messaging service is called Ping Pong. It allows text messages of up to 140 characters to be sent and received between wireless devices using only an e-mail address.

Summary

In the same way instant messaging (IM) on the Internet has become a standard method of communication among U.S. teenagers, SMS will provide a choice that will grow the total messaging market and, in some cases, replace voice traffic and IM. Desktop-to-desktop IM is not conducive to mobility. Even desktop-to-mobile IM inhibits mobility because a cellular subscriber using a limited screen and keyboard cannot keep up with messaging traffic generated from a fully featured desktop. Unlike SMS, IM clients operating on wireless devices have proven complex, cumbersome, and difficult to use.

See also

Paging

Wireless Messaging

SOFTWARE-DEFINED RADIO

Software-defined radios can be reprogrammed quickly to transmit and receive on multiple frequencies in different transmission formats. This reprogramming capability could change the way users traditionally communicate across wireless services and promote more efficient use of radio spectrum. In a software-defined radio, functions that were formerly carried out solely in hardware, such as generation of the transmitted radio signal and tuning of the received radio signal, are performed by software. Because these functions are carried out in software, the radio is programmable, allowing it to transmit and receive over a wide range of frequencies and to emulate virtually any desired transmission format.

The concept of software-defined radio originated with the military, where it was used quickly for electronic warfare applications. Now the cellular/wireless industries in the United States and Europe have begun work to adapt the technology to commercial communications services in the hope of realizing its long-term economic benefits. If all goes according to plan, future radio services will provide seamless access across cordless telephone, wireless local loop, Personal Communications Services (PCS), mobile cellular, and satellite modes of communication, including integrated data and paging.

Generations of Radio Systems

First-generation hardware-based radio systems are built to receive a specific modulation scheme. A handset would be built to work over a specific type of analog network or a specific type of digital network. The handset worked on one network or the other, but not both, and it could certainly not cross between analog and digital domains.

Second-generation radio systems also are based in hardware. Miniaturization enables two sets of components to be packaged into a single, compact handset. This enables the unit to operate in dual mode—for example, switching between Advanced Mobile Phone Service (AMPS) or TDMA modulation as necessary. Such handsets are implemented using snap-in components: Two existing chip sets—one for AMPS and one for TDMA, for example—are used together. Building such handsets typically costs only 25 to 50 percent more than a single-mode handset but offers network operators and users far more flexibility.

Handsets that work across four or more modes/bands entail far more complexity and processing power and call for a different architecture altogether. The architecture is based in software and programmable digital signal processors (DSPs). This architecture is referred to as "software-defined radio" or just "software radio." It represents the third generation of radio systems.

As new technologies are placed onto existing networks and wireless standards become more fragmented—particularly in the United States—the need for a single radio unit that can operate in different modes and bands becomes more urgent. A software radio handset could, for example, operate in a GSM-based PCS network, a legacy AMPS network, and a future satellite mobile network.

Operation

As noted, a software radio is one in which channel modulation waveforms are defined in software. Waveforms are generated as sampled digital signals, converted from digital to analog via a wideband digital-to-analog converter (DAC), and then up-converted from an intermediate frequency (IF) to the desired radiofrequency (RF).

In similar fashion, the receiver employs a wideband analog-to-digital converter (ADC) that captures all the channels of the software radio node. The receiver then extracts, down-converts, and demodulates the channel waveform using the software loaded on a general-purpose processor.

Multimode/Multiband

As competing technologies for wireless networks emerged in the early 1990s, it became apparent that subscribers would have to make a choice: The newer digital technologies offered more advanced features, but coverage would be spotty for some years to come. The older analog technologies offered wider coverage but did not support the advanced features. A compromise was offered in the form of wireless multimode/multiband systems that offered subscribers the best of both worlds.

At the same time, wireless multimode/multiband systems allow operators to economically grow their networks to support new services where the demand is highest. With multimode/multiband handsets, subscribers can access new digital services as they become available while retaining the capability to communicate over existing analog networks. The wireless system gives users access to digital channels wherever digital service is available while providing a transparent handoff when users roam between cells alternately served by various digital and analog technologies. As long as subscribers stay within cells served by advanced digital technologies, they will continue to enjoy the advantages provided by these technologies. When they reach a cell that is supported by analog technology, they will have access only to the features supported by that technology. The intelligent roaming capability of multimode/multiband systems automatically chooses the best system for the subscriber to use at any given time.

Third-generation radio systems are frequency-agile and extend this flexibility even further by supporting more modes and bands. It is important to remember, however, that software radio systems may never catch up to encompass all the modes and bands that are available today and that may become available in the future. Users will always be confronted by choices. Making the right choice will depend on calling patterns, the features associated with the different technologies and standards, and the type of systems in use at international locations visited most frequently.

Multimode and multiband handsets have been available from several manufacturers since 1995. These handsets support more than one technology for their mode of operation and more than one frequency band.

An example of a multimode wireless system is one that supports both AMPS and Narrowband AMPS (N-AMPS). Narrowband AMPS is a system-overlay technology that offers enhanced digital-like features, such as Digital Messaging Service, to phones operating in a traditional analog-based AMPS network. Among the vendors offering dual-mode AMPS/N-AMPS handsets is Nokia, the world's second-largest manufacturer of cellular phones.

An example of a multiband wireless system is one that supports GSM at both 900 and 1800 MHz in Europe. Among the vendors offering dual-band GSM handsets is Motorola. The company's International 8800 Cellular Telephone allows GSM 1800 subscribers to roam on either their home or GSM 900 networks (where roaming agreements are in place) using a single cellular telephone.

Of course, handsets can be both multimode and multiband. Ericsson, for example, offers dual-band/dual-mode handsets that support communication over both 800-MHz AMPS/Digital AMPS (D-AMPS) and 1900-MHz D-AMPS networks. Subscribers on a D-AMPS 1900 channel can hand off both to and from a D-AMPS channel on 800 MHz as well as to and from an analog AMPS channel.

Multimode and multiband wireless systems allow operators to expand their networks to support new services where they are needed most, expanding to full coverage at a pace that makes economic sense. From the subscribers' perspective, multimode and multiband wireless systems allow them to take advantage of new digital services that are initially deployed in large cities while still being able to communicate in areas served by the older analog technologies.

With its multimode capabilities, the wireless system preferentially selects a digital channel wherever digital service is available. If the subscriber roams out of the cell served by digital technology—from one served by CDMA to one served by AMPS, for example—a handoff occurs transparently. As long as subscribers stay within CDMA cells, they will continue to enjoy the advantages the technology provides, such as better voice quality and soft handoff, which virtually eliminates dropped calls. When subscribers reach a cell that supports only AMPS, voice quality diminishes and the chances for dropped calls increases.

However, these multimode/multiband handsets are not software-programmable. They rely instead on packaging dual sets of hardware in the same handset. Miniaturization of the various components makes this both practical and economical, but this approach has its limitations when the number of modes and frequencies that must be supported goes beyond two or three. Beyond that point, a totally new approach is required that relies more on programmable components.

Regulation

Despite the promising concept of software-defined radios, the rollout of consumer products that use the technology has been slow. In September 2001, the FCC adopted rule changes to accommodate the authorization and deployment of software-defined radios. Under the previous rules, if a manufacturer wanted to make changes to the frequency, power, or type of modulation for an approved transmitter, a new approval was required, and the equipment had to be relabeled with a new identification number. Because software-defined radios have the capability of being reprogrammed in the field, these requirements could be overly burdensome and hinder the deployment of software-defined radios to consumers.

Under the new rules, software modifications in a software-defined radio can be made through a "permissive change" that has a streamlined filing process. The FCC identification number will not have to be changed, so equipment in the field will not have to be relabeled. These permissive changes can be obtained only by the original grantee of the equipment authorization. To allow for changes to equipment by other parties such as software developers, the FCC will permit an optional "electronic label" for software-defined radios, in which the FCC identification number could be displayed on a liquid-crystal display (LCD) or similar screen. It will allow another party to obtain an equipment approval in its name and become the party responsible for compliance instead of the original grantee.

The FCC also requires that a grantee take adequate steps to prevent unauthorized software modifications to radios, but it declined to set specific security requirements at this time. This will allow manufacturers flexibility to develop innovative equipment while at the same time provide for oversight of the adequacy of such steps through the equipment authorization process.

Summary

Software radio architectures not only reduce the complexity and expense of serving a diverse customer base, they also simplify the integration and management of rapidly emerging standards. With software-based radio systems, access points, cell sites, and wireless data network hubs can be reprogrammed to meet changing standards requirements instead of replacing them or maintaining them in parallel with a newer infrastructure.

From the perspective of users, the same hardware would continue to be used—only the software gets upgraded. This could signal the end of outdated cellular telephones. Consumers will be able to upgrade their phones with new applications—much as they would purchase new programs to add new capabilities to their computers. Although the benefits are clear, commercial software-defined radio systems are still a few years away. Until they become available, users will have to make do with the current generation of multimode/multiband handsets.

See also

Cellular Telephones

SPECIALIZED MOBILE RADIO

Specialized Mobile Radio (SMR) is used to provide two-way radio dispatch service for the public safety, construction, and transportation industries. In 1979, the FCC established SMR service in the 800-MHz band and, in 1986, established SMR service in the 900-MHz band. Although SMR is used primarily for voice communications, such systems also support data and facsimile services. Generally, SMR systems provide dispatch services for companies with multiple vehicles using the "push to talk" method of communication.

A traditional SMR system consists of one or more base station transmitters, one or more antennas, and mobile radio units obtained from the SMR operator for a fee or purchased from a retail source.

SMR has limited roaming capabilities, but its range may be extended through interconnection with the Public Switched Telephone Network (PSTN), as if the user were a cellular subscriber. Both types of services operate over different assigned frequencies within the range of 800 to 900 MHz. Cellular services are assigned to bands between 824 and 849 MHz and 869 and 894 MHz.

SMR networks traditionally used one large transmitter to cover a wide geographic area. This limited the number of subscribers because only one subscriber could talk on one frequency at any given moment. The number of frequencies allocated to SMR is smaller than for cellular, and there have been several operators in each market. Because dispatch messages are short, SMR services were able to work reasonably well.

Types of Systems

The 800-MHz SMR systems operate on two 25-kHz channels paired, while the 900-MHz systems operate on two 12.5-kHz channels paired. Because of the different sizes of the channel bandwidths allocated for 800- and 900-MHz systems, the radio equipment used for 800-MHz SMR is not compatible with the equipment used for 900-MHz SMR systems.

SMR systems consist of two distinct types: conventional and trunked systems. A conventional system allows an end user the use of only one channel. If someone else is already using that end user's assigned channel, that user must wait until the channel is available.

In contrast, a trunked system combines channels and contains processing capabilities that automatically search for an open channel. This search capability allows more users to be served at any one time. A majority of the current SMR systems are trunked systems.

Summary

In 1993, Congress reclassified most SMR licensees as Commercial Radio Service (CMRS) providers and established the authority to use competitive bidding to issue new licenses. With the development of digital systems, the SMR marketplace now offers new services such as acknowledgment paging and inventory tracking, credit card authorization, automatic vehicle location, fleet management, inventory tracking, remote database access, and voice mail.

See also

Cellular Voice Communications

SPECTRUM AUCTIONS

In recent years, the FCC has assigned licenses for wireless spectrum by putting it up for auction. The idea behind the auction process is that is encourages companies to roll out new services as soon as possible to recover their investments in the licenses. In getting spectrum into the hands of those who initially value it the most, competitive bidding also facilitates efficient spectrum aggregation rather than fragmented secondary markets.

In the past, the FCC often relied on comparative hearings, in which the qualifications of competing applicants were examined to award licenses in cases where two or more applicants filed applications for the same spectrum in the same market. This process was time-consuming and resource-intensive. The FCC also used lotteries to award licenses, but this created an incentive for companies to acquire licenses on a speculative basis and resell them.

Of the three methods of assigning spectrum, competitive bidding has proved to be the most effective way to ensure that licenses are assigned quickly and to the companies that value them the most while recovering the value of the spectrum resource for the public.[1] In addition, auctions avoid the perception of the government making decisions that are biased toward or against individual industry players. The rules and procedures of the auctions are clearly established, and the outcomes are definitive.

---

[1]The revenues generated from spectrum auctions go to the U.S. Treasury, not the FCC.

Auction Process

The FCC's auctions of electromagnetic spectrum assign licenses using a unique methodology called "electronic simultaneous multiple-round auctions." This methodology is similar to a traditional auction, except that instead of licenses being sold one at a time, a large set of related licenses is auctioned simultaneously and bidders can bid on any license offered. The auction closes when all bidding activity has stopped on all licenses.

Another characteristic of the auction process is that it is automated. When the FCC began to design auctions for the airwaves, it became apparent that manual auction methods could not adequately allocate large numbers of licenses when thousands of interdependent licenses were being auctioned to hundreds of bidders at the same time. The FCC's Automated Auction System (AAS) provides the necessary tools to conduct large auctions very efficiently. The system accommodates the needs of bidders by allowing them to bid from their offices using a PC and a modem through a private and secure wide area network (WAN). The system also can accommodate on-site bidders and telephonic bidding.

Bidders and other interested parties are able to track the progress of the auctions through the Auction Tracking Tool (ATT), a stand-alone application that allows a user to track detailed information on an auction. During an auction, the FCC releases result files after every round, with details on all the activity that occurred in that round. Users can use the ATT to import these round result files into a master database file and then view a number of different tables containing a large amount of data in a spreadsheet view. Users can sort, filter, and query the tables to track the activity of an auction in virtually any way they desire. There are also canned tables containing simple summary data to allow more casual observers to track the progress of the auction in general.

The FCC also provides the capability to plot maps of auction winners, high bidders by round, and more general auction activity. Through a Geographic Information System (GIS), interested parties can use a Web browser–based application to construct queries against the database for a particular auction and have the results displayed in a map format. The GIS presents its query results primarily in maps, which the user can export to easily transportable graphical formats. The GIS also allows the user to display data in tabular format. Currently, there are three queries that can be executed against any closed or open auction in the GIS:

- *Market analysis by number of bids* Allows users to see which licenses received a bid in a given round and how many bids each market received.
- *Round results summary* Provides a high-level summary of activity for the selected round, depicting markers for which a new high bid was received, markets for which a bid was withdrawn, and markets that had no new activity.
- *Bidder activity* Allows users to query the database to generate a map showing all the licenses for which a particular area has a high bid in a given round.

Companies interested in participating in spectrum auctions must submit an electronic application to the FCC disclosing their ownership structure and identifying the markets/licenses on which they intend to bid. Approximately 2 weeks after the filing deadline and 2 weeks before the start of the auction, potential bidders must submit a refundable deposit that is used to purchase the bidding units required to place bids in the auction. This deposit is not refundable after the auction closes.

At a minimum, an applicant's total up-front payment must be enough to establish eligibility to bid on at least one of the licenses applied for, or else the applicant will not be eligible to participate in the auction. In calculating the up-front payment amount, an applicant should determine the maximum number of bidding units it may wish to bid on in any single round and submit a payment that covers that number of bidding units. Bidders have to check their calculations carefully because there is no provision for increasing a bidder's maximum eligibility after the up-front payment deadline.

About 10 days before the auction, qualified bidders receive their confidential bidding access codes, Automated Auction System software, telephonic bidding phone number, and other documents necessary to participate in the auction. Five days before the start of the auction, the FCC sponsors a mock auction that allows bidders to work with the software, become comfortable with the rules and the conduct of a simultaneous multipleround auction, and familiarize themselves with the telephonic bidding process.

When the auction starts, it continues until all bidding activity has stopped on all licenses. To ensure the competitiveness and integrity of the auction process, the rules prohibit applicants for the same geographic license area from communicating with each other during the auction about bids, bidding strategies, or settlements.

The winning bidders for spectrum in each market are awarded licenses. Within 10 business days, each winning bidder must submit sufficient funds (in addition to its up-front payment) to bring its total amount of money on deposit to 20 percent of its net winning bids (actual bids less any applicable bidding credits). Up-front payments are applied first to satisfy the penalty for any withdrawn bid before being applied toward down payments.

If a company fails to pay on time, the FCC takes back the licenses and holds them for a future auction. The licenses are granted for a 10-year period, after which the FCC can take them back if the holder fails to provide service over that spectrum.

Summary

The FCC's simultaneous multiple-round auction methodology and the AAS software have generated interest worldwide. The FCC has demonstrated the system to representatives of many countries, including Argentina, Brazil, Canada, Hungary, Peru, Russia, South Africa, and Vietnam. Mexico licensed the FCC's copyrighted system and has used it successfully in a spectrum auction. In addition, in 1997, the FCC was awarded a bronze medal from the Smithsonian Institution for recognition of the visionary use of information technology.

See also

Federal Communications Commission

Spectrum Planning

SPECTRUM PLANNING

Radio spectrum is managed as a scarce natural resource in the United States. It is scarce because at any given time and place one use of a portion of the spectrum precludes any other use of that portion. In the broadcasting service alone, the broadcaster must know where the station's signal can be received in order to meet the needs of advertisers. Interference is unacceptable because it unnaturally limits the broadcaster's market. Similarly, a taxi company or a police department must be able to reliably determine their coverage areas and know that they will be able to operate without interference in that area. Consequently, for the public good, the use of the radio spectrum is regulated, access is controlled, and rules for its implementation are enforced because of the possibility for interference between uncoordinated uses.

Spectrum planning in the United States is the shared responsibility of the National Telecommunications and Information Administration (NTIA) and the Federal Communications Commission (FCC). The NTIA is an agency within the U.S. Department of Commerce and is responsible for ensuring the spectrum requirements of federal government users. The FCC is an independent regulatory agency headed by five members who are appointed for 5-year terms by the President, with the advice and consent of the Senate. It is responsible for ensuring the spectrum needs of commercial operators and that the public interest is served by a competitive environment.

Both the NTIA and the FCC work with other executive branch agencies to allocate portions of the spectrum for specific radio services. Any spectrum shared by federal and nonfederal users is jointly managed by the NTIA and the FCC. Because 93.1 percent of the spectrum below 30 GHz is shared, with only 5.5 and 1.4 percent allocated, respectively, to the private sector and the government on exclusive bases, effective coordination between the NTIA and the FCC is essential.

The NTIA and the FCC establish their individual spectrum requirements and then coordinate these requirements with one another through the Interdepartment Radio Advisory Committee (IRAC). If difficulties arise in the coordination process, the IRAC typically facilitates the negotiation efforts.

The current approach to wireless spectrum allocation is viewed by experts as seriously fractured and in need of revision, given the rapid evolution of technology. While progress has been made in the FCC's methods of assigning spectrum, primarily through auctions, allocation policy has not kept pace with the increasing demand, with the result that the allocation system is not moving spectrum to its best uses in a timely manner.

To assist it in identifying and evaluating changes in spectrum policy that will increase the public benefits derived from the use of the radio spectrum, the FCC has established a Spectrum Policy Task Force. The responsibilities of the task force are to provide specific recommendations to the FCC for ways in which to evolve the current "command and control" approach to spectrum policy into a more integrated, market-oriented approach that provides greater regulatory certainty while minimizing regulatory intervention. The task force also assists the FCC in addressing ubiquitous spectrum issues, including interference protection, spectral efficiency, effective public safety communications, and implications of international spectrum policies.

Congress also can influence spectrum allocations through legislation. Once spectrum is allocated, the frequencies become available for managed federal use under the authority of NTIA and for state and local government and commercial use under the authority of the FCC.

Federal courts also have something to say about spectrum matters—specifically about FCC enforcement. For example, NextWave Telecom, a small wireless carrier embroiled in bankruptcy, won back the rights to airwave licenses worth billions of dollars when a U.S. Appeals Court ruled in mid-2001 that the FCC had stripped the company's spectrum improperly when it defaulted on its payments. The 90 licenses at issue, which NextWave picked up in a 1996 spectrum auction for more than $4.7 billion, provide as much as 30 MHz in major cities. Ultimately, NextWave sold the licenses to AT&T, Cingular, Verizon, and VoiceStream for $16 billion, with NextWave getting $5 billion and the U.S. Treasury getting $11 billion.

International spectrum allocations typically influence allocation decisions within the United States; however, the interests of the federal government and commercial spectrum users usually take precedence. Therefore, domestic and international spectrum allocations do not always coincide, as in the case of third-generation (3G) networks, which use different frequency bands—one set for the United States and a different set for the rest of the world.

Summary

Radio spectrum is managed as a scarce natural resource. It is scarce because at any given time and place one use of a portion of the spectrum precludes any other use of that portion. In the broadcasting service alone, the broadcaster must know where the station's signal can be received in order to meet the needs of advertisers. Interference is unacceptable because it unnaturally limits the broadcaster's market. Similarly, a taxi company or a police department must be able to reliably determine its coverage area and know that it will be able to operate without interference in that area. Consequently, for the public good, the use of the radio spectrum is regulated, access is controlled, and rules for its use are enforced because of the possibility for interference between uncoordinated uses.

See also

Federal Communications Commission

Spectrum Auctions

SPREAD-SPECTRUM RADIO

Spread spectrum is a digital coding technique in which the signal is taken apart or "spread" so that it sounds more like noise to the casual listener, allowing many more users to share the available bandwidth while affording each conversation a high degree of privacy. Actress Hedy Lamarr (Figure S-5) and composer George Antheil share the patent for spread-spectrum technology. Their patent for a "Secret Communication System," issued in 1942, was based on the frequency-hopping concept, with the keys on a piano representing the different frequencies and frequency shifts.[2]

Lamarr had become intrigued with radio-controlled missiles and the problem of how easy it was to jam the guidance signal. She realized that if the signal jumped from

---

[2]In 1942, the technology did not exist for a practical implementation of spread spectrum. When the transistor finally did become available, the Navy used the idea in secure military communications. When transistors became really cheap, the idea was used in cellular phone technology to keep conversations private. By the time the Navy used the idea, the original patent had expired, and Lamarr and Antheil never received any royalty payments for their idea.

Figure S-5   Actress Hedy Lamarr (1914–2000), codeveloper of spread-spectrum technology.

frequency to frequency quickly—like changing stations on a radio—and both sender and receiver changed in the same order at the same time, then the signal could never be blocked without knowing exactly how and when the frequency changed. Although the frequency-hopping idea could not be implemented at that time because of technology limitations, it eventually became the basis for cellular communication based on Code Division Multiple Access (CDMA) and wireless Ethernet LANs based on infrared technology.

Frequency Assignment

Spread spectrum uses the industrial, scientific, and medical (ISM) bands of the electromagnetic spectrum. The ISM bands include the frequency ranges at 902 to 928 MHz and 2.4 to 2.484 GHz, which do not require a site license from the FCC.

Spread spectrum is a highly robust wireless data transmission technology that offers substantial performance advantages over conventional narrowband radio systems. As noted, the digital coding technique used in spread spectrum takes the signal apart and spreads it over the available bandwidth, making it appear as random noise. The coding operation increases the number of bits transmitted and expands the bandwidth used. Noise has a flat, uniform spectrum with no coherent peaks and generally can be removed by filtering. The spread signal has a much lower power density but the same total power.

This low power density, spread over the expanded transmitter bandwidth, provides resistance to a variety of conditions that can plague narrowband radio systems, including

- *Interference* A condition in which a transmission is being disrupted by external sources, such as the noise emitted by various electromechanical devices, or internal sources such as cross-talk.

- *Jamming* A condition in which a stronger signal overwhelms a weaker signal, causing a disruption to data communications.

- *Multipath* A condition in which the original signal is distorted after being reflected off a solid object.

- *Interception* A condition in which unauthorized users capture signals in an attempt to determine its content.

Non-spread-spectrum narrowband radio systems transmit and receive on a specific frequency that is just wide enough to pass the information, whether voice or data. In assigning users different channel frequencies, confining the signals to specified bandwidth limits, and restricting the power that can be used to modulate the signals, undesirable cross-talk—interference between different users—can be avoided. These rules are necessary because any increase in the modulation rate widens the radio signal bandwidth, which increases the chance for cross-talk.

The main advantage of spread-spectrum radio waves is that the signals can be manipulated to propagate fairly well through the air, despite electromagnetic interference, to virtually eliminate cross-talk. In spread-spectrum modulation, a signal's power is spread over a larger band of frequencies. This results in a more robust signal that is less susceptible to interference from similar radio-based systems, since, although they too are spreading their signals, they use different spreading algorithms.

Spreading

Spread spectrum is a digital coding technique in which a narrowband signal is taken apart and "spread" over a spectrum of frequencies (Figure S-6). The coding operation increases the number of bits transmitted and expands the amount of bandwidth used. With the signal's power spread over a larger band of frequencies, the result is a more robust signal that is less susceptible to impairment from electromechanical noise and other sources of interference. It also makes voice and data communications more secure.

Using the same spreading code as the transmitter, the receiver correlates and collapses the spread signal back down to its original form. The result is a highly robust wireless data transmission technology that offers substantial performance advantages over conventional narrowband radio systems.

There are two spreading techniques in common use today: direct sequence and frequency hopping.

Direct Sequence In direct sequence spreading—the most common implementation of spread-spectrum technology—the

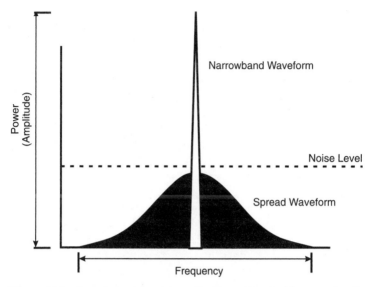

Figure S-6 Spread spectrum transmits the entire signal over a bandwidth that is much greater than that required for standard narrowband transmission. Increasing the frequency range allows more signal components to be transmitted, which results in a more accurate reconstruction of the original signal at the receiving device.

radio energy is spread across a larger portion of the band than is actually necessary for the data. Each data bit is broken into multiple subbits called "chips." The higher modulation rate is achieved by multiplying the digital signal with a chip sequence. If the chip sequence is 10, for example, and it is applied to a signal carrying data at 300 kbps, then the resulting bandwidth will be 10 times wider. The amount of spreading depends on the ratio of chips to each bit of information.

Because data modulation widens the radio carrier to increasingly larger bandwidths as the data rate increases, this chip rate of 10 times the data rate spreads the radio carrier to 10 times wider than it would otherwise be for data alone. The rationale behind this technique is that a spread-spectrum signal with a unique spread code cannot create the exact spectral characteristics as another spread-coded signal.

Using the same code as the transmitter, the receiver can correlate and collapse the spread signal back down to its original form, while other receivers using different codes cannot.

This feature of spread spectrum makes it possible to build and operate multiple networks in the same location. When each network is assigned its own unique spreading code, all transmissions can use the same frequency range yet remain independent of each other. The transmissions of one network appear to the other as random noise and are filtered out because the spreading codes do not match.

This spreading technique would appear to result in a weaker signal-to-noise ratio, since the spreading process lowers the signal power at any one frequency. Normally, a low signal-to-noise ratio would result in damaged data packets that would require retransmission. However, the processing gain of the despreading correlator recovers the loss in power when the signal is collapsed back down to the original data bandwidth but is not strengthened beyond what would have been received had the signal not been spread.

The FCC has set rules for direct sequence transmitters. Each signal must have 10 or more chips. This rule limits the practical raw data throughput of transmitters to 2 Mbps in the 902-MHz band and 8 Mbps in the 2.4-GHz band. The number of chips is directly related to a signal's immunity to interference. In an area with a lot of radio interference, users will have to give up throughput to successfully limit interference.

Frequency Hopping In frequency hopping, the transmitter jumps from one frequency to the next at a specific hopping rate in accordance with a pseudo-random code sequence. The order of frequencies selected by the transmitter is taken from a predetermined set as dictated by the code sequence. For example, the transmitter may have a hopping pattern of going from Channel 3 to Channel 12 to Channel 6 to Channel 11 to Channel 5, and so on (Figure S-7). The receiver tracks these changes. Since only the intended receiver is aware of

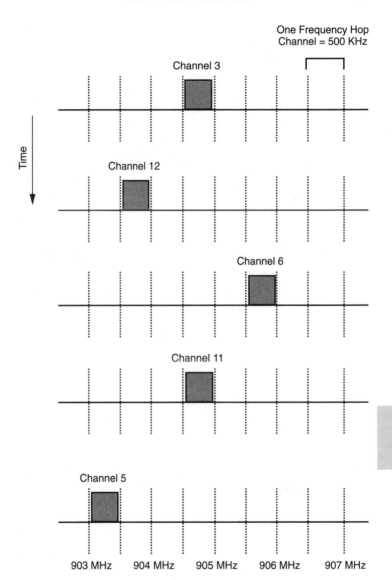

Figure S-7 Frequency-hopping spread spectrum.

the transmitter's hopping pattern, then only that receiver can make sense of the data being transmitted.

Other frequency-hopping transmitters will be using different hopping patterns that usually will be on noninterfering frequencies. Should different transmitters coincidentally attempt to use the same frequency and the data of one or both become garbled at that point, retransmission of the affected data packets is required. Those data packets will be sent again on the next hopping frequency of each transmitter.

The FCC mandates that frequency-hopped systems must not spend more than 0.4 second on any one channel each 20 seconds, or 30 seconds in the 2.4-GHz band. Furthermore, they must hop through at least 50 channels in the 900-MHz band or 75 channels in the 2.4-GHz band. These rules reduce the chance of repeated packet collisions in areas with multiple transmitters.

Summary

Direct sequence spread spectrum offers better performance, but frequency-hopping spread spectrum is more resistant to interference and is preferable in environments with electromechanical noise and more stringent security requirements. Direct sequence is more expensive than frequency hopping and uses more power. Although spread spectrum generally provides more secure data transmission than conventional narrowband radio systems, this does not mean the transmissions are immune from interception and decoding by knowledgeable intruders with sophisticated tapping equipment. For this reason, many vendors provide optional encryption for added security.

See also

Code Division Multiple Access

Frequency Division Multiple Access

Time Division Multiple Access

T

TELEGRAPHY

Telegraphy is a form of data communication that is based on the use of a signal code. The word *telegraphy* comes from Greek *tele* meaning "distant" and *graphein* meaning "to write"—writing at a distance.

The inventor of the first electric telegraph was Samuel Finley Breese Morse, an American inventor and painter (Figure T-1). On a trip home from Italy, Morse became acquainted with the many attempts to create usable telegraphs for long-distance telecommunications. He was fascinated by this problem and studied books on physics for 2 years to acquire the necessary scientific knowledge.

Early Attempts

Morse focused his research on the characteristics of electromagnets, whereby they became magnets only while the current flows. The intermittence of the current produced two states—magnet and no magnet—from which he developed a code for representing characters, which eventually became known as Morse Code. (The International Morse Code is a

Figure T-1 Samuel Finley Breese Morse (1791–1872), *c.* 1865.

system of dots and dashes that can be used to send messages by a flash lamp, telegraph key, or other rhythmic device such as a tapping finger.)

His first attempts at building a telegraph failed, but he eventually succeeded with the help of some friends who were more technically knowledgeable. The signaling device was very simple. It consisted of a transmitter containing a battery and a key, a small buzzer as a receiver, and a pair of wires connecting the two. Later, Morse improved it by adding a second switch and a second buzzer to enable transmission in the opposite direction as well.

In 1837, Morse succeeded in a public demonstration of his first telegraph. Although he received a patent for the device in 1838, he worked for 6 more years in his studio at New York University to perfect his invention. Finally, on May 24, 1844, with a $30,000 grant from Congress, Morse unveiled the results of his work. Over a line strung from Washington, D.C., to Baltimore, Morse tapped out the message, "What hath God wrought." The message reached Morse's collaborator, Alfred Lewis Vail, in Baltimore, who immediately sent it back to Morse.

With the success of the telegraph assured, the line was expanded to Philadelphia, New York, Boston, and other major cities and towns. The telegraph lines tended to follow the rights-of-way of railroads, and as the railroads expanded westward, the nation's communications network expanded as well.

Morse Code

Morse Code uses a system of dots and dashes that are tapped out by an operator using a telegraph key. (It also can be used to communicate via radio and flash lamp.) Various combinations of dots and dashes represent characters, numbers, and symbols separated by spaces. The Morse Code for letters, numbers, and symbols used in the United States is described in Table T-1.

Morse Code is the basis of today's digital communication. Although it has virtually disappeared in the world of professional communication, it is still used in the world of amateur radio (HAM) and is kept alive by history buffs. There are even pages on the Web that teach telegraphy and perform translations of text into Morse Code.

Wireless Telegraphy

An Italian inventor and electrical engineer, Guglielmo Marconi (1874–1937), pioneered the use of wireless telegraphy. Telegraph signals previously had been sent through electrical wires. In experiments he conducted in 1894,

TABLE T-1   Morse Code

Letters	Numbers
A . —	1 . — — — —
B — . . .	2 . . — — —
C — . — .	3 . . . —
D — . .	4 . . . . —
E .	5 . . . . .
F . . — .	6 — . . . .
G — — .	7 — — . . .
H	8 — — — . .
I . .	9 — — — — .
J . — — —	0 — — — — —
K — . —	. . — . — . —
L . — . .	? . . — — . .
M — —	/ — . . — .
N — .	= — . . . —
O — — —	
P . — — .	
Q — — . —	
R . — .	
S . . .	
T —	
U . . —	
V . . . —	
W . — —	
X — . . —	
Y — . — —	
Z — — . .	

Marconi (Figure T-2) demonstrated that telegraph signals also could be sent through the air.

A few years earlier, Heinrich Hertz had produced and detected the waves across his laboratory. Marconi's achievement was in producing and detecting the waves over long distances, laying the groundwork for what today we know as radio. So-called Hertzian waves were produced by sparks in one circuit and detected in another circuit a few meters away. By continuously refining his techniques, Marconi could soon detect signals over several kilometers, demon-

Figure T-2 Guglielmo Marconi, at 22 years of age, behind his first patented wireless receiver (1896).

strating that Hertzian waves could be used as a medium for communication.

The results of these experiments led Marconi to approach the Italian Ministry of Posts and Telegraphs for permission to set up the first wireless telegraph service. He was unsuccessful, but in 1896, his cousin, Henry Jameson-Davis, arranged an introduction to Nyilliam Preece, engineer-in-chief of the British Post Office. Encouraging demonstrations in London and on Salisbury Plain followed, and in 1897, Marconi obtained a patent and established the Wireless Telegraph and Signal Company, Ltd, which opened the world's first radio factory at Chelmsford, England, in 1898.

Experiments and demonstrations continued. Queen Victoria at Osborne House received bulletins by radio about the health of the Prince of Wales, convalescing on the royal yacht off Cowes. Radio transmission was pushed to greater and greater lengths, and by 1899, Marconi had sent a signal 9 miles across the Bristol Channel and then 31 miles across the English Channel to France. Most people believed that the curvature of the earth would prevent sending a signal much farther than 200 miles, so when Marconi was able to transmit across the Atlantic in 1901, it opened the door to a rapidly developing wireless industry. Commercial broadcasting was still in the future—the British Broadcasting Company (BBC) was established in 1922—but Marconi had achieved his aim of turning Hertz's laboratory demonstration into a practical means of communication.

Summary

By Morse's death in 1872, the telegraph was being used worldwide and would pave the way for the invention of the telephone. Western Union had the monopoly on commercial telegraph service but spurned Alexander Graham Bell, who approached the company with an improvement that would convey voice over the same wires. Bell had to form his own company, American Bell Telephone Company, to offer a commercial voice communication service. Since then, voice and data technologies have progressed through separate evolutionary paths. Only in recent years has voice-data integration been pursued as a means of containing the cost of telecommunication services.

See also

Hertz

Cellular Telephones

TELEMETRY

Telemetry is the monitoring and control of remote devices from a central location via wire-line or wireless links. Applications include utility meter reading, load management, environmental monitoring, vending machine management, and security alarm monitoring. Companies are deploying telemetry systems to reduce the cost of manually reading and checking remote devices. For example, vending machines need not be visited daily to check for proper operation or out-of-stock conditions. Instead, this information can be reported via modem to a central control station so that a repair technician or supply person can be dispatched as appropriate. Such telemetry systems greatly reduce service costs.

When telemetry applications use wireless technology, additional benefits accrue. The use of wireless technology enables systems to be located virtually anywhere without depending on the telephone company for line installation. For instance, a kiosk equipped with a wireless modem can be located anywhere in a shopping mall without incurring line installation costs. Via wireless modems, data are collected from all the area kiosks at the end of the day for batch processing at a data center. The kiosk also runs continuous diagnostics to ensure proper operation. If a malfunction occurs, problems are reported via the wireless link to a central control facility, which can diagnose and fix the problem remotely or dispatch a technician if necessary.

Security Application

A number of wireless security systems are available for commercial and residential use (Figure T-3). Wireless technology provides installation flexibility, since sensors can be placed anywhere without proximity to phone jacks. A number of sensors are available to detect such things as temperature, frequency, or motion. The system can be programmed to automatically call monitoring station personnel, police, or designated friends and neighbors when an alert is triggered. Such systems can be set to randomly turn lights on and off at designated times to give the appearance of occupancy. Depending on vendor, the system may even perform continuous diagnostics to report low battery power or tampering.

In a typical implementation, the security system console monitors the sensors placed at various potential points of entry. The console expects the sensor to send a confirmation

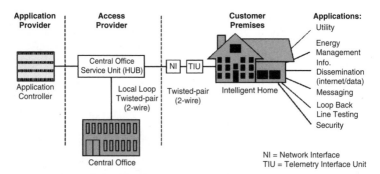

Figure T-3   Telemetry applications for residential users.

signal at preset intervals, say, every 90 seconds. If the console does not get the signal, it knows that something is wrong. For example, a sensor attached to a corner of the window or other glass panel is specially tuned to vibrations caused by breaking glass. When it detects the glass breaking, the sensor opens its contact and sends a wireless signal to an audio alarm located on the premises, at a police station, or at a private security firm.

The use of wireless technology for security applications actually can improve service reliability. A security service that does not require a dedicated phone line is not susceptible to intentional or accidental outages when phone lines are down or there is bad weather. Wireless links offer more immunity to such problems.

Traffic Monitoring

Another common application of wireless technology is its use in traffic monitoring. For example, throughout the traffic signal control industry there has been a serious effort to find a substitute for the underground hard-wired inductive loop that is in common use today to detect the presence of vehicles at stoplights. Although the vehicle detection loop is inherently simple, it has many disadvantages:

- Slot cutting for loop and lead-in wire is time-consuming and expensive.
- Traffic is disrupted during installation.
- Reliability depends on geographic conditions.
- Maintenance costs are high, especially in cold climates.

A wireless proximity detector can overcome these problems. Its signals activate traffic lights in the prescribed sequence. The proximity detector, usually mounted on a nearby pole, focuses the wireless signal very narrowly on the road to represent a standard loop. The microprocessor-based detector provides real-time information while screening out such environmental variations as temperature, humidity, and barometric pressure. By tuning out environmental variations, the detectors provide consistent output. This increases the reliability of traffic control systems. Using a laptop computer with a Windows-compatible setup package, information can be exchanged with the detector via an infrared link. From the laptop, the pole-mounted detector can be set up remotely, calibrated, and put through various diagnostic routines to verify proper operation.

Role of Cellular Carriers

Cellular carriers are well positioned to offer wireless telemetry services. The cellular telephone system has a total of 832 channels, half of which are assigned to each of the two competing cellular carriers in each market. Each cellular carrier uses 21 of its 416 channels as control channels. Each control channel set consists of a Forward Control Channel (FOCC) and a Reverse Control Channel (RECC).

The FOCC is used to send general information from the cellular base station to the cellular telephone. The RECC is used to send information from the cellular telephone to the base station. The control channels are used to initiate a cellular telephone call. Once the call is initiated, the cellular system directs the cellular telephone to a voice channel.

After the cellular telephone has established service on a voice channel, it never goes back to a control channel. All information concerning handoff to other voice channels and termination of the telephone call is handled via communication over the voice channels.

This leaves the control channels free to provide other services such as telemetry, which is achieved by connecting a gateway to a port at the local mobile switching center (MSC) or regional facility. The gateway can process the telemetry messages according to the specific needs of the applications.

For instance, if telemetry is used to convey a message from an alarm panel, the gateway will process the message on a real-time, immediate basis and pass the message to the central alarm monitoring service. On the other hand, if a soft drink vending company uses telemetry to poll its machines at night for their stock status, the gateway will accumulate all the data from the individual vending machines and process them in batch mode so that the management reports can be ready for review the next morning.

Individual applications can have different responses from the same telemetry radio. While the vending machine uses batch processing for its stock status, it could have an alarm message conveyed to the vending company on an immediate basis, indicating a malfunction. A similar scenario is applicable for utility meter reading. Normal meter readings can be obtained on a batch basis during the night and delivered to the utility company the following morning. However, real-time meter readings can be made any time during the day for customers who desire to close out or open service and require an immediate, current meter reading. Telemetry can even be used to turn on or turn off utility service remotely by the customer service representative.

Web-Enabled Telemetry

The next big market for telemetry systems is for those which distribute data through the Internet for access by a Web browser, which lowers the cost of implementing telemetry applications. A growing number of companies offer Web-enabled telemetry solutions for such applications as ground station telemetry processing as well as remote and wide area monitoring.

With such systems, a host device acts as a network server that plugs into the local area network (LAN) with standard Category 5 cabling and RJ-45 connectors. The host device is connected to remote units over phone lines or wireless links to the Internet. Data are sent and received between the host and remote units in standard Transmission Control Protocol/Internet Protocol (TCP/IP) packets. Client computers connected anywhere on the LAN or the Internet can use a standard Web browser to display the collected data with no requirement for additional software. In some cases, integrated Java applets provide real-time telemetry display.

With accumulated data published as Web pages over an Ethernet LAN or over the Internet, it possible to monitor and control a process through a browser from anywhere in the world. For example, using a remote unit attached to a heater with a wireless link to an access point, an engineer can monitor the temperature, change set points or alarm points, turn the heater on and off, or make other modifications from anywhere on the local network, or anywhere on the Internet, simply by using a Web browser. The remote unit also can send an e-mail message to the engineer alerting him or her to an alarm condition or updating the status of the remote device. Leveraging the technology of the Internet even further, the engineer could receive a message from the remote unit on an Internet-capable pager or cell phone.

Through the Web browser an administrator can set monitoring and measurement parameters from any computer on the network and configure and change all communication parameters of the remote units in the field. Access to configurations and data is protected through passwords. Selectable local disk archiving protects data for future reference should the primary storage disk fail.

Summary

Telemetry services, once implemented by large companies over private networks, are becoming more widely available for a variety of mainstream business and consumer applications. Wireless technology permits more flexibility in the implementation of telemetry systems and can save on line installation costs. Telemetry systems are inherently more reliable when wireless links are used to convey status and control information, since they are less susceptible to outages due to tampering and severe weather. As data communications technology advances and companies continue to exploit the global ubiquity of the Internet, telemetry solutions will become more pervasive.

See also

Access Points

Cellular Data Communications

TIME DIVISION MULTIPLE ACCESS

Time Division Multiple Access (TDMA) technology is used in digital cellular telephone communication. It divides each cellular channel into three time slots to increase the amount of conversations that can be carried. TDMA improves the bandwidth utilization and overall system capacity offered by older FM radio systems by dividing the 30-kHz channel into three narrower channels of 10 kHz each. Newer forms of TDMA allow even more users to be supported by the same channel.

TDMA systems have been providing commercial digital cellular service since mid-1992. Versions of the technology are used to provide Digital American Mobile Phone Service (D-AMPS), Global System for Mobile (GSM) communications, Personal Digital Cellular (PDC), and Digital Enhanced Cordless Telecommunications (DECT). Originally, the TDMA specification was described in EIA/TIA Interim Standard 54 (IS-54). An evolved version of that standard is IS-136, which is used in the United States for both cellular and Personal Communications Services (PCS) in the 850-MHz and 1.9-GHz frequency bands, respectively. The difference is that IS-136 makes use of a control channel to provide advanced call features and messaging services.

Time Slots

As noted, TDMA divide ` the original 30-kHz channel into three time slots. Users are assigned their own time slot into which voice or data are inserted for transmission via synchronized timed bursts. The bursts are reassembled at the receiving end and appear to provide continuous, smooth communication because the process is very fast. The digital bit streams that correspond to the three distinct voice conversations are encoded, interleaved, and transmitted using a digital modulation scheme called Differential Quadrature Phase-Shift Keying (DQPSK). Together these manipulations reduce the effects of most common radio transmission impairments.

If one side of the conversation is silent, however, the time slot goes unused. Enhancements to TDMA use dynamic time slot allocation to avoid the wasted bandwidth when one side of the conversation is silent. This technique almost doubles the bandwidth efficiency of TDMA over the original analog systems. Frequency reuse further enhances network capacity. In nonadjacent cells, the same frequency sets are used as in other cells, but the cells with the same frequency sets are spaced many miles apart to reduce interference (Figure T-4).

Framing

In a TDMA system, the digitized voice conversations are separated in time, with the bit stream organized into frames,

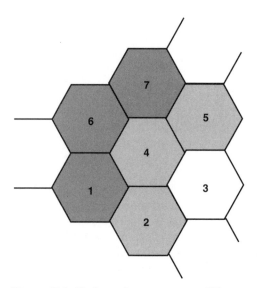

Figure T-4   Each number represents a different set of channels or paired frequencies used by a cell. A cellular system separates each cell that shares the same channel set, which minimizes interference while allowing the same frequencies to be used in another part of the system.

typically on the order of several milliseconds. A 6-millisecond frame, for example, is divided into six 1-millisecond time slots, with each time slot assigned to a specific user. Each time slot consists of a header and a packet of user data for the call assigned to it (Figure T-5). The header generally contains synchronization and addressing information for the user data.

If the data in the header become corrupted as a result of a transmission problem—signal fade, for example—the entire slot can be wasted, in which case no more data will be transmitted for that call until the next frame. The loss of an entire data packet is called "frame erasure." If the transmission problem is prolonged (i.e., deep fade), several frames in sequence can be lost, causing clipped speech or forcing the retransmis-

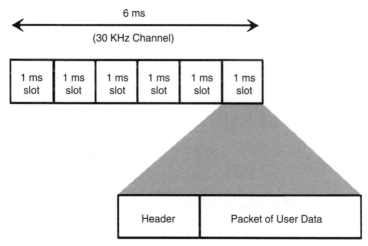

Figure T-5   Six TDMA time slots.

sion of data. Most transmission problems, however, will not be severe enough to cause frame erasure. Instead, only a few bits in the header and user data will become corrupted, a condition referred to as "single-bit errors."

Network Functions

The IS-54 standard defines the TDMA radio interface between the mobile station and the cell site radio. The radio downlink from the cell site to the mobile phone and the radio uplink from the mobile phone to the cell site are functionally similar. The TDMA cell site radio is responsible for speech coding, channel coding, signaling, modulation/demodulation, channel equalization, signal strength measurement, and communication with the cell site controller.

The TDMA system's speech encoder uses a linear predictive coding technique and transfers Pulse Code Modulated (PCM) speech at 64 kbps to and from the network. The channel coder performs channel encoding and decoding, error correction, and bit interleaving and deinterleaving. It processes speech and signaling information, builds the time slots for the channels, and communicates with the main controller, modulator, and demodulator.

The modulator receives the coded information and signaling bits for each time slot from the channel coders. It performs DQPSK modulation to produce the necessary digital components of the transmitter waveform. These waveform samples are converted from digital to analog signals. The analog signals are then sent to the transceiver, which transmits and receives digitally modulated radiofrequency (RF) control and information signals to and from the cellular phones.

The modulator/equalizer receives signals from the transceiver. It performs filtering, automatic gain control, receive signal strength estimation, adaptive equalization, and demodulation. The demodulated data for each of the time slots is then sent to the channel coder for decoding.

Newer voice coding technology is available that produces near-landline speech quality in wireless networks based on the IS-136 TDMA standard. One technology uses an algebraic code excited linear predictive (ACELP) algorithm, an enhanced internationally accepted code for dividing waves of sound into binary bits of data. The ACELP coders can be integrated easily into current wireless base station radios as well as new telephones. ACELP-capable phones enable users to take advantage of the improved digital clarity over both North American frequency bands—850 MHz for cellular and 1.9 GHz for PCS.

Call Handoff

Receive signal strength estimation is used in the call handoff process. The traditional handoff process involves the cell site currently serving the call, the switch, and the neighboring cells that potentially can continue the call. The neighboring cells measure the signal strength of the potential call to be handed off and report that measurement to the serving cell, which uses it to determine which neighboring cell can best handle the call. TDMA systems, on the other hand, reduce the time needed and the overhead required to complete the handoff by assigning some of this signal strength data gathering to the cellular phone, relieving the neighboring cells of this task and reducing the handoff interval.

Digital Control Channel

Digital Control Channel (DCCH), described in IS-136, gives TDMA features that can be added to the existing platform through software updates. Among these new features are

- *Over-the-air activation* Allows new subscribers to activate cellular or PCS service with just a phone call to the service provider's customer service center.
- *Messaging* Allows users to receive visual messages up to the maximum length allowed by industry standards (200 alphanumeric characters). Transmission of messages permits a mobile unit to function as a pager.
- *Sleep mode* Extends the battery life of mobile phones and allows subscribers to leave their portables powered on throughout the day, ready to receive calls.
- *Fraud prevention* The cellular system is capable of identifying legitimate mobile phones and blocking access to invalid ones.

A number of other advanced features also can be supported over the DCCH, such as voice encryption and secure data transmission, caller ID, and voice mail notification.

Summary

Many vendors and service providers have committed to supporting either TDMA or CDMA. Those who have committed to CDMA claim that they did so because they consider TDMA to be too limited in meeting the requirements of the next generation of cellular systems. Although TDMA provides service providers with a significant increase in capacity over AMPS, the standard was written to fit into the existing AMPS channel structure for easy migration.

Proponents of TDMA, however, note that the inherent compatibility between AMPS and TDMA, coupled with the deployment of dual-mode/dual-band terminals, offers full mobility to subscribers with seamless handoff between PCS and cellular networks. They also note that the technology is operational in many of the world's largest wireless networks and is providing reliable, high-quality service without additional development or redesign. These TDMA systems can be easily and cost-effectively integrated with existing wireless and landline systems, and the technology is evolving to meet the quality and service requirements of the global third-generation (3G) wireless infrastructure.

See also

Code Division Multiple Access
Frequency Division Multiple Access

U

ULTRA WIDEBAND

Ultra wideband (UWB) offers the promise of new radar and imaging services that can save lives by helping to rescue hostages, locate disaster victims trapped under the rubble of a collapsed building, detect hidden flaws in the construction of highways or airport runways, secure homes and businesses, and possibly even provide short-range high-speed Internet access to the classroom. UWB devices operate by employing very narrow or short-duration pulses that result in very large or wideband transmission bandwidths. Its ultrawide disbursement of ultra-low power bursts presents novel interference questions that must be addressed, including how to ensure that existing services are not adversely impacted—especially those services which support public safety—and whether widespread deployment would have any appreciable effect on the noise floor. With appropriate technical standards, however, UWB devices can operate using spectrum occupied by existing radio services without causing interference, thereby permitting scarce spectrum resources to be used more efficiently.

In early 2002, the Federal Communications Commission (FCC) issued standards designed to ensure that existing and planned radio services, particularly safety services, are adequately protected from UWB users. The FCC will enforce the rules and act quickly on any reports of interference. The standards are based in large measure on standards that the National Telecommunications and Information administration (NTIA) believes are necessary to protect against interference to vital federal government operations.

On an ongoing basis, the FCC intends to review the standards for UWB devices, explore more flexible standards, and address the operation of additional types of UWB operations and technology. Since there is no production UWB equipment available at this writing and there is little operational experience with the impact of UWB on other radio services, the FCC chose to err on the side of conservatism in setting emission limits when there are unresolved interference issues.

The FCC establishes different technical standards and operating restrictions for three types of UWB devices based on their potential to cause interference. These three types of UWB devices are imaging systems, including ground-penetrating radars (GPRs), wall, through-wall, medical imaging, and surveillance devices; vehicular radar systems; and communications and measurement systems.

- *Imaging systems* Provides for the operation of GPRs and other imaging devices subject to certain frequency and power limitations. The operators of imaging devices must be eligible for licensing, except that medical imaging devices may be operated by a licensed health care practitioner. At the request of NTIA, the FCC will notify or coordinate with NTIA prior to the operation of all imaging systems. Imaging systems include

    Ground-penetrating radar systems GPRs must be operated below 960 MHz or in the frequency band 3.1 to 10.6 GHz. GPRs operate only when in contact with or within close proximity of the ground for the purpose of detecting or obtaining the images of buried objects.

The energy from the GPR is intentionally directed down into the ground for this purpose. Operation is restricted to law enforcement, fire and rescue organizations, scientific research institutions, commercial mining companies, and construction companies.

*Wall-imaging system*s Wall-imaging systems must be operated below 960 MHz or in the frequency band 3.1 to 10.6 GHz. Wall-imaging systems are designed to detect the location of objects contained within a "wall," such as a concrete structure, the side of a bridge, or the wall of a mine. Operation is restricted to law enforcement, fire and rescue organizations, scientific research institutions, commercial mining companies, and construction companies.

Through-wall imaging systems These systems must be operated below 960 MHz or in the frequency band 1.99 to 10.6 GHz. Through-wall imaging systems detect the location or movement of persons or objects that are on the other side of a structure such as a wall. Operation is limited to law enforcement and fire and rescue organizations.

Medical systems These devices must be operated in the frequency band 3.1 to 10.6 GHz. A medical imaging system may be used for a variety of health applications to "see" inside the body of a person or animal. Operation must be at the direction of or under the supervision of a licensed health care practitioner.

*Surveillance system*s Although technically these devices are not imaging systems, for regulatory purposes they are treated in the same way as through-wall imaging and are permitted to operate in the frequency band 1.99 to 10.6 GHz. Surveillance systems operate as "security fences" by establishing a stationary radio frequency (RF) perimeter field and detecting the intrusion of persons or objects in that field. Operation is limited to law enforcement, fire and rescue organizations, public utilities, and industrial entities.

- *Vehicular radar systems* Provides for the operation of
  vehicular radar systems in the 24-GHz band using directional antennas on terrestrial transportation vehicles provided the center frequency of the emission and the
  frequency at which the highest radiated emission occurs
  are greater than 24.075 GHz. These devices are able to
  detect the location and movement of objects near a vehicle, enabling features such as near collision avoidance,
  improved airbag activation, and suspension systems that
  better respond to road conditions.

- *Communications and measurement systems* Provides for
  use of a wide variety of other UWB devices, such as high-
  speed home and business networking devices as well as
  storage tank measurement devices under Part 15 of the
  FCC's rules subject to certain frequency and power limitations. The devices must operate in the frequency band
  3.1 to 10.6 GHz. The equipment must be designed to
  ensure that operation can only occur indoors, or it must
  consist of handheld devices that may be employed for such
  activities as peer-to-peer operation.

Summary

UWB technologies are destined to play a significant public
safety role. UWB devices will save the lives of firefighters
and police officers, prevent automobile accidents, assist
search-and-rescue crews in seeing through the rubble of disaster sites, enable broadband connections between home
electronics, and allow new forms of communications in the
years ahead. The U.S. government already uses UWB extensively to make soldiers, airport runways, and highway
bridges safer. But opinion differs greatly on the interference
effect of the widespread use of UWB technologies by the public. If interference does occur, it conceivably could affect critical government and nongovernment spectrum users.
National defense and several safety-of-life systems depend
on bands that have the potential to be impacted by UWB

devices. For this reason, the FCC and NTIA will cooperate in managing the use of UWB technology.

See also

Federal Communications Commission

Spectrum Planning

UNIVERSAL MOBILE TELEPHONE SERVICE

One of the major new third-generation (3G) mobile systems being developed within the global IMT-2000 framework is the Universal Mobile Telecommunications System (UMTS), which has been standardized by the European Telecommunications Standards Institute (ETSI). UMTS makes use of UMTS Terrestrial Radio Access (UTRA) as the basis for a global terrestrial radio access network. Europe and Japan are implementing UTRA in the paired bands 1920–1980 MHz and 2110–2170 MHz. Europe also has decided to implement UTRA in the unpaired bands 1900–1920 MHz and 2010–2025 MHz.

UMTS combines key elements of Time Division Multiple Access (TDMA)—about 80 percent of today's digital mobile market is TDMA-based—and Code Division Multiple Access (CDMA) technologies with an integrated satellite component to deliver wideband multimedia capabilities over mobile communications networks. The transmission rate capability of UTRA will provide at least 144 kbps for full-mobility applications in all environments, 384 kbps for limited-mobility applications in the macro- and microcellular environments, and 2.048 Mbps for low-mobility applications particularly in the micro- and picocellular environments. The 2.048-Mbps rate also may be available for short-range or packet applications in the macrocellular environment.

Because the UMTS incorporates the best elements of TDMA and CDMA, this 3G system provides a glimpse of how future wireless networks will be deployed and what possible services may be offered within the IMT-2000 family of systems.

UMTS Objectives

UMTS makes possible a wide variety of mobile services ranging from messaging to speech, data and video communications, Internet and intranet access, and high-bit-rate communication up to 2 Mbps. As such, UMTS is expected to take mobile communications well beyond the current range of wireline and wireless telephony, providing a platform that will be ready for implementation and operation in the year 2002.

UMTS is intended to provide globally available, personalized, and high-quality mobile communication services. Its objectives include

- Integration of residential, office, and cellular services into a single system, requiring one user terminal.
- Speech and service quality at least comparable to current fixed networks.
- Service capability up to multimedia.
- Separation of service provisioning and network operation.
- Number portability independent of network or service provider.
- The capacity and capability to serve over 50 percent of the population.
- Seamless and global radio coverage and radio bearer capabilities up to 144 kbps and further to 2 Mbps.
- Radio resource flexibility to allow for competition within a frequency band.

Description

UMTS separates the roles of service provider, network operator, subscriber, and user. This separation of roles makes possible innovative new services without requiring additional network investment from service providers. Each UMTS user has a unique network-independent identification number, and several users and terminals can be associated with the same subscription, enabling one subscription and bill per household to include all members of the family as users with their own terminals. This arrangement would give children access to various communications services under their parents' account. This application also would be attractive for businesses that require cost-efficient system operation— from subscriber/user management down to radio system—as well as adequate subscriber control over the user services.

UMTS supports the creation of a flexible service rather than standardizing the implementations of services in detail. The provision of services is left to service providers and network operators to decide according to the market demand. The subscriber—or the user when authorized by the subscriber— selects services into individual user service profiles, either with the subscription or interactively with the terminal.

UMTS supports its services with networking, broadcasting, directory, localization, and other system facilities, giving UMTS a clear competitive edge over mobile speech and restricted data services of earlier-generation networks. Being adept at providing new services, UMTS is also competitive in the cost of speech services and as a platform for new applications.

UMTS offers a high-quality radio connection that is capable of supporting several alternative speech codecs at 2 to 64 kbps, as well as image, video, and data codecs. Also supported are advanced data protocols covering a large portion of those used in Integrated Services Digital Network (ISDN). The concept includes variable and high bit rates up to 2 Mbps.

Functional Model

The UMTS functional model relies on distributed databases and processing, leaving room for service innovations without the need to alter implemented UMTS networks or existing UMTS terminals. This service-oriented model provides three main functions: management and operation of services, mobility and connection control, and network management.

- *Management and operation of services* A service data function (SDF) handles storage and access to service-related data. A service control function (SCF) contains overall service and mobility control logic and service-related data processing. A service switching function (SSF) invokes service logic—to request routing information, for example. A call control function (CCF) analyzes and processes service requests in addition to establishing, maintaining, and releasing calls.

- *Mobility and connection control* Drawing on the contents of distributed databases, UMTS will provide for the real-time matching of user service profiles to the available network services, radio capabilities, and terminal functions. This function will handle mobile subscriber registration, authentication, location updating, handoffs, and call routing to a roaming subscriber.

- *Network management* Under UMTS, the administration and processing of subscriber data, maintenance of the network, and charging, billing, and traffic statistics will remain within the traditional telecommunications management network (TMN).

TMN consists of a series of interrelated national and international standards and agreements that provide for the surveillance and control of telecommunications service provider networks on a worldwide scale. The result is the ability to achieve higher service quality, reduced costs, and faster product integration. TMN is also applicable in wireless communications, CATV networks, private overlay networks, and other large-scale, high-bandwidth communications networks. With regard to UTMS (and other

3G wireless networks), TMN will be enhanced to accommodate new requirements. In areas such as service profile management, routing, and radio resource management between UMTS services, networks, and terminal capabilities, new TMN elements will be developed.

Bearer Services

Under UMTS, four kinds of bearer services will be provided to support virtually any current and future application:

- *Class A* This bearer service offers constant-bit-rate (CBR) connections for isochronous (real-time) speech transmission. This service provides a steady supply of bandwidth to ensure the highest quality speech.

- *Class B* This bearer service offers variable-bit-rate connections that are suited for bursty traffic, such as transaction-processing applications.

- *Class C* This bearer service is a connection-oriented packet protocol that can be used support time-sensitive legacy data applications such as those based on IBM's Systems Network Architecture (SNA).

- *Class D* This is a connectionless packet bearer service. This is suitable for accessing data on the public Internet or private intranets.

Summary

By harnessing the best in cellular, terrestrial, and satellite wideband technology, UMTS will guarantee access to all communications, from simple voice telephony to high-speed, high-quality multimedia services. It will deliver information directly to users and provide them with access to new and innovative services and applications. It will offer mobile personalized communications to the mass market regardless of location, network, or terminal used. Users will be provided with adaptive multimode/multiband phones or terminals with a flexible air interface to enable global roaming across locations and with backward compatibility with second-generation (2G) systems.

See also

Cellular Telephones

International Mobile Telecommunications

V

VOICE CLONING

Voice cloning is a technology that promises human-sounding synthetic speech that can be used to support existing applications and encourage the development of new applications, particularly for use in mobile phone voice mail, announcement messages, and voice-activated features. Although synthesized speech systems go back to 1939, today's technology offers voice quality that is so realistic that it justifies being called "cloning."

Voice cloning is based on technology developed by AT&T Labs and has two components. The first is a text-to-speech engine that turns written words into natural-sounding speech. The second includes a library of voices and the ability to custom-develop a voice, perhaps duplicating a celebrity spokesperson. The English-speaking voice, male or female, can be used to read text on a computer, cell phone, or personal digital assistant (PDA). The technology can even be added to a car's computer system to recite driving directions, provide city and restaurant guides, and report on the performance of key subsystems.

The speech software is so good at reproducing the sounds, inflections, and intonations of a human voice that it can recreate voices and even bring the voices of long-dead celebrities back to life. The software, which turns printed text into synthesized speech, makes it possible for a company to use recordings of a person's voice to utter new things that the person never actually said. The software, called Natural Voices, is not flawless—the synthesized speech may contain a few robotic tones and unnatural inflections—but this is the first text-to-speech software to raise the specter of voice cloning, replicating a person's voice so perfectly that the human ear cannot tell the difference.

The product itself is provided as a text-to-speech server engine and client software development kit (SDK) that is an integrated collection of C++ classes to help developers integrate text to speech into their applications. The SDK includes a sample application that can be used to explore potential uses of the SDK and text-to-speech server. Both the text-to-speech server engine and SDK run on popular computer and development platforms, including Linux, Solaris, and Windows NT and 2000. An installation package installs the AT&T Labs Natural Voices TTS engine, documentation, tools, class libraries, sample applications, and demo applications onto the target system.

AT&T Labs also offers a custom voice product that entails a person going to a studio where staff record 10 to 40 hours of readings. Texts range from business and news reports to outright babble. The recordings are then chopped into the smallest number of units possible and sorted into databases. When the software processes text, it retrieves the sounds and reassembles them to form new sentences. In the case of long-dead celebrities, archival recordings can be used in the same way.

Applications

Potential customers for the software, which is priced in the thousands of dollars, include telephone call centers, companies that make software that reads digital files aloud, and makers of automated voice devices. Businesses could use the software to

- Create new revenue-generating applications and services for cell phone users.
- Improve customer relationships by putting a pleasant-sounding voice interface on applications, products, or services.
- Realize mobility and "access anywhere/anytime/any device" strategies by making computer-based information accessible by voice.
- Facilitate international expansion plans through a wide variety of text-to-speech languages.

Third-party developers can use voice cloning technology to add significant enhancements to existing applications and services, drive new revenue opportunities, and add "stickiness" to applications or services. Voice on an e-commerce Web site, for example, can make content easily accessible to the visually impaired, which would keep them coming back for future purchases.

The software also can be used by publishers of video games and books on tape. In the near future, people will want high-end speech technology that enables them to interact at length with their cell phones and Palm organizers instead of typing entries and squinting at a tiny screen.

Ownership Issues

Voice cloning technology raises ownership issues. For example, who owns the rights to a celebrity's voice? This and related issues can be addressed in contracts that include voice-licensing clauses. Current technical limitations may alleviate any worries that a person's voice could be cloned without permission.

Although the technology is not yet good enough to carry out fraud, synthesized voices eventually may be capable of tricking people into thinking that they are getting phone calls from people they know—such as a politician during an election campaign. Politicians already make use of machines that perfectly mimic their signatures and handwritten postscript messages, making it appear that they are sending personal letters to constituents. In the not too distant future, we can expect voice cloning to add another personal touch to campaigning.

Summary

What is unique about voice cloning is the ability to recreate custom voices. AT&T has previously licensed speech technology, such as SpeechWorks, to other companies but contends that the latest version represents a huge technological leap forward. Despite the technical breakthroughs by AT&T Labs, many engineers are skeptical that a completely simulated voice can be indistinguishable from that of a human. With the pressure on to perfect the technology, however, it is too soon to rule out this possibility. Already industry analysts are predicting that the market for text-to-speech software will reach more than $1 billion in the next 5 years, providing ample incentive to fine-tune the technology.

See also

Voice Compression

VOICE COMPRESSION

Voice compression entails the application of various algorithms to the voice stream to reduce bandwidth requirements while preserving the quality or audibility of the voice transmission. Numerous compression standards for voice have emerged over the years that allow businesses to achieve substantial savings on leased lines with only a modest cost for additional hardware. Using these standards, the normal 64-kbps voice channel can be reduced to 32, 16, or 8 kbps, or even as little as 6.3 and 5.3 kbps, for sending voice over the Internet or cellular phone networks. As the compression ratio increases, however, voice quality diminishes.

In the 1960s, the CCITT standardized the use of Pulse Code Modulation (PCM) as the internationally accepted coding standard (G.711) for toll-quality voice transmission. Under this standard, a single voice channel requires 64 kbps when transmitted over the telephone network, which is based on Time Division Multiplexing (TDM). The 64-kbps PCM time slot—or payload bit rate—forms the basic building block for today's public telephone services and equipment, such that 24 time slots or channels of 64 kbps each that can be supported on a T1 line.

Pulse Code Modulation

A voice signal takes the shape of a wave, with the top and the bottom of the wave constituting the signal's frequency level, or amplitude. The voice is converted into digital form by an encoding technique called Pulse Code Modulation (PCM). Under PCM, voice signals are sampled at the minimum rate of two times the highest voice frequency level of 4000 Hertz (Hz), which equates to 8000 times per second. The amplitudes of the samples are encoded into binary form using enough bits per sample to maintain a high signal-to-noise ratio. For quality reproduction, the required digital transmission speed for 4-kHz voice signals works out to 8000 samples per second × 8 bits per sample = 64,000 bps (64 kbps).

The conversion of analog voice signals to and from digital is performed by a coder-decoder, or codec, which is a key component of D4 channel banks and multiplexers. The codec translates amplitudes into binary values and performs mu-law quantizing. The mu-law process (North America only) is an encoding-decoding scheme for improving the signal-to-noise ratio. This is similar in concept to Dolby noise reduction, which ensures quality sound reproduction.

Other components in the channel bank or multiplexer interleave the digital signals representing as many as 24 channels to form a 1.544-Mbps bit stream (including 8 kbps for control) suitable for transmission over a T1 line. PCM exhibits high quality, is robust enough for switching through the public network without suffering noticeable degradation, and is simple to implement. But PCM allows for only 24 voice channels over a T1 line. Digital compression techniques can be applied to multiply the number of channels on a T1 line, several of which are described in Table V-1.

Compression Basics

Among the most popular compression methods is Adaptive Differential Pulse Code Modulation (ADPCM), which has

TABLE V-1 Description of Commonly Used Digital Compression Techniques

Digital Encoding Method	Standard	Bit Rate, kbps	Mean Opinion Score
Pulse Code Modulation (PCM)	G.711	64	4.4
Adaptive Differential Pulse Code Modulation (ADPCM)	G.726	40, 32, 24, 16	4.2
Low Delay Code Excited Linear Prediction (LDCELP)	G.728	16	4.2
Conjugate Structure Algebraic Code Excited Linear Prediction (CSACELP)	G.729A	8	4.2
Multipulse Maximum Likelihood Quantization (MPMLQ)	G.723	6.3	3.9
Algebraic Code Excited Linear Prediction (ACELP)	G.723	5.3	3.5

Note: The mean opinion score (MOS) is the accepted measure of voice quality, determined through a statistical sample of user opinions.

been a worldwide standard since 1984. It is used primarily on private T-carrier networks to double the channel capacity of the available bandwidth from 24 to 48 channels, but it can be applied to microwave and satellite links as well.

ADPCM is also used on some cellular networks such as those based on the Personal Handyphone System (PHS) and Personal Air Communications Systems (PACS). Both employ 32-kbps ADPCM waveform encoding, which provides near landline voice quality. ADPCM has demonstrated a high degree of tolerance to the cascading of voice encoders (vocoders), as experienced when a mobile subscriber calls a voice-mail system and the mailbox owner retrieves the message from a mobile phone. With other mobile technologies, the playback quality is noticeably diminished, but with PHS and PACS, it is very clear.

The ADPCM device accepts the 8000-sample-per-second rate of PCM and uses a special algorithm to reduce the 8-bit samples to 4-bit words. These 4-bit words, however, no longer represent sample amplitudes but only the difference between successive samples. This is all that is necessary for a like device at the other end of the line to reconstruct the original amplitudes.

Integral to the ADPCM device is circuitry called the "adaptive predictor" that predicts the value of the next signal based only on the level of the previously sampled signal. Since the human voice does not usually change significantly from one sampling interval to the next, prediction accuracy can be very high. A feedback loop used by the predictor ensures that voice variations are followed with minimal deviation.

Consequently, the high accuracy of the prediction means that the difference in the predicted and actual signal is very small and can be encoded with only 4 bits rather than the 8 bits used in PCM. In the event that successive samples vary widely, the algorithm adapts by increasing the range represented by the 4 bits. However, this adaptation will decrease the signal-to-noise ratio and reduce the accuracy of voice frequency reproduction.

At the other end of the digital facility is another compression device (Figure V-1), in which an identical predictor performs the process in reverse to reinsert the predicted signal and restore the original 8-bit code.

By halving the number of bits to accurately encode a voice signal, T1 transmission capacity is doubled from the original 24 channels to 48 channels, providing the user with a 2 for 1 cost savings on monthly charges for leased T1 lines.

It is also possible for ADPCM to compress voice to 16 kbps by encoding voice signals with only 2 bits instead of 4

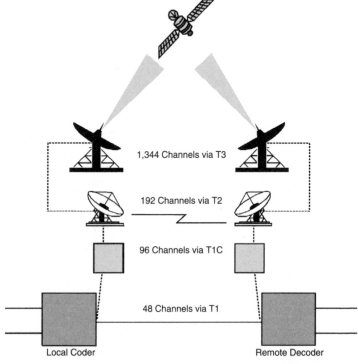

Figure V-1 Some basic network configurations employing Adaptive Differential Pulse Code Modulation (ADPCM) to double the number of channels on the available bandwidth of various digital facilities.

bits, as discussed above. This 4 to 1 level of compression provides 96 channels on a T1 line without significantly reducing signal quality.

Although other compression techniques are available for use on wire and wireless networks, ADPCM offers several advantages. ADPCM holds up well in the multinode environment, where it may undergo compression and decompression several times before arriving at its final destination. And unlike many other compression methods, ADPCM does not distort the distinguishing characteristics of a person's voice during transmission.

Variable-Rate ADPCM

Some vendors have designed ADPCM processors that not only compress voice but also accommodate 64-kbps pass-through as well. The use of very compact codes allows several different algorithms to be handled by the same ADPCM processor. The selection of algorithm is controlled in software and is done by the network manager. Variable-rate ADPCM offers several advantages.

Compressed voice is more susceptible to distortion than uncompressed voice—16 kbps more so than 32 kbps. When line conditions deteriorate to the point where voice compression is not possible without seriously disrupting communications, a lesser compression ratio may be invoked to compensate for the distortion. If line conditions do not permit compression even at 32 kbps, 64-kbps pass-through may be invoked to maintain quality voice communication. Of course, channel availability is greatly reduced, but the ability to communicate with the outside world becomes the overriding concern at this point rather than the number of channels.

Variable-rate ADPCM provides opportunities to allocate channel quality according to the needs of different classes of users. For example, all intracompany voice links may operate at 16 kbps, while those used to communicate externally may be configured to operate at 32 kbps.

The number of channels may be increased temporarily by compressing voice to 16 kbps instead of 32 kbps until new facilities can be ordered, installed, and put into service. As new links are added to keep up with the demand for more channels, the other links may be returned to operation at 32 kbps. Variable-rate ADPCM, then, offers much more channel configuration flexibility than products that offer voice compression at only 32 kbps.

Other Compression Techniques

Other compression schemes can be used over T-carrier facilities, such as Continuously Variable Slope Delta (CVSD) modulation and Time Assigned Speech Interpolation (TASI).

CVSD The higher the sampling rate, the smaller is the average difference between amplitudes. At a high enough sampling rate—32,000 times a second in the case of 32-kbps voice—the average difference is small enough to be represented by only 1 bit. This is the concept behind CVSD modulation, where the 1 bit represents the change in the slope of the analog curve. Successive 1s or 0s indicate that the slope should get steeper and steeper. This technique can result in very good voice quality if the sampling rate is fast enough.

Like ADPCM, CVSD will yield 48 voice channels at 32 kbps on a T1 line. But CVSD is more flexible than ADPCM in that it can provide 64 voice channels at 24 kbps or 96 voice channels at 16 kbps. This is so because the single-bit words are sampled at the signaling rate. Thus, to achieve 64 voice channels, the sampling rate is 24,000 times a second, while 96 voice channels takes only 16,000 samples per second. In reducing the sampling rate to obtain more channels, however, the average difference between amplitudes becomes greater. And since the greater difference between amplitudes is still represented by only 1 bit, there is a noticeable drop in voice quality. Thus the flexibility of CVSD comes at the expense of quality. It is even possible for CVSD to provide 192 voice channels at 8 kbps.

TASI Since people are not normally able to talk and listen simultaneously, network efficiency at best is only 50 percent. And since all human speech contains pauses that constitute wasted time, network efficiency is further reduced by as much as 10 percent, putting maximum network efficiency at only 40 percent.

Statistical voice compression techniques, such as Time Assignment Speech Interpolation (TASI), take advantage of this quiet time by interleaving various other conversation segments together over the same channel. TASI-based systems actually seek out and detect the active speech on any line and assign only active talkers to the T1 facility. Thus TASI makes more efficient utilization of "time" to double T1 capacity. At the distant end, the TASI system sorts out and reassembles the interwoven conversations on the line to which they were originally intended.

The drawback to statistical compression methods is that they have trouble maintaining consistent quality. This is so because such techniques require a high number of channels, at least 100, from which a good statistical probability of usable quiet periods may be gleaned. However, with as few as 72 channels, a channel gain ratio of 1.5 to 1 may be achieved. If the number of input channels is too few, a condition known as "clipping" may occur, in which speech signals are deformed by the cutting off of initial or final syllables.

A related problem with statistical compression techniques is freeze-out, which usually occurs when all trunks are in use during periods of heavy traffic. In such cases, a sudden burst in speech can completely overwhelm the total available bandwidth, resulting in loss of entire strings of syllables. Another liability inherent in statistical compression techniques, even for large T1 users, is that they are not suitable for transmissions having too few quiet periods, such as when facsimile and music on-hold is used. Statistical compression techniques, then, work better in large configurations than in small ones.

Summary

Adding lines and equipment is one way that organizations can keep pace with increases in traffic. But even when funds are immediately available for such network upgrades, communications managers must contend with the delays inherent in ordering, installing, and putting new facilities into service. To accommodate the demand for bandwidth in a timely manner, communications managers can apply an appropriate level of voice compression to obtain more channels out of the available bandwidth. Depending on the compression technique selected, there need not be a noticeable decrease in voice quality.

See also

Data Compression

Personal Access Communications Systems

Personal Handyphone System

W

WIRED EQUIVALENT PRIVACY

Wired Equivalent Privacy (WEP) is the security protocol specified in the IEEE 802.11b Standard for Wireless Fidelity (Wi-Fi) networks. A key point of vulnerability exists on the wireless link between client devices and access points. Here, WEP provides a level of security and privacy ostensibly comparable to what is expected of a wired local area network (LAN). Since it was not intended as an end-to-end security solution, however, users must implement additional safeguards to fully protect their information.

WEP relies on a secret key that is shared between a mobile station such as a notebook equipped with a wireless Ethernet card and an access point that provides a wired connection to the LAN. The secret key is used to encrypt packets before they are transmitted, and an integrity check is used to ensure that packets are not modified in transit.

The IEEE standard does not discuss how the shared key is established. In practice, most installations use a single key that is shared between all mobile stations and access points. Commercial products offer more sophisticated key management techniques that can be used to help defend against hacker attacks, but products for the residential market generally lack these features because they require a more technical understanding of security concepts.

WEP uses the RC4 encryption algorithm, which is known as a "stream cipher." A stream cipher operates by expanding a short key into an infinite pseudo-random key stream. The sending device implements the XOR (exclusive or) operation on the key stream with the plaintext to produce ciphertext. The receiving device has a copy of the same key and uses it to generate an identical key stream. By implementing XOR on the key stream with the ciphertext, the original plaintext is recovered.

Researchers from the University of California at Berkeley have found that this mode of operation makes stream ciphers vulnerable to several attacks. If an attacker flips a bit in the ciphertext, then on decryption the corresponding bit in the plaintext will be flipped. Also, if an eavesdropper intercepts two ciphertexts encrypted with the same key stream, it is possible to obtain the XOR of the two plaintexts. According to the Berkeley researchers, knowledge of this XOR can enable statistical attacks to recover the plaintexts. The statistical attacks become increasingly effective as more ciphertexts using the same key stream become known. Once one of the plaintexts becomes known, it is a relatively simple matter to recover all of the others. These attack methods work equally well on both 64- and 128-bit versions of WEP.

WEP has defenses against both of these attacks. To ensure that a packet has not been modified in transit, it uses an integrity check (IC) field in the packet. To avoid encrypting two ciphertexts with the same key stream, an initialization vector (IV) is used to augment the shared secret key and produce a different RC4 key for each packet. The IV is also included in the packet. However, the Berkeley researchers contend that both these measures are implemented incorrectly, resulting in poor security.

The integrity check field is implemented as a CRC-32 checksum, which is part of the encrypted payload of the packet. However, CRC-32 is *linear*, which means that it is possible to compute the bit difference of two CRCs based on the bit difference of the messages over which they are taken. In other words, flipping bit X in the message results in a deterministic set of bits in the CRC that must be flipped to produce a correct checksum on the modified message. Because flipping bits carries through after an RC4 decryption, this allows the attacker to flip arbitrary bits in an encrypted message and correctly adjust the checksum so that the resulting message appears valid.

Vendors that offer business-class products that use WEP for security on wireless links have dealt with these problems by adding features to WEP. Cisco Systems, for example, offers Dynamic WEP Key Management, which allows network administrators to set time increments in which WEP keys are exchanged per user per session. Increasing the frequency in which keys are exchanged helps systems mitigate the possibility of successful attacks.

Although WEP will be refined continually to increase the security of wireless links, even Cisco recognizes that no single security scheme works for all customers. Accordingly, in addition to WEP, Cisco also offers virtual private network (VPN), firewall, and other features to enhance the end-to-end security of corporate networks.

Summary

WEP seeks to establish a similar level of protection as that offered by the wired network's physical security measures by encrypting data transmitted over the wireless LAN. Data encryption protects the vulnerable wireless link between clients and access points. Once this measure has been taken, other typical LAN security mechanisms such as password protection, end-to-end encryption, VPNs, client firewall software, and authentication can be put in place to further ensure privacy.

See also

Access Points

Wireless Fidelity

Wireless LAN Security

WIRELESS E911

The number 911 is the designated universal emergency number in North America for both wireline and wireless telephone service. Dialing 911 puts the caller in immediate contact with a public safety answering point (PSAP) operator who arranges for the dispatch of appropriate emergency services—ambulance, fire, police, rescue—based on the nature of the reported problem. Since its inception in 1968, this concept has amply demonstrated its value by saving countless lives in thousands of cities and towns across the United States and Canada.

In a series of orders since 1996, the Federal Communications Commission (FCC) has taken action to improve the quality and reliability of 911 emergency services for wireless phone users by adopting rules to govern the availability of basic 911 services and the implementation of enhanced 911 (E911) for wireless services. To further these goals, the agency has required wireless carriers to implement E911 service, subject to certain conditions and schedules. The wireless 911 rules apply to all cellular, broadband Personal Communications Service (PCS) and certain Specialized Mobile Radio (SMR) service providers.

These carriers are required to provide to the PSAP the telephone number of the originator of a 911 call and the location of the cell site or base station receiving a 911 call. This information assists in the provision of timely emergency responses both by providing some information about the general location from which the call is being received and by permitting emergency call takers to reestablish a connection with the caller if the call is disconnected.

All mobile phones manufactured for sale in the United States after February 13, 2000, that are capable of operating in an analog mode, including dual-mode and multimode handsets, must include a special method for processing 911 calls. When a 911 call is made, the handset must override any programming that determines the handling of ordinary calls and must permit the call to be handled by any available carrier, regardless of whether the carrier is the customer's preferred service provider.

As of October 2001, wireless carriers were required to begin providing automatic location identification (ALI) as part of E911 service implementation, according to the following schedule:

1. Begin selling and activating ALI-capable handsets no later than October 1, 2001.

2. Ensure that at least 25 percent of all new handsets activated are ALI-capable no later than December 31, 2001.

3. Ensure that at least 50 percent of all new handsets activated are ALI-capable no later than June 30, 2002.

4. Ensure that 100 percent of all new digital handset activated are ALI-capable no later than December 31, 2002 and thereafter.

5. By December 31, 2005, achieve 95 percent penetration of ALI-capable handsets among its subscribers.

Originally, the FCC envisioned that carriers would need to deploy network-based technologies to provide ALI. However, there have been significant advances in location technologies that employ new or upgraded handsets and that are based on the Global Positioning System (GPS). These methods are approved for implementing enhanced 911 services as well.

Summary

Emergency 911 services have become valuable tools in rendering prompt and appropriate assistance to people in critical need. Most states have laws that mandate prompt action on all calls received by a PSAP operator. Unfortunately, 911 systems are so taken for granted that many calls are not for emergencies at all, and expensive resources end up being expended needlessly on trivial pursuits. PSAP operators now receive calls on such matters as garbage collection dates, late mail delivery, a leaky faucet or heater the landlord won't fix, directions to stores and restaurants, and whether or not to see a lawyer for this or that problem. The 911 systems in some communities have become so bogged down with nonemergency calls that the subject is frequently addressed by public awareness campaigns in the print and broadcast media.

See also

Global Positioning System

Personal Communications Service

Specialized Mobile Radio

WIRELESS APPLICATION PROTOCOL

The Wireless Application Protocol (WAP) is a specification developed by the WAP Forum for sending and reading Internet content and messages on small wireless devices, such as cellular phones equipped with text displays. Common WAP-enabled information services are news, stock quotes, weather reports, flight schedules, and corporate announcements. Special Web pages called "WAP portals" are specifically formatted to offer information and services. CNN and Reuters are among the content providers that offer news for delivery to cell phones, wireless personal digital assistants (PDAs), and handheld computers. Electronic commerce and e-mail are among the WAP-enabled services that can be accessed from these devices as well.

Typically, these devices will have very small screens, so content must be delivered in a "no frills" format. In addition, the bandwidth constraints of today's cellular services mean that the content must be optimized for delivery to handheld devices. To get the information in this form, Web sites are built with a light version of the HyperText Markup Language (HTML) called the Wireless Markup Language (WML).

The strength of WAP is that it spans multiple air link standards and, in the true Internet tradition, allows content publishers and application developers to be unconcerned about the specific delivery mechanism. Like the Internet, the WAP architecture is defined primarily in terms of network protocols, content formats, and shared services. This approach leads to a flexible client-server architecture that can be implemented in a variety of ways but which also provides interoperability and portability at the network interfaces. The WAP protocol stack is depicted in Figure W-1.

WAP solves the problem of using Internet standards such as HTML, HyperText Transfer Protocol (HTTP), TLS, and Transmission Control Protocol (TCP) over mobile networks. These protocols are inefficient, requiring large amounts of mainly text-based data to be sent. Web content written with HTML generally cannot be displayed in an effective way on the small-sized screens of pocket-sized mobile phones and pagers, and navigation around and between screens is not easy with one hand.

Furthermore, HTTP and TCP are not optimized for the intermittent coverage, long latencies, and limited bandwidth associated with wireless networks. HTTP sends its headers and commands in an inefficient text format instead of compressed binary format. Wireless services using these protocols are often slow, costly, and difficult to use. The TLS security standard, too, is problematic, since many messages need to be exchanged between client and server. With wireless transmission latencies, this back-and-forth traffic flow results in a very slow response for the user.

WAP has been optimized to solve all these problems. It makes use of binary transmission for greater compression of data and is optimized for long latency and low to medium bandwidth. WAP sessions cope with intermittent coverage and can operate over a wide variety of wireless transports using the Internet Protocol (IP) where possible and other

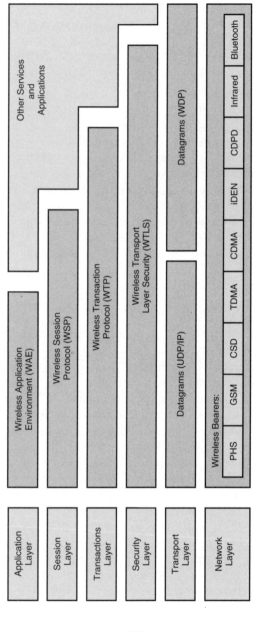

Figure W-1 The Wireless Application Protocol (WAP) stack.

396

optimized protocols where IP is impossible. The WML used for WAP content makes optimal use of small screens, allows easy navigation with one hand without a full keyboard, and has built-in scalability from two-line text displays through to the full graphic screens on smart phones and communicators.

WAP Applications Environment

WAP applications are built within the Wireless Application Environment (WAE), which closely follows the Web content delivery model, but with the addition of gateway functions. Figure W-2 contrasts the conventional Web model with the WAE model. All content is specified in formats that are similar to the standard Internet formats and is transported using standard protocols on the Web while using an optimized

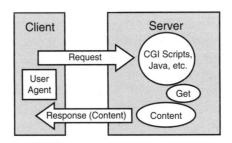

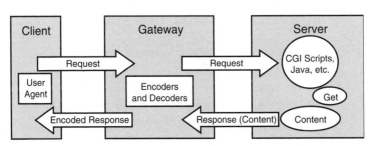

Figure W-2 The standard Web content delivery model (*top*) and the Wireless Application Environment mode (*bottom*).

HTTP-like protocol in the wireless domain (i.e., WAP). The architecture is designed for the memory and CPU processing constraints that are found in mobile terminals. Support for low-bandwidth and high-latency networks is also included in the architecture as well. Where existing standards were not appropriate due to the unique requirements of small wireless devices, WAE has modified the standards without losing the benefits of Internet technology.

The major elements of the WAE model include

- *WAE user agents* These client-side software components provide specific functionality to the end user. An example of a user agent is a browser that displays content downloaded from the Web. In this case, the user agent interprets network content referenced by a Uniform Resource Locator (URL). WAE includes user agents for the two primary standard content types: encoded WML and compiled WML Script.

- *Content generators* Applications or services on servers may take the form of Common Gateway Interface (CGI) scripts that produce standard content formats in response to requests from user agents in the mobile terminal. WAE does not specify any particular content generator, since many more are expected to become available in the future.

- *Standard content encoding* A well-defined content encoding, allowing a WAE user agent (e.g., a browser) to conveniently navigate Web content. Standard content encoding includes compressed encoding for WML, bytecode encoding for WMLScript, standard image formats, a multipart container format, and adopted business and calendar data formats (i.e., vCard and vCalendar).

- *Wireless Telephony Application (WTA)* This collection of telephony specific extensions provides call and feature control mechanisms, allowing users to access and interact with mobile telephones for phonebooks and calendar applications.

WMLScript is a lightweight procedural scripting language based on JavaScript. It enhances the standard browsing and presentation facilities of WML with behavioral capabilities. For example, an application programmer can use WMLScript to check the validity of user input before it is sent to the network server, provide users with access to device facilities and peripherals, and interact with the user without a round-trip to the network server (e.g., display an error message).

WAP 2.0

The latest version of the Wireless Application Protocol is WAP 2.0, which continues the convergence of WAP with the evolving Internet, merging the work of the WAP Forum, the World Wide Web Consortium (W3C), and the Internet Engineering Task Force (IETF) and enabling more rapid development of new mobile Internet applications.

New technologies of WAP 2.0 that will improve the user experience are data synchronization, multimedia messaging service (MMS), persistent storage interface, provisioning, and pictograms. Additionally, WTA, Push, and User Agent Profile (UAPROF) use more advanced features in WAP 2.0 than in previous versions.

- *Data synchronization* adopts the SyncML protocol to ensure a common solution framework with a multitude of devices. The SyncML messages are supported over both the Wireless Session Protocol (WSP) and the HTTP/1.1 protocols.

- *Multimedia messaging service* provides the framework to develop applications that support feature-rich messaging solutions, permitting delivery of varied types of content in order to tailor the user experience.

- *Persistent storage interface* provides a set of storage services that allows the user to organize, access, store, and retrieve data on wireless devices.

- *Provisioning* permits the network operator to manage the devices on its network with a common set of tools.

- *Pictogram* permits the use of a set of tiny images, allowing users to quickly convey concepts in a small amount of space while transcending traditional language boundaries.

- *Wireless Telephony Application* provides a range of advanced telephony services within the application environment, enabling a host of call handling functions such as making and answering calls, placing them on hold, and redirecting them even while performing data-centric tasks. The availability of these services enables operators to offer customers a unique user interface to control complex network features, such as call forwarding options.

- *Push technology* allows trusted application servers to proactively send personalized content to the end user, such as a sales offer for a product a person might be interested in buying, a new e-mail notification, or a location-dependent promotion. Push technology complements the traditional "pull" model of the Internet, where users request specific information from a Web site.

- *User agent profile* enables application servers to send the appropriate content to the user and to recognize the capabilities of devices, such as screen size and color, to maximize performance potential, bringing the user increased satisfaction.

WAP 2.0 is a next-generation specification that addresses the needs of all players in the wireless industry who plan on incorporating the platform-agnostic specification in their products and services to grow the wireless market by offering value-added features.

Summary

WAP is an open global specification that empowers mobile users with wireless devices to easily access and interact with information and services instantly. It is designed to work with most wireless networks, including Bluetooth,

Infrared, CDPD, CDMA, GSM, PDC, PHS, TDMA, iDEN, DECT, and GRPS. It can be built to run on any operating system, including PalmOS, EPOC, Windows CE, FLEXOS, OS/9, and JavaOS. WAP offers the additional advantage of providing service interoperability between different device families.

See also

Bluetooth

Cellular Data Communications

Infrared Networking

Wireless IP

WIRELESS APPLICATION SERVICE PROVIDERS

Application service providers (ASPs) host business-class applications in their data centers and make them available to customers on a subscription basis over the network.[1] Among the business functions commonly outsourced in this way are customer relationship management (CRM), financial management, human resources, procurement, and enterprise resource planning (ERP). The ASP owns the applications, and subscribers are charged a fixed monthly fee for use of the applications over secure network connections. Wireless ASPs (WASPs) provide hosted wireless applications so that companies will not have to build their own sophisticated wireless infrastructures.

---

[1]The difference between Web hosting or e-commerce hosting and the kind of hosting performed by ASPs is that in the former case the customer owns the applications and merely runs them on the shared or dedicated servers of an Internet service provider (ISP). In the latter case the ASP has strategic alliances with third-party software providers for licensed use of the applications over the network. The ASP pays fees to the software firms based on factors such as the number and type of customers' users.

An ASP enables customers to avoid many of the significant and unpredictable ongoing application management challenges and costs. Following the implementation of software applications, performed for a fixed fee or on a time and materials basis, customers pay a monthly service fee based largely on the number of applications used, total users, the level of service required, and other factors. By providing application implementation, integration, management, and various upgrade services and related hardware and network infrastructure, the ASP reduces information technology (IT) burdens of its customers, enabling them to focus on their core businesses and react quickly to dynamic market conditions.

Traditionally, organizations have installed, operated, and maintained enterprise software applications internally. The implementation of enterprise software applications often takes twice as long as planned. Moreover, the ongoing costs of operating these applications, including patching, upgrading, training, and management expenses, are often significant, unpredictable, and inconsistent and may increase over time. The emergence of the Internet, the increased communications bandwidth, and the rewriting of enterprise software to be delivered over IP networks are transforming the way enterprise software applications are being provided to companies.

Instead of in-house installations, these applications are beginning to be hosted by third parties, in which the hosting company maintains the applications on an off-site server, typically in a data center, and delivers the applications to customers over the Internet as a service. In addition, competitive pressures have led to a renewed focus on core competencies, with many businesses concluding that building and maintaining IT capabilities across their entire set of applications are not core competencies. In response to these factors, companies are adopting hosted applications rather than managing them in-house.

An ASP typically can complete a standard implementation of its services in 2 to 14 weeks. This allows customers to avoid the longer implementation times frequently experienced with installing and integrating customized, sophisticated applications. This enables customers to achieve the desired benefits quickly by reducing the time required to establish or augment IT capabilities with wire or wireless infrastructures of their own.

To address this market, many types of companies are setting themselves up as ASPs in this relatively new market, including long-distance carriers, telephone companies, computer firms, Internet service providers (ISPs), software vendors, integrators, and business management consultants. Intel Corp., for example, has built data centers around the world to be ready to host electronic business sites for millions of businesses that will embrace the Internet within 5 years.

Another ASP, Corio, enables businesses to obtain best-of-breed applications at an affordable cost. Corio is responsible for maintaining and managing the applications and ensuring their availability to its customers from its data centers. For a fixed monthly fee for the suite of integrated business applications and services, businesses can achieve a 70 percent reduction on average of total cost of ownership (TCO) in the first year versus traditional models and a 30 to 50 percent TCO reduction over a 5-year period.

Among the WASPs are Etrieve and Wireless Knowledge. Etrieve combines the reach of wireless, information processing technology, and voice recognition to provide mobile professionals with the right information, at the right time, in the right format. As an extension of the desktop, Etrieve enables mobile professionals to manage their critical office information—e-mail, calendar, and address book—by voice, by text, and by importance. Wireless Knowledge, a subsidiary of Qualcomm, provides mobile access to Microsoft Exchange or Lotus Domino groupware.

Past Attempts

Application outsourcing has been around for nearly 30 years under the concept of the service bureau. In the service bureau arrangement, business users rented applications running the gamut from rudimentary data processing to high-end proprietary payroll. Companies such as EDS and IBM hosted the applications at centralized sites for a monthly fee and typically provided access via low-speed private-line connections.

In an early 1990s incarnation of the service-bureau model, AT&T rolled out hosted Lotus Notes and Novell NetWare services, complete with 24×7 monitoring and management. Users typically accessed the applications over a Frame Relay service or dedicated private lines. AT&T's Notes hosting effort failed and was discontinued in early 1996. The carrier lacked the expertise needed to provide application-focused services and did not offer broad enough access to these applications. The lesson: Large telecommunications companies are focused on building networks, which is quite different from implementing and managing enterprise applications.

In 1998, there emerged renewed interest in this type of service with a new twist—that of providing an array of standardized services to numerous business customers. Economies of scale could be achieved in this "one to many" model by cost reductions incurred in service delivery; specifically by relying on managed IP networks. Further cost reductions could be achieved by developing implementation templates, innovative application management tools, and integration models that can be used for numerous applications across a variety of companies and industries.

To help sell the benefits of applications outsourcing, 25 companies have formed the Applications Service Provider Industry Consortium. The consortium includes a wide range of companies, including AT&T on the service-provider side. Compaq Computer, IBM, and Sun Microsystems are representative of the systems and software vendors. Interconnect companies such as Cisco Systems are also members. The consortium's goals include education, common definitions, research, standards, and best practices.

Summary

Several trends have come together to rekindle the market for applications outsourcing. The rise of the Internet as an essential business tool, the increasing complexity of enterprise software programs, and the shortage of IT expertise have created a ready environment for carriers and other companies entering the applications hosting business. The economics of outsourcing are compelling, and new companies are being created to deal with customers' emerging outsourcing requirements. As wireless technologies become popular, WASPs have emerged to bring applications to mobile devices, saving companies the trouble and expense of doing it themselves.

See also

Wireless Internet Service Providers

WIRELESS CENTREX

Centrex—short for "central office exchange"—is a service that handles business calls at the telephone company's switch rather than through a customer-owned, premises-based Private Branch Exchange (PBX). Centrex provides a full complement of station features, remote switching, and network interfaces that provides an economical alternative to owning a PBX. Centrex offers remote options for businesses with multiple locations, providing features that appear to users and the outside world as if the remote sites and the host switch are one system. Centrex-capable switches now support wireless links, which extends the boundaries of business services.

Centrex users have access to direct inward dialing (DID) features, as well as station identification on outgoing calls. Each station has a unique line appearance in the central office, in a manner similar to residential telecommunications subscriber connections. A Centrex call to an outside line exits the switch in the same manner as a toll call exits a local exchange. Users dial a four- or five-digit number without a prefix to call internal extensions and dial a prefix (usually 9) to access outside numbers.

The telephone company operates, administers, and maintains all Centrex switching equipment for the customers. It also supplies the necessary operating power for the switching equipment, including backup power to ensure uninterrupted service during commercial power failures.

Centrex may be offered under different brand names. BellSouth calls it Essex, and SBC Communications calls it Plexar, while Verizon calls it CentraNet. Centrex is also offered through resellers that buy Centrex lines in bulk from the local exchange carrier. Using its own or commercially purchased software, the reseller packages an offering of Centrex and perhaps other basic and enhanced telecommunications services to meet the needs of a particular business. The customer gets a single bill for all local, long-distance, 800, 900, and calling-card services at a fee that is less than the customer would otherwise pay.

Centrex Features

Centrex service offerings typically include direct inward dialing (DID), direct outward dialing (DOD), and automatic identification of outward dialed calls (AIOD). Advanced digital Centrex service provides all the basic and enhanced features of the latest PBXs in the areas of voice communications, data communications, networking, and Integrated Services Digital Network (ISDN) access. Commonly available features include voice mail, electronic mail, message center support, and modem pooling.

For large networks, the Centrex switch can act as a tandem switch, linking a company's PBXs through an electronic tandem network. Centrex is also compatible with most private switched network applications, including the Federal

Telecommunications System (FTS) and the Defense Switched Network (DSN).

Many organizations subscribe to Centrex service primarily because of its networking capabilities, particularly for setting up a virtual citywide network without major cost or management concerns. With city-wide Centrex, a business can set up a network of business locations with a uniform dialing plan, a single published telephone number, centralized attendant service, and full feature transparency for only an incremental cost per month over what a single Centrex site would cost.

Wireless Service

A number of value-added services are available through wireless Centrex. Pacific Bell's Wireless Centrex, for example, combines an existing Centrex service with Ericsson's Freeset Business Wireless Telephone system to create a private wireless environment at a company's business location without incurring expensive cellular airtime charges. At the corporate facility multiple, overlapping cells cover assigned areas. The number of cells required is determined by traffic density at a given location.

An on-premises system called the "radio exchange" handles such functions as powering, control, and facilities for connection to Centrex. Base stations relay calls from the radio exchange to the portable telephones. Each base station provides multiple simultaneous speech channels. The coverage of each base station depends on the character of the environment, but it is typically between 8000 and 15,000 square feet. Portable telephones contain the intelligence needed to accommodate roaming and cell-to-cell handover.

When the radio exchange receives an incoming call, it transmits the identification signal of a portable telephone to all base stations. Because the portable telephone communicates with the nearest base station, even in standby mode, it receives the signal and starts ringing. When the call is answered, the portable telephone selects the channel with the best quality transmission.

Pacific Bell's Wireless Centrex service gives business users full wireless mobility plus the following options:

- Account codes
- Authorization codes
- Automatic callback
- Automatic recall
- Call diversion/call forwarding
- Call diversion override
- Call hold
- Call transfer
- Call waiting
- Call pickup
- Speed calling
- Call park
- Conference/three-way calling
- Remote access to network services
- Distinctive/priority ringing
- Do not disturb
- External call forwarding
- Executive intrusion/executive busy override
- Individual abbreviated dialing/single digit dialing
- Speed dialing
- Last number redial
- Loudspeaker paging
- Message waiting indication
- Remote access to subscriber features
- Select call forwarding

Summary

Centrex offers high-quality, dependable, feature-rich telephone service that supports a variety of applications. For many organizations, Centrex offers distinct advantages over on-premises PBX or key/hybrid systems. Centrex can save money over the short term because there is no outlay of cash for an on-premises system. If the service is leased on a month-to-month basis, there is little commitment and no penalty for discontinuing the service. A company can pick up and move without worrying about reinstalling the system, which may not be right for the new location. With wireless capabilities, businesses get the advantage of mobility without the expense of investing in wireless infrastructure of their own.

See also

Wireless PBX

WIRELESS COMMUNICATIONS SERVICES

Wireless Communications Services (WCS) is a category of service that operates in the 2.3-GHz band of the electromagnetic spectrum from 2305 to 2320 MHz and 2345 to 2360 MHz. The FCC issued licenses for WCS as the result of a spectrum auction held in April 1997. Licensees are permitted—within their assigned spectrum and geographic areas—to provide any fixed, mobile, radiolocation, or broadcast-satellite service. One use for the WCS spectrum is for services that adhere to the Personal Access Communications System (PACS) standard. This standard is applied to consumer-oriented products, such as personal cordless devices.

There are two 10-MHz WCS licenses for each of 52 major economic areas (MAEs) and two 5-MHz WCS licenses for each of 12 regional economic area groupings (REAGs).

WCS licensees are permitted to partition their service areas into smaller geographic service areas and to disaggregate their spectrum into smaller blocks without limitation. Licenses are good for a term of 10 years and are renewable just like PCS and cellular licensees. In addition, WCS licensees will be required to provide "substantial service" within their 10-year license term.

WCS is implemented through small relay stations, which may interface with the Public Switched Telephone Network (PSTN). Where WCS poses interference problems with existing Multipoint Distribution Service (MDS) or Instructional Television Fixed Service (ITFS) operations, the WCS licensees must bear the full financial obligation for the remedy. WCS licensees must notify potentially affected MDS/ITFS licensees at least 30 days before commencing operations from any new WCS transmission site or increasing power from an existing site of the technical parameters of the WCS transmission facility. The FCC expects WCS and MDS/ITFS licensees to coordinate voluntarily and in good faith to avoid interference problems, which will result in the greatest operational flexibility in each of these types of operations.

Summary

In establishing WCS, the FCC believed that the flexible use of the 2305- to 2320-MHz and 2345- to 2360-MHz frequency bands would help ensure that new technologies are developed and deployed, such as a wireless system tailored to provide portable Internet access over wide areas at data rates comparable to an ISDN-type connection. Because the technical characteristics of such a system would differ significantly from those for some other systems that might use this band (e.g., PCS), the FCC neither restricted the services provided in this band nor dictated technical standards for operation beyond those required to avoid interference and protect the public interest. In fact, WCS licensees are not constrained to a single use of this spectrum and, therefore, may offer a mix of services and technologies to their customers.

See also

Personal Access Communications Systems

Spectrum Auctions

WIRELESS FIDELITY

Wireless Fidelity (Wi-Fi) refers to a type of Ethernet specified under the IEEE 802.11a and IEEE 802.11b Standards
for LANs operating in the 5- and 2.4-GHz unlicensed frequency bands respectively. Wi-Fi is equally suited to residential users and businesses, and equipment is available
that allows both bands to be used to support separate networks simultaneously.

The IEEE 802.11 Standard makes the wireless network a
straightforward extension of the wired network. This has
allowed for a very uncomplicated implementation of wireless
communication with obvious benefits—they can be installed
using the existing network infrastructure with minimal
retraining or system changes. Notebook users can roam
throughout their sites while remaining in contact with the
network via strategically placed access points that are
plugged into the wired network.

Wireless users can run the same network applications
they use on an Ethernet LAN. Wireless adapter cards used
on laptop and desktop systems support the same protocols as
Ethernet adapter cards. For most users, there is no noticeable functional difference between a wired Ethernet desktop
computer and a wireless computer equipped with a wireless
adapter other than the added benefit of the ability to roam
within the wireless cell. Under many circumstances, it may
be desirable for mobile network devices to link to a conventional Ethernet LAN in order to use servers, printers, or an
Internet connection supplied through the wired LAN. A wireless access point (AP) is a device used to provide this link.

The IEEE 802.11b Standard designates devices that operate in the 2.4-GHz band to provide a data rate of up to 11

Mbps at a range of up to 300 feet (100 meters) using direct-sequence spread-spectrum technology. Some vendors have implemented proprietary extensions to the IEEE 802.11b Standard, allowing applications to burst beyond 11 Mbps to reach as much as 22 Mbps. Users can share files and applications, exchange e-mail, access printers, share access to the Internet, and perform any other task as if they were directly cabled to the network.

The IEEE 802.11a Standard designates devices that operate in the 5-GHz band to provide a data rate of up to 54 Mbps at a range of up to 900 feet (300 meters). Sometimes called "Wi-Fi5," this amount of bandwidth allows users to transfer large files quickly or even watch a movie in MPEG format over the network without noticeable delays. This technology works by transmitting high-speed digital data over a radio wave using Orthogonal Frequency Division Multiplexing (OFDM) technology.

OFDM works by splitting the radio signal into multiple smaller subsignals that are then transmitted simultaneously at different frequencies to the receiver. OFDM reduces the amount of interference in signal transmissions, which results in a high-quality connection. Wi-Fi5 products automatically sense the best possible connection speed to ensure the greatest speed and range possible with the technology. Some vendors have implemented proprietary extensions to the IEEE 802.11a Standard allowing applications to burst beyond 54 Mbps to reach as much as 72 Mbps.

IEEE 802.11 wireless networks can be implemented in infrastructure mode or ad-hoc mode. In infrastructure mode—referred to in the IEEE specification as the "basic service set"—each wireless client computer associates with an AP via a radio link. The AP connects to the 10/100-Mbps Ethernet enterprise network using a standard Ethernet cable and provides the wireless client computer with access to the wired Ethernet network. Ad-hoc mode is the peer-to-peer network mode, which is suitable for very small installations. Ad-hoc mode is referred to in the IEEE 802.11b spec-ification as the "independent basic service set."

Security for Wi-Fi networks is handled by the IEEE stan-dard called Wired Equivalent Privacy (WEP), which is avail-able in 64- and 128-bit versions. The more bits in the encryption key, the more difficult it is for hackers to decode the data. It was originally believed that 128-bit encryption would be virtually impossible to break due to the large num-ber of possible encryption keys. However, hackers have since developed methods to break 128-bit WEP without having to try each key combination, proving that this system is not totally secure. These methods are based on the ability to gather enough packets off the network using special eaves-dropping equipment to then determine the encryption key. Although WEP can be broken, it does take considerable effort and expertise to do so. To help thwart hackers, WEP should be enabled and the keys rotated on a frequent basis.

The wireless LAN industry has recognized that WEP is not as secure as once thought and is responding by develop-ing another standard, known as IEEE 802.11i, that will allow WEP to use the Advanced Encryption Algorithm (AES) to make the encryption key even more difficult to determine. AES replaces the older 56-bit Digital Encryption Standard (DES), which had been in use since the 1970s. AES can be implemented in 128-, 192-, and 256-bit versions. Assuming a computer with enough processing power to test 255 keys per second, it would take 149 trillion years to crack AES.

Summary

Wi-Fi is a certification of interoperability for IEEE 802.11b systems awarded by the Wireless Ethernet Compatibility Alliance (WECA), now known as the Wi-Fi Alliance. The Wi-Fi seal indicates that a device has passed independent tests and will reliably interoperate with all other Wi-Fi certified equipment. Customers benefit from this standard by avoiding becoming locked into one vendor's solution—they can purchase Wi-Fi certified access points and client devices from different vendors and still expect them to work together.

See also

> Access Points
>
> Wired Equivalent Privacy
>
> Wireless Internet Service Providers
>
> Wireless LANs
>
> Wireless Security

WIRELESS INTERNET ACCESS

In most organizations today, Internet access via the LAN is the norm. Workers share applications, data, and services through their hub- or switch-based LAN connection. When they require Internet access to do research on the Web or send e-mail to someone on another network, for example, the traffic goes through a router connected to the hub or switch and then out to the ISP via a dedicated, "always on" connection such as a Digital Subscriber Line (DSL) or T1 link or even cable. Telecommuters may have to rely on dialup services that offer no more than 56 kbps.

However, there are two trends that make the case for wireless Internet access. One trend has to do with the workforce itself; more and more workers are becoming mobile and require a flexible method of Internet access from wherever they happen to be. This greatly improves flexibility in that the user does not have to find an available phone line in order to dial into the Internet. Wireless access to the Internet also increases productivity in that the user can accomplish work-related tasks at the most opportune time, even while traveling or taking a "vacation."

Second, more companies require high-speed Internet access and cannot always get ISDN, DSL, or cable in their areas or afford the price of one or more T1 lines. Larger companies that require broadband Internet access in the multi-megabit range via T3 or OC-3 may not be able to wait for fiber installation to their locations. These companies are looking to fixed wireless providers for fast installation as well as cheap bandwidth. Whereas a new fiber build may take months to complete under a best-case scenario, wireless broadband connectivity from the customer premises to the service provider's hub antenna can be accomplished in a matter of days.

Business users have a growing number of wireless Internet access services to choose from, and vendors and service providers are emphasizing technologies designed to let users take the Internet with them wherever they go. Already, wireless offerings run the gamut from Wireless Application Protocol (WAP)–enabled mobile phones to broadband architectures that offer fixed wireless access to IP-based networks at megabit-per-second speeds. In addition, Internet connectivity is now available with all major mobile phone protocols, including Code Division Multiple Access (CDMA), Time Division Multiple Access (TDMA), and Global System for Mobile (GSM) telecommunications.

Wireless Access Methods

There are several methods to wirelessly access the Internet, including analog cellular and cellular digital packet data networks, packet radio services, and satellite.

Cellular Networks Analog cellular phone subscribers can send files and e-mail wirelessly via the Internet by hooking up a modem to their phones. Of course, this method of access requires that the user have a cell phone in the first place, as well as an adapter cable to connect the phone to the modem.

Cellular modems work anywhere a cell phone does. The problem is that they work only as well as the cell phone does at any given moment. Checking e-mail on the Internet can be slow, and connection quality varies from network to network.

To access Web content on the Internet, however, really requires a digital cellular service [also called Personal Communication Service (PCS)] and a special phone equipped with a liquid-crystal display (LCD) screen. AT&T's PocketNet phone, for example, lets users access the Internet in areas served by its Cellular Digital Packet Data (CDPD) network.

With this type of cellular service, standard analog cell phones will not work; a digital CDPD phone is required. An alternative is to use a dual-mode phone that operates in standard analog cellular mode for voice conversations and CDPD mode for access to the Internet at 19.2 kbps. AT&T's PocketNet service includes a personal information manager (PIM) that contains an address book, calendar, and to-do list that are maintained on AT&T's Web site and accessed through the phone. The personal address book is tied to the PocketNet phone's "easy dialing" feature for fast, convenient calling.

Another feature offered by CDPD is mobility management, which routes messages to users regardless of the location or the technology. Gateways provide the cellular network with the capability to recognize when subscribers move out of the CDPD coverage area and transfer messages to them via circuit-switched cellular.

Packet Radio Networks An example of a radio network is provided by the Ricochet service developed by Metricom, now a subsidiary of Aerie Networks, that provides network solutions and wireless data communications for industrial and PC applications. The Ricochet service comprises radios, wired access points (AP), and network interconnection facilities (NIF) that enable data to be sent across a network of intelligent radio nodes at speeds of over 176 kbps, with bursts of up to 400 kbps.

The network uses frequency-hopping spread-spectrum packet radios. A large number of these low-power radios are installed throughout a geographic region in a mesh topology, usually placed on top of streetlights or utility poles. These radio receivers, about the size of a shoebox, are also referred to as "microcell radios." Only a small amount of power is required for the radio, which is received by connecting a special adapter to the streetlight. No special wiring is required. These radios are placed about every quarter to a half mile and take only about 5 minutes each to install. Using 162 frequency-hopping channels in a random pattern accommodates many users at the same time, along with providing a high degree of security.

Distributed among the radios are APs, which are used to route the wireless packets on to the wired backbone. Gathering and converting the packets so that they can be transmitted to the wired backbone is accomplished via a T1-based Frame Relay connection. The packets can then be sent to another AP, the Internet, or the appropriate service provider. If the packet is sent to another WAP, it is being used as an alternate route through the mesh. The number of paths a packet can take through the network enhances speed throughput, since network blockage is not as frequent and many possible repeaters exist for the packet.

The Ricochet modem weighs only 8 ounces and thus is very portable. It can be plugged into the serial port of a computer and can be connected to online services and networks just like a standard phone modem. The modem works with most communications software, as well as with Intel-based and Macintosh hardware platforms and operating systems.

Ricochet's wired backbone is based on standard Internet Protocol (IP) technology, routing data via a metropolitan service area. If a data packet has to move across the country, Ricochet's NIF system is used. The NIF functions as a router, collecting packets from the WAPs and using leased lines to connect with NIFs situated in the different metropolitan areas. A name server is part of the Ricochet network backbone, providing security by validating all connection requests.

Metricom had been testing a new wireless data technology that would have provided data rates equal to that provided by wired ISDN 128-kbps service. This technology, called Ricochet II, uses two bands of unlicensed spectrum: the 900-MHz band and the 2.4-GHz band. The new network is compatible with the company's existing Ricochet network and modems. After investing $1.3 billion in infrastructure, Metricom declared bankruptcy and discontinued operations in August 2001. Under new management, the company is in the process of reactivating the existing network and expanding Ricochet to areas where high-speed broadband access is currently unavailable.

The outlook for Ricochet is unclear. Newer, cheaper Wi-Fi networks are being set up that greatly surpasses the speed of Ricochet. In fact, a new type of wireless Internet service provider (WISP) has emerged, such as Boingo Wireless, which uses the 2.4-GHz frequency band to offer data rates of up to 11 Mbps in public places.

Fixed Wireless Access As an alternative to traditional wire-based local telephone service, fixed wireless access technology provides a wireless link to the Public Switched Telephone Network (PSTN). Unlike cellular technologies, however, which provide services to mobile users, fixed wireless services require a rooftop antenna to an office building or home that has a line of sight with a service provider's hub antenna.

Fixed wireless access systems come in two varieties: narrowband and broadband. A narrowband fixed wireless access service can provide bandwidth up to 128 kbps, which can support one voice conversation and a data session such as Internet access or fax transmission. A broadband fixed wireless access service can provide bandwidth in the multimegabit-per-second range, which is enough to support telephone calls, television programming, and broadband Internet access.

A narrowband fixed wireless service requires a wireless access unit that is installed on the exterior of a home or business to allow customers to originate and receive calls with no change to their existing analog telephones. Voice and data calls are transmitted from the transceiver at the customer's location to the base station equipment, which relays the call through carrier's existing network facilities to the appropriate destination. No investment in special phones or facsimile machines is required; customers use all their existing equipment.

Narrowband fixed wireless systems use the licensed 3.5-GHz radio band with 100-MHz spacing between uplink and downlink frequencies. Subscribers receive network access over a radio link within a range of 200 meters (600 feet) to 40 kilometers (25 miles) of the carrier's hub antenna. About 2000 subscribers can be supported per cell site.

Broadband fixed wireless access systems are based on microwave technology. Multichannel Multipoint Distribution Service (MMDS) operates in the licensed 2- to 3-GHz frequency range, while Local Multipoint Distribution Service (LMDS) operates in the licensed 28- to 31-GHz frequency range. Both services are used by Competitive Local exchange Carriers (CLECs) primarily to offer broadband Internet access. These technologies are used to bring data traffic to the fiberoptic networks of Interexchange Carriers (IXCs) and nationwide CLECs, bypassing the local loops of the Incumbent Local Exchange Carriers (ILECs).

Fixed wireless access technology originated out of the need to contain carriers' operating costs in rural areas, where pole and cable installation and maintenance are more expensive than in urban and suburban areas. However, wireless access technology also can be used in urban areas to bypass the local exchange carrier for long-distance calls. Since the IXC or CLEC avoids having to pay the ILEC's local loop interconnection charges, the savings can be passed back to the customer. This arrangement is also referred to as a "wireless local loop."

Satellite Services Satellites are another solution for the rapidly growing demand to transmit Internet and other networking traffic because they offer reliable connections to virtually anywhere in the world.

CyberStar, for example, offers a portfolio of business services, including a global broadband IP multicasting service that allows business users to send high-bandwidth voice, video, and data files to branch offices. By integrating IP, applications, video and audio streaming, IP multicast, Webcasting, and high-speed delivery onto an independent, satellite-based platform, CyberStar gives the IT manager a scalable broadband network.

The satellite service enhances, not replaces, existing enterprise communications infrastructures. The CyberStar system is intended as an overlay that fits with whatever companies are currently running. They can keep mission-critical applications on the current network and run data-intensive and media-rich applications over a reliable and cost-efficient satellite network.

Customers pay for the bandwidth they use rather than signing up for a flat-rate service. Pricing is based on the amount of traffic that is sent each month and the number of sites that are receiving that traffic. Customers are charged initial installation costs per site, which includes antennas, satellite receiver cards, and service activation fees.

Other satellite service providers also support Internet traffic. For homes and businesses, Hughes Network Systems (HNS) has given has been offering its one-way DirecPC Internet access service for several years as well as a newer one- and two-way broadband service, called DirecWAY, which offers 400 kbps.

In the near future, low-earth-orbit (LEO) satellites such as Teledesic will provide Internet access and support other broadband applications with bandwidth on demand ranging from 64 kbps to multimegabit-per-second speeds.

Service Caveats

The biggest and most obvious concern to those interested in wireless technology is that it is more expensive than its wired counterpart and requires an investment in special equipment. As portability becomes more of a factor with regard to Internet access, corporations will become more willing to pay the higher cost of wireless technology. As history has shown, once prices drop due to corporate participation, consumers also will be able to reap the benefits of wireless technology.

Another concern is how network managers are going to integrate the wireless and wired worlds. This involves not only two skill sets, but also two network management systems and two separate application development paths. Nevertheless, wireless technology is becoming more accepted because it can now be economically integrated with wired networks. This can be done with wireless LAN access points that are connected to the hub or switch with Category 5 cable.

Performance is another issue that needs to be improved in wireless Internet access, particularly for real-time applications. The perception is that a wireless connection should have the same throughput and latency as a wired connection, but this is usually not possible. This tends to give the notion that there is not a reasonable response time for interactive applications with wireless connections. While good response time can be had with fixed wireless systems that use MMDS and LMDS, these technologies have other issues that bear consideration, such as limited distance and a line-of-sight requirement. These technologies also falter during heavy rain, dense fog, and dust storms. While the service is out periodically, wired links for backup may be required.

As noted, there are some wireless Internet access providers that are starting to use the 2.4-GHz ISM (industrial, scientific, medical) frequency band to offer commercial Wi-Fi services. But this is unlicensed spectrum, which means that there is no guarantee against interference. MMDS and LMDS, on the other hand, are services provided over licensed spectrum. Since the FCC controls the use of this spectrum, business users can expect interference-free wireless connections for Internet access.

Summary

There are a variety of technologies available for wireless Internet access. Although wide-scale deployment of wireless Internet access services is still somewhere in the future, the long-term prospects for those services may be brighter than ever, due in large part to the number of carriers getting involved with the various technologies and the increasing investments being made toward further development. In many parts of the United States, there is the growing realization that obtaining broadband Internet access through cable and DSL will not be available anytime soon, if ever. Wireless technologies have an important role to play on filling this niche.

See also

Cellular Data Communications

Direct Broadcast Satellite

Fixed Wireless Access

Local Multipoint Distribution Service

Multichannel Multipoint Distribution Service

Wireless Fidelity

Wireless Internet Service Providers

Wireless Local Loops

WIRELESS INTERNET SERVICE PROVIDERS

Wireless Internet service providers (WISPs) use Wi-Fi technology to offer access to the Internet at speeds of up to 11 Mbps when users are in range of an antenna. Based on the IEEE 802.11b Standard, Wi-Fi technology is used by WISPs to provide wireless Internet access at airports, hotels, convention centers, coffee shops, and other public places. Users download free software from the WISP's Web page and use it to find an available signal (Figure W-3). On powering up a notebook computer equipped with a Wi-Fi-compatible IEEE 802.11b card, the software searches all available networks and establishes a wireless connection within seconds.

Service Plans

There are usually several service plans to chose from, according to the number of days the user expects to be online:

- *10-day connect package* A package of 10 connect days a month, which includes unlimited access in a WISP's service location for up to 24 hours. The user can even disconnect and reconnect within each 24-hour period from the same location with no additional charge. Each additional connect day is charged separately.

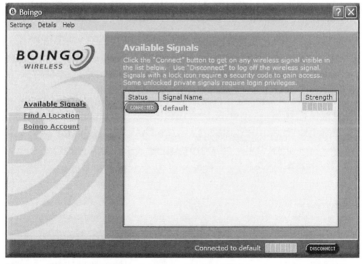

Figure W-3 Free software from wireless Internet service providers (WISPs) such as Boingo Wireless lets users know when they are in range of a high-speed wireless signal so that they can make a connection.

- *Unlimited usage* Unlimited usage allows the user to stay connected to the wireless network all day every day for a fixed monthly charge.

- *Pay-as-you-go* For users who are not sure how much they will use the service, they can sign up for the service and pay a daily charge instead of a monthly fee. A connect day includes unlimited access in a WISP's service location for up to 24-hours. The user can disconnect and reconnect within each 24-hour period from the same location with no additional charge.

In addition to being able to connect to hundreds of hot spot locations, a monthly service plan also may include

- Wi-Fi "sniffer" software that checks the airwaves for available wireless networks.

- Location directory to find service locations.

- Web-based account management, allowing the user to manage his or her own account. Online and 800 number customer support is available.

- Save and manage security keys and network settings.

- Built-in personal virtual private network (VPN) to ensure secure connections to corporate networks.

Wi-Fi services are also offered by traditional ISPs such as Earthlink, giving customers another way to access their services while at public places where they can use the precious little time they have catching up on e-mail or connecting to the Internet.

Summary

Some cellular service providers are offering wireless data access via an integrated GPRS/EDGE/IEEE 802.11b service offering. By combining the benefits of their existing 2.5/3G and Wi-Fi networks, they expect to give customers what they want most from wireless data services: ubiquitous coverage and high speed. Customers will have seamless service between the two wireless networks via a combo PC card for notebook computers that provides access to both GPRS/EDGE and IEEE 802.11b networks, giving them the ability to move between the two environments without having to change cards. The existing GPRS and upcoming EDGE networks provide wide area coverage for applications where customers need constant access to such applications as e-mail and calendar, whereas Wi-Fi networks available in convenient public locations allow them to spend time accessing larger data files and multimedia messages or browse the Web.

See also

Enhanced Data Rates for Global Evolution

General Packet Radio Service

Wireless Fidelity

WIRELESS INTERNETWORKING

Corporations are making greater use of wireless technologies for extending the reach of LANs where a wired infrastructure is absent, impractical, or too costly to install. Wireless bridges and routers can extend data communications between buildings in a campus environment or between buildings in a metropolitan area. A variety of technologies may be used for extending the reach of LANs to remote locations, including microwave, laser, and spread spectrum. All rely on directional antennas at each end and a clear line of site between locations.

Wireless Bridges

Short-haul microwave bridges, for example, provide an economical alternative to leased lines or underground cabling.

Because they operate over very short distances—less than a mile—and are less crowded, they are less stringently regulated and have the additional advantage of not requiring an FCC license.

The range of a bridge is determined by the type of directional antenna. A four-element antenna, for example, provides a wireless connection of up to 1 mile. A ten-element antenna provides a wireless connection of up to 3 miles.

Directional antennas require a clear line of sight. To ensure accurate alignment of the directional antennas at each end, menu-driven diagnostic software is used. Once the antennas are aligned and the system ID and channel are selected with the aid of configuration software, the system is operational. Front-panel light-emitting diodes (LEDs) provide a visual indication of link status and traffic activity. The bridge unit has a diagnostic port, allowing performance monitoring and troubleshooting through a locally attached terminal or remote computer connected via modem.

Because they are fully compatible with the IEEE 802.3 Ethernet Standard, microwave bridges support all Ethernet functionality and applications without the need for any special software or network configuration changes. For Ethernet connections, the interface between the microwave equipment and the network is virtually identical to that between the LAN and any cable medium, where retiming devices and transceivers at each end of the cable combine to extend the Ethernet cable segments. Typically, microwave bridges support all Ethernet media types via AUI connectors for thick Ethernet (10Base5), 10Base2 connectors for thin Ethernet, and twisted-pair connectors for 10BaseT Ethernet. These connections allow microwave bridges to function as an access point to wired LANs.

Like conventional Ethernet bridges, microwave bridges perform packet forwarding and filtering to reduce the amount of traffic over the wireless segment. The microwave bridges contain Ethernet address filter tables that help to reduce the level of traffic through the system by passing only the Ethernet packets bound for an inter- or intrabuilding destination over the wireless link. Since the bridges are "self-learning," the filter tables are automatically filled with Ethernet addresses as the bridge learns which devices reside on its side of the link. In this way, Ethernet packets not destined for a remote address remain local. The table is dynamically updated to account for equipment added or deleted from the network. The size of the filter table can be 1000 entries or more depending on vendor.

With additional hardware, microwave bridges have the added advantage of pulling double duty as a backup to local T1/E1 facilities. When a facility degrades to a preestablished error-rate threshold or is knocked out of service entirely, the traffic can be switched over to a wireless link to avoid loss of data. When line quality improves or the facility is restored to service, the traffic is switched from the wireless link to the wireline link.

Wireless Routers

Wireless remote access routers scale wireless connect geographically disbursed LANs by creating a wireless wide area network (WAN) over which network traffic is routed at distances of 30 miles or more using microwave, laser, or spread-spectrum technology.

Applications of wireless routers include remote-site LAN connectivity and network service distribution. Organizations with remote offices such as banks, health care entities, government agencies, schools, and other service organizations can connect their distributed computing resources with wireless routers. Industrial and manufacturing companies can reliably and cost-effectively connect factories, warehouses, and research facilities. Network service providers can distribute Internet, Very Small Aperture Terminal (VSAT), and other network services to their customers. Performance is comparable to commonly used wired WAN connections, approaching T1 speeds with a 1.3-Mbps data rate.

Unlike wireless bridges, which simply connect LAN segments into a single logical network, wireless routers function at the network layer with IP/IPX routing, permitting the network designer to build large, high-performance, manageable networks. Wireless routers are capable of supporting star, mesh, and point-to-point topologies that are implemented with efficient Media Access Control (MAC) protocols. These topologies can even be combined in an internetwork.

A polled protocol (star topology) provides efficient shared access to the channel even under heavy loading (Figure W-4). In the star topology, remote stations interconnect with the central base station and with other remote stations through the base station. Only one location needs line of site to the remotes. Networks and workstations at each location tie into a common internetwork. The maximum range between the central base station and remote stations is approximately 15 miles.

For small-scale networks, a CSMA/CA protocol supports a mesh topology (Figure W-5). In the mesh topology,

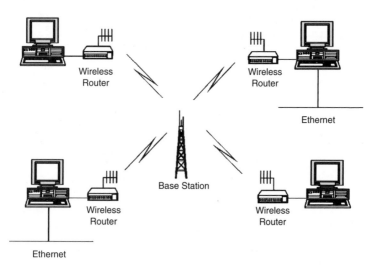

Figure W-4   Star topology of a router-based wireless WAN.

each site must be line of sight to every other. The CSMA/CA protocol ensures efficient sharing of the radio channel. The range with omnidirectional antennas is up to 3.5 miles.

The point-to-point topology is useful where there are only a few sites (Figure W-6). A router also can function as a repeater link between sites. The single-hop node-to-node range is up to 30 miles depending on such factors as terrain and antennas, with a multiple hop range extending on the order of a hundred miles. Clusters of nodes also can be connected using a point-to-point protocol when building large-scale internetworks.

Summary

Interconnecting LANs with bridges and routers that use wireless technologies is an economical alternative to leased lines and carrier-provided services. Installation, setup, and maintenance are fairly easy with the graphical management tools provided by vendors.

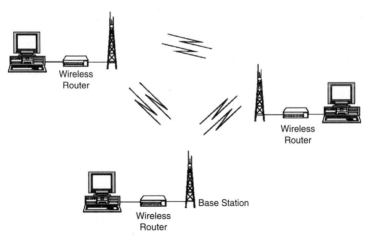

Figure W-5   Mesh topology of a router-based wireless WAN.

Figure W-6   A point-to-point topology of a router-based wireless WAN.

See also

Access Points

Bridges

Routers

Wireless Management Tools

WIRELESS INTRANET ACCESS

For many businesses today, mobile staff require immediate access to up-to-date company information without having to physically plug into a LAN to access the corporate intranet. The advantages of a wireless connection are obvious: Newly gathered information can be entered directly into the corporate information system, and in the opposite direction, the most current information, such as product inventory, documentation, and pricing, can be retrieved immediately from the corporate information system—regardless of the employee's location in a building or campus, which may change continuously throughout the day.

An intranet is a private, secure version of the public Internet that enables employees to access corporate databases, applications, and services over a TCP/IP network through a Web browser. These resources reside on local or remote servers that can be accessed when users log onto the corporate network with their user name and password.

After the authentication process, users click the Web browser icon displayed on the desktop, which opens the home page of the intranet. From there they can go to department Web pages to obtain specific information, access a directory to find the address of another employee, go to the order entry system to add a customer order, check the inventory database for product availability, or launch a query to the decision support system (DSS) that returns a breakdown of sales by region.

Virtual Private Networks

Using a portable wireless device, such as a laptop or PDA equipped with a Web browser, gives employees instant access to resources on the corporate intranet to do their job, including messaging applications and productivity tools. For remote users on the move, connectivity to the corporate intranet can be accomplished through a commercial wireless service that provides a means of accessing a company's VPN gateway, where authentication and encryption are applied to ensure secure wireless access to intranet resources.

WAP Portals

For employees in the field, such as salespeople and maintenance crews, service providers offer wireless connectivity to corporate VPN gateways that are cabled to a LAN that provides intranet access. But because of bandwidth constraints on wireless WANs, data reduction techniques must be used to facilitate information retrieval from the intranet. Overcoming this throughput limitation can be accomplished by formatting selected intranet content in an abbreviated form using the Wireless Markup Language (WML), which can be accessed even by such memory-constrained devices as PDAs and Internet-enabled cell phones. The WML-formatted content can be accessed via the company's Wireless Application Protocol (WAP) portal.

Wi-Fi

By installing IEEE 802.11b (Wi-Fi) wireless access points throughout the building and equipping laptop computers with wireless network cards, companies can provide employees with wireless connectivity to the LAN at speeds of up to 11 Mbps over the unlicensed 2.4-GHz frequency band. The laptops communicate wirelessly with the access point, which is wired to the LAN hub with Category 5 cable. Once employees turn on their laptops, log onto the network, and pass authentication, they can do anything their colleagues can do with their wired desktop computers, including access the intranet with a Web browser.

Bluetooth

Another way to access the corporate intranet is via a Bluetooth-enabled access point cabled to the LAN. Like Wi-Fi, Bluetooth operates in the 2.4-GHz band but at the comparatively slow rate of 30 to 400 kbps across a range of only 30 feet. Bluetooth supports "piconets" that link laptops, PDAs, mobile phones, and other devices on an as-needed basis. It improves on infrared in that it does not require a line of sight between the devices and has greater range than infrared's 3 to 10 feet.

While Bluetooth does not have the power and range of a full-fledged LAN, its master-slave architecture does permit the devices to face different piconets, in effect, extending the range of the signals beyond 30 feet.

Like Wi-Fi, Bluetooth uses spread-spectrum technology, but since it is a frequency hopper, there is little chance that 2.4-GHz Wi-Fi devices will interfere with Bluetooth. The Bluetooth standard specifies a very fast frequency hop rate—1600 times a second among 79 frequencies—so it will be the first to sense problems and act to steer clear of interference from other 2.4-GHz devices.

Packaged Solutions

Packaged wireless intranet solutions are becoming available. Accessible through cell phones and PDAs such as Palm's products, these solutions lets users access their intranets to collaborate on projects, documents, schedules, and contacts. Users automatically receive a company Web mail address when the system administrator provides them with access to the corporate intranet. This allows users to send and receive e-mail and includes functionality that enables employees to create folders and rules, attach files, filter e-mail, send sort messages, and import contacts from the contact manager.

Portal vendors offer wireless services that let mobile users have unlimited access to enterprise information portals and resources. Plumtree, for example, offers its Wireless Device Server, an add-on to its Internet Device Server portal package that enables wireless access to portal e-mail, intranet databases, the Web, and other resources. Wireless Device Server lowers the cost of wireless, enabling a corporate intranet by automatically detecting devices and presenting content in the appropriate WML or HTML format. This dispenses with the need for a company to maintain separate sites for different devices.

For building intranet content intended specifically for wireless access, companies can even turn to their database vendor. Oracle, for example, offers a wireless edition of its Oracle 9i Application Server that provides tools for creating wireless Internet/intranet content, services, and applications.

Summary

As wireless technologies and services become more ubiquitous, vendors are addressing mobile users' needs for products that allow them to access resources on their company's intranet. Among other things, these products adapt

Web-based applications to handheld devices, manage wireless connectivity, and deliver end-to-end security.

See also

> Access Points
>
> Bluetooth
>
> Wireless Fidelity
>
> Wireless Internets

WIRELESS IP

For a long time, the term *wireless IP* was interchangeable with Cellular Digital Packet Data (CDPD), which supports data over analog cellular networks at 19.2 kbps. As cellular technologies progressed, however, describing wireless IP merely as CDPD is wholly inadequate. Today, wireless IP refers to a new class of wireless data platform that enables wireless carriers for the first time to offer premium-pay, business-quality IP services such as mobile VPNs, enhanced security, location-based services in General Packet Radio Service (GPRS)/Enhanced Data for Global Evolution (EDGE), Universal Mobile Telecommunication System (UMTS) core networks, and Code Division Multiple Access (CDMA) 3G networks.

Wireless IP switches enable mobile operators to bring the service richness of data services to the mobile environment, offering a whole new class of mobile data services that promise to increase mobile worker productivity. These switches also seamlessly transport user traffic from the mobile data network onto public data networks such as the Internet. In this role, wireless IP switches perform the function of the Packet Data Serving Node/Home Agent (PDSN/HA) that supports CDMA 2000 wireless networks and a Gateway GPRS Support Node (GGSN), which supports GPRS and UMTS.

The tunnel switching capability of these new switches supports mobile VPNs in both GPRS and CDMA 3G environments. Once traffic is offered to the network from a customer location, the wireless IP switch can mediate between the different tunnel types and decide whether or not to convert them into other tunnel types and provide access to private networks or terminate the tunnel and provide plain Internet access.

Advanced IP address management is available for mobile customers who want to access the private network from wireless carrier radio access network. The wireless IP switch uses network address translation (NAT) and port address translation (PAT) to switch that private address to a public address. Under the administrative control of the service provider, a customer is allowed access to certain actions (i.e., monitor, configure, change, create, etc.) to customize its services.

Wireless IP switches offer compression support, which is essential to mobile VPNs and Point-to-Point Protocol (PPP) performance because bandwidth is usually limited. Compression reduces traffic volume, which helps boost access performance while reducing packet fragmentation, thus decreasing delay time.

To ensure that priority applications get the treatment they deserve, wireless IP switches provide class of service (CoS) and quality of service (QoS) on IP networks, as well as Multiprotocol Label Switching (MPLS) path mapping and differentiated services marking.

End-to-end security is important for establishing user confidence in network-based intranets and extranets. Wireless IP switches are capable of supporting triple-DES (3DES) encryption for tens of thousands of flows simultaneously while maintaining full line rate throughput without any delays.

Virtual routing is the ability to create individual instances of routers for customers and providers. This functionality greatly simplifies the routing configuration for the service provider because the service provider obtains access to all these customers through a single router interface. With the virtual router capability and the ability to interconnect virtual routers within the chassis configuration of the wireless IP switch, a user can create very complex wireless IP core network topologies within the same platform.

Security is handled by the wireless IP switch's policy-driven stateful firewall, which provides granular firewall policies with "follow me" characteristics. Thus, wherever a customer accesses the network, the wireless IP switch will gather and implement the specific policies associated with that user. The wireless IP switch also supports other firewall capabilities, including intrusion detection and denial of service protection.

The wireless IP switch's element management system (EMS) allows subscribers to select their service providers as well as their services from a Web browser. This reduces service provider operations costs, reduces service turn-on times, and improves end-customer satisfaction by allowing subscribers to provision their own services.

With policy-based provisioning, the wireless IP switch delivers individualized services based on directory profiles. This eliminates the need for tedious manual router configuration and provides single-point IP service delivery and control while decreasing the number of devices to manage and reducing the cost of operations. Policies follow users throughout network.

Summary

Businesses have long been skeptical of using IP for critical and real-time applications and have been wary about the security of IP-based networks as well, especially over wireless networks. Vendors such as Lucent and Nortel, however, have taken all these concerns into account with their wireless IP switches. With flexible configuration options for service modules, interface cards, switch fabrics, and control modules, today's wireless IP switches provide a wide range of service choices. High service availability is ensured with redundant power supplies, switch fabrics, service modules, and processors. Instant failover functionality is designed into the hardware so that if a primary component fails, the backup component takes over immediately without any interruption of service. Interface cards support high-availability features such as port failover, rerouting, and SONET automatic protection switching (APS). Software provides full-service failover of virtual routers (including forwarding tables), firewall, QoS, and VPN policies.

See also

Cellular Data Communications

WIRELESS LAN SECURITY

Wireless links are inherently less secure than copper or fiber media. In copper media, the wires are twisted together to minimize the amount of signal radiation that can be picked up by eavesdroppers. Some types of copper media used in data centers include shielding, which prevents any signal radiation. In fiber media, the difference in the density of the core and surrounding cladding prevents light from radiating away and being picked up by eavesdroppers. Of course, it is still possible to intercept communications over copper and fiber media, but it is much easier to do so when the communication occurs over unprotected wireless links.

The biggest threat to the security of a wireless LAN is the failure to use any form of security in the belief that the information traversing the wireless link is not important enough to safeguard. But leaving the wireless link unprotected can have unintended consequences. For example, a hacker might not be interested in intercepting a telecommuter's communications at all but may see the wireless link as an opportunity to access a VPN connection to the telecommuter's corporate network, where sensitive information is stored on application servers and distributed databases. Some companies are diligent in this regard and do not allow telecommuters to use wireless technologies at home, considering it a willful security breach if they do so.

With Wi-Fi systems, the first line of defense in securing a wireless link is to enable Wired Equivalent Privacy (WEP). This will deter the casual intruder, but experienced hackers have been known to break into WEP-enabled systems in 15 minutes or less. Although WEP is being improved continually, additional measures must be taken to strengthen security. The Wi-Fi Alliance recommends one or more of the following:

- Turn WEP on and manage the WEP key by changing the default key and, subsequently, changing the WEP key daily to weekly.

- Turn windows sharing off on sensitive files and directories.

- Protect access to drives and folders with passwords.

- Change the default Service Set Identifier (SSID) that comes with the product.

- Use session keys if available in the product.

- Use MAC address filtering if available in the product.

- Use a VPN system. Although it would require a VPN server, the VPN client is already included in many operating systems such as Windows 98 Second Edition, Windows 2000, and Windows XP.

These steps alone, however, are not enough to completely protect the wireless network. The SSID, for example, is a two-edged sword. This is a plaintext identifier of no more than 32 characters that gets attached to the header of packets sent over a wireless LAN. The SSID differentiates one wireless LAN from another, so all access points and all devices attempting to connect to a specific WLAN must use the same SSID, as if it were a password. A device will not be permitted to join a wireless LAN unless it can provide the unique SSID. But because hacker tools can sniff the SSID in plaintext from a packet, it does not really enhance network security.

Encryption Enhancements

As noted, Wi-Fi systems implement an encryption scheme known as WEP that protects data transmitted through the air, making it difficult (but no impossible) for wireless stations that do not operate on the same encryption key to obtain the transmitted data. The IEEE 802.11 Standard defines the encryption key to be 64 bits in length, 40 bits of which can be defined by the user as the selected secret key. The other 24 bits are generated by the system and change on each successive transmission to make sure that the actual encryption key changes constantly. This cryptography scheme is known as RC4.

Some vendors have added to this standard encryption scheme the option to use longer encryption keys. For example, they may use 128-bit keys, 104 bits of which are the user-defined secret key. The longer the key, the more time it takes for a hacker to obtain the key using brute force methods. To prevent comprising the encryption key if the notebook's PC card is stolen, some Wi-Fi vendors even avoid storing the keys in the wireless PC card's flash ROM (read-only memory).

Without the correct encryption key, a wireless client station cannot communicate with an access point and, therefore, cannot get onto the network, much less monitor the traffic from a neighboring station.

But not even these safeguards are entirely hacker-proof. There are now tools freely available on the Internet specifically designed for WEP key hacking, eliminating the need to use brute-force methods. These tools exploit weaknesses in the key-scheduling algorithm of RC4. Once the tool captures the data traffic, the key can be derived from it.

The weakness of WEP lies in the predictability of the so-called initialization vector, which is the 24 bits of the encryption key generated by the system that changes on each successive transmission—ostensibly to improve security. But some values of this initialization vector can be predicted for generating "weak keys" that can be used to gain access to the wireless network. All IEEE 802.11–compliant wireless networks are vulnerable to this kind of attack, since they all implement WEP security in a similar manner.

Authentication via RADIUS

Some access points have the added capability to restrict access to the network to those stations whose MAC address is included in a Remote Authentication Dial-In User Service (RADIUS) database. To enable this feature, the access points need to be configured to communicate to the RADIUS server each time a wireless station makes the initial contact.

To add to the availability of the service, the access points can connect to two servers, a primary and a backup, in case the primary is down. The network administrator builds the database at the RADIUS server by including the MAC addresses of all the stations that are allowed access to the network. Stations with MAC addresses that do not appear in this table are not granted access, and the traffic generated by these stations will be filtered out.

To ensure the effectiveness of RADIUS, the network administrator must make sure that all access points are configured to use the RADIUS database for MAC address authentication. To do so, the database has to be populated with the MAC addresses of PC cards. New cards have to be added to the database, and cards that have been reported as stolen or no longer in use must be removed from the database.

Key Change Administration

Regular key changes are important to maintain the integrity of the security system. Some Wi-Fi systems allow for the use of multiple keys in support of dynamic key rollover. Multiple keys active at the same time can cover the period required for all users to rotate to a new key.

To support frequent key change procedures, vendors offer software tools that allow changing the WEP encryption keys remotely. In other words, a network or security administrator is able to transmit a new WEP key or set of keys to client stations and to have them active the next time the PC card driver is loaded. The end-user will not be aware that this key change has taken place and does not even know the exact value of the WEP key.

To allow visitors and guests to use a network where RADIUS-based MAC authentication and WEP encryption are effective, local procedures can be implemented where IT staff can loan PC cards with MAC addresses that are registered in the RADIUS database. These adapters stay on the premises, and a procedure is implemented to enforce the return of the PC card by the guest/visitor to the IT staff. PC cards should be used that do not hold the WEP key in ROM; that way, if a card is not returned, network security is not compromised.

However, if a laptop used for wireless LAN communication is stolen, a key change must be implemented immediately. This is so because if the key is not stored in the PC card's ROM, then it is maintained in the registry of the operating system. In changing the keys as soon as the theft is reported, the risk of unauthorized access to the wireless network is minimized.

Closed Systems

Wi-Fi access points have the capability to filter wireless client stations on the network name, which is a parameter that the wireless client needs to supply in order to attach to a wireless network. If a wireless client station is not configured with the correct network name, this station will not be able to attach to the access point.

However, the Wi-Fi standard does allow wireless clients to attach to any network if no value for this parameter is entered. This is called an "empty string," and it is typically used to access a public access point, such as those available Starbucks cafés. To ensure a reasonable level of security, some products like Agere's ORiNOCO systems are equipped with a so-called closed system option, which is a user-selectable option to reject wireless stations that try to attach to the network using the empty string option. With "Closed System" selected, wireless clients that are not configured with the correct network name cannot access the network.

Firewalls

To protect communications over wireless LANs requires the use of more advanced security techniques, such as a firewall. Firewalls can be standalone devices that are dedicated to safeguarding the enterprise network. Similar functionality can be added to wireless routers, in which case the security features are programmed through the router's operating system. Internet appliances can have firewall capabilities as well, such a DSL routers and cable modems. In addition, there is client software that can be loaded into notebooks and desktop computers that gives users personal control over security (Figure W-7).

A packet-filtering firewall examines all the packets passed to it and then forwards them or drops them based on predefined rules (Figure W-8). The network administrator can control how packet filtering is performed, permitting or denying connections using criteria based on the source and destination host or network and the type of network service.

Summary

By itself, WEP is offers no guarantee of security against hackers. A combination of steps must be taken to create the highest degree of security for a wireless network, including the proper selection of equipment, the implementation of

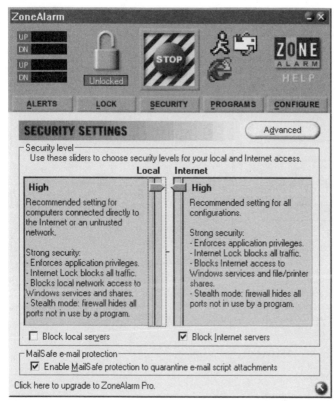

Figure W-7 Firewall software for individual computers allows users to control their own level of security. Shown is ZoneAlarm from Zone Labs, which is available free at the company's Web site for personal use.

RADIUS, and the use of firewalls. With the right security provisions in place, wireless systems will have equal or more privacy than can be expected from existing wired stations. It is essential, however, that proper operational procedures be defined to implement and activate the security options and that there be corporate enforcement of these procedures.

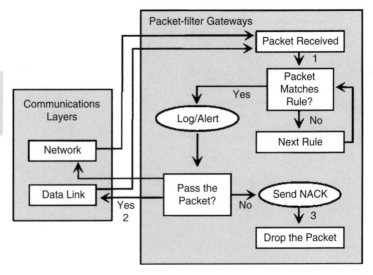

Figure W-8 Operation of a packet-filtering firewall: (1) inbound/outbound packets are examined for compliance with company-defined security rules; (2) packets found to be in compliance are allowed to pass into the network; (3) packets that are not in compliance are dropped.

See also

    Wired Equivalent Privacy
    Wireless Fidelity

WIRELESS LANS

A wireless local area network (WLAN) is a data communications system implemented as an extension—or as an alternative—to a wired LAN. Using a variety of technologies including narrowband radio, spread spectrum, and infrared, wireless LANs transmit and receive data through the air, minimizing the need for wired connections.

Applications

Wireless LANs have become popular in a number of vertical markets, including health care, retail, manufacturing, and warehousing. These industries have profited from the productivity gains of using handheld terminals and notebook computers to transmit real-time information to centralized hosts for processing. Wireless LANs allow users to go where wires cannot always go. Specific uses of wireless LANs include

- Hospital staff members can become more productive when using handheld or notebook computers with a wireless LAN capability to deliver patient information, regardless of their location.

- Consulting or accounting audit teams, small workgroups, or temporary office staff can use wireless LANs to quickly set up for ad-hoc projects and become immediately productive.

- Network managers in dynamic enterprise environments can minimize the overhead cost of moves, adds, and changes with wireless LANs, since the need to install or extend wiring is eliminated.

- Warehouse workers can use wireless LANs to exchange information with central databases, thereby increasing productivity.

- Branch office workers can minimize setup requirements by installing preconfigured wireless LANs.

- Wireless LANs are an alternative to cabling multiple computers in the home.

While the initial investment required for wireless LAN hardware can be higher than the cost of conventional LAN hardware, overall installation expenses and life-cycle costs can be significantly lower. Long-term cost savings are greatest in dynamic environments requiring frequent moves, adds, and changes. Wireless LANs can be configured in a variety of topologies to meet the needs of specific applications and installations. They can grow by adding access points and extension points to accommodate virtually any number of users.

Technologies

There are several technologies to choose from when selecting a wireless LAN solution, each with advantages and limitations. Most wireless LANs use spread spectrum, a wideband radio frequency technique developed by the military for use in reliable, secure, mission-critical communications systems. To achieve these advantages, the signal is spread out over the available bandwidth and resembles background noise that is virtually immune from interception.

There are two types of spread-spectrum radio: frequency hopping and direct sequence. Frequency-hopping spread spectrum (FHSS) uses a narrowband carrier that changes frequency in a pattern known only to the transmitter and receiver. Properly synchronized, the net effect is to maintain a single logical channel. To an unintended receiver, FHSS appears to be short-duration impulse noise.

Direct-sequence spread spectrum (DSSS) generates a redundant bit pattern for each bit to be transmitted and requires more bandwidth for implementation. This bit pattern, called a "chip" (or "chipping code"), is used by the receiver to recover the original signal. Even if one or more bits in the chip are damaged during transmission, statistical techniques embedded in the radio can recover the original data without the need for retransmission. To an unintended receiver, DSSS appears as low-power wideband noise.

Another technology used for wireless LANs is infrared (IR), which uses very high frequencies that are just below visible light in the electromagnetic spectrum. Like light, IR cannot penetrate opaque objects—to reach the target system, the waves carrying data are sent in either directed (line-of-sight) or diffuse (reflected) fashion. Inexpensive directed systems provide very limited range of not more than 3 feet and typically are used for personal area networks but occasionally are used in specific wireless LAN applications. High-performance directed IR is impractical for mobile users and therefore is used only to implement fixed subnetworks. Diffuse IR wireless LAN systems do not require line-of-sight transmission, but cells are limited to individual rooms. As with spread-spectrum LANs, IR LANs can be extended by connecting the wireless access points to a conventional wired LAN.

Operation

As noted, wireless LANs use electromagnetic waves (radio or infrared) to communicate information from one point to another without relying on a wired connection. Radio waves are often referred to as "radio carriers" because they simply perform the function of delivering energy to a remote receiver. The data being transmitted are superimposed on the radio carrier so that they can be extracted accurately at the receiving end. This process is generally referred to as "carrier modulation." Once data are modulated onto the radio carrier, the radio signal occupies more than a single frequency, since the frequency or bit rate of the modulating information adds to the carrier.

Multiple radio carriers can exist in the same space at the same time without interfering with each other if the radio waves are transmitted on different frequencies. To extract data, a radio receiver tunes into one radio frequency while rejecting all other frequencies.

In a typical wireless LAN configuration, a transmitter/receiver (transceiver) device, called an "access point," connects to the wired network from a fixed location using standard cabling. At a minimum, the access point receives, buffers and transmits data between the wireless LAN and the wired network infrastructure. A single access point can support a small group of users and can function within a range of less than 100 to several hundred feet. The access point (or the antenna attached to the access point) is usually mounted high but may be mounted essentially anywhere that is practical as long as the desired radio coverage is obtained.

Users access the wireless LAN through wireless LAN adapters. These adapters provide an interface between the client network operating system (NOS) and the airwaves via an antenna. The nature of the wireless connection is transparent to the NOS.

Configurations

Wireless LANs can be simple or complex. The simplest configuration consists of two PCs equipped with wireless adapter cards, which form a network whenever they are within range of one another (Figure W-9). This peer-to-peer network requires no administration. In this case, each client would only have access to the resources of the other client and not to a central server.

Installing an access point can extend the operating range of the wireless network, effectively doubling the range at which the devices can communicate. Since the access point is connected to the wired network, each client would have access to the server's resources as well as to other clients (Figure W-10). Each access point can support many clients—the specific number depends on the nature of the transmissions involved. In some cases, a single access point can support up to 50 clients.

Depending on the manufacturer and the frequency band used in its products, access points have an operating range of about 500 feet indoors and 1000 feet outdoors. In a very large

Figure W-9 A wireless peer-to-peer network created between two notebook computers equipped with external wireless adapters.

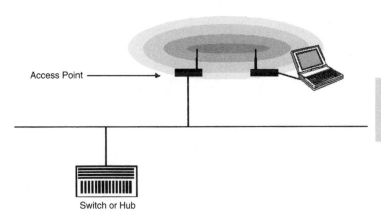

Access Point

Switch or Hub

Figure W-10 A wireless client connected to the wired LAN via an access point.

facility such as a warehouse or on a college campus, it probably will be necessary to install more than one access point
(Figure W-11). Access point positioning is determined by a site
survey. The goal is to blanket the coverage area with overlapping coverage cells so that clients can roam throughout the
area without ever losing network contact. Access points hand
the client off from one to another in a way that is invisible to
the client, ensuring uninterrupted connectivity.

To solve particular problems of topology, the network
designer might choose to use extension points (EPs) to augment
the network of access points (Figure W-12). These devices look
and function like access points (APs), but they are not tethered
to the wired network, as are APs. EPs function as repeaters by
boosting signal strength to extend the range of the network by
relaying signals from a client to an AP or another EP.

Another component of wireless LANs is the directional
antenna. If a wireless LAN in one building must be connected
to a wireless LAN in another building a mile away, one solution might be to install a directional antenna on the two
buildings—each antenna targeting the other and connected
to its own wired network via an access point (Figure W-13).

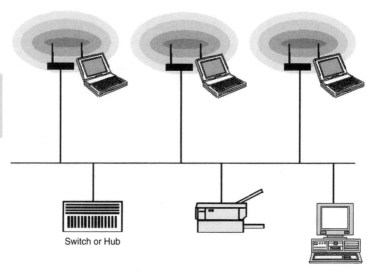

Figure W-11 Multiple access points extend wireless coverage and enable roaming.

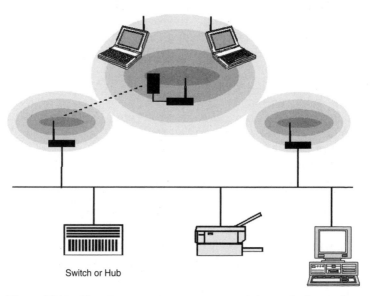

Figure W-12 Use of an extension point to extend the reach of a wireless network.

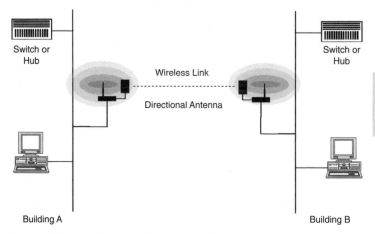

Figure W-13 A directional antenna can be used to interconnect wireless LANs in different buildings.

Wireless LAN Standards

There are several wireless LAN standards, each suited for a particular environment: IEEE 802.11a and 802.11b, HomeRF, and Bluetooth.

For the residential and office environments, the IEEE 802.11a offers a data transfer rate of up to 11 Mbps at a range of up to 300 feet from the base station. It operates in the 2.4-GHz band and transmits via the direct-sequence spread-spectrum method. Multiple base stations can be linked to increase that distance as needed, with support for multiple clients per access point. IEEE 802.11a specifies the 5-GHz frequency band, offering a data transfer rate of up to 54 Mbps.

The HomeRF 2.0 Standard draws from IEEE 802.11b and Digital Enhanced Cordless Telecommunication (DECT), a popular standard for portable phones worldwide. Operating in the 2.4-GHz band, HomeRF was designed from the ground up for the home market for both voice and data. It offers throughput rates comparable to IEEE 802.11b and supports the same kinds of terminal devices in both point-to-point and multipoint configurations. HomeRF transmits at up to 10 Mbps over a range of about 150 feet from the base station, which makes it suitable for the average home. HomeRF transmits using spread-spectrum frequency hopping; that is, it hops around constantly within its prescribed bandwidth. When it encounters interference, like a microwave oven or an adjacent wireless LAN, it adapts by moving to another frequency.

The key advantage that HomeRF has over IEEE 802.11b in the home environment is its superior ability to adapt to interference from devices like portable phones and microwaves. As a frequency hopper, it coexists well with other frequency-hopping devices that proliferate in the home. Another advantage of HomeRF is that it continuously reserves a chunk of bandwidth via "isochronous channels" for voice services. Speech quality is high; there is no clipping while the protocol deals with interference.

The IEEE 802.11b Standard does not include frequency hopping. In response to interference, IEEE 802.11b simply retransmits or waits for the higher-level TCP/IP protocol to sort out signal from noise. This works well for data but can result in voice transmissions sounding choppy. Voice and data are treated the same way, converting voice into data packets but offering no priority to voice. This results in unacceptable voice quality. Another problem with IEEE 802.11b is that its Wired Equivalent Protocol (WEP) encryption, designed to safeguard privacy, has had problems living up to its claim.

Bluetooth also operates in the 2.4-GHz band but was not created originally to support wireless LANs; it was intended as a replacement for cable between desktop computers, peripherals, and handheld devices. Operating at the comparatively slow rate of 30 to 400 kbps across a range of only 30 feet, Bluetooth supports "piconets" that link laptops, PDAs, mobile phones, and other portable devices on an as-needed basis. It improves on infrared in that it does not require a line of sight between the devices and has greater range than infrared's 3 to 10 feet. Bluetooth also supports voice channels.

While Bluetooth does not have the power and range of a full-fledged LAN, its master-slave architecture does permit the devices to face different piconets, in effect, extending the range of the signals beyond 30 feet. Like HomeRF, Bluetooth is a frequency hopper, so devices that use these two standards can coexist by hopping out of each other's way. Bluetooth has the faster hop rate, so it will be the first to sense problems and act to steer clear of interference from HomeRF devices.

The three standards each have particular strengths that make them ideal for certain situations, as well as specific shortcomings that render them inadequate for use beyond their intended purpose:

- While suited for the office environment, IEEE 802.11b is not designed to provide adequate interference adaptation and voice quality for the home. Data collisions force packet retransmissions, which is fine for file transfers and print jobs but not for voice or multimedia that cannot tolerate the resulting delay.

- HomeRF delivers an adequate range for the home market but not for many small businesses. It is better suited than IEEE 802.11b for streaming multimedia and telephony, applications that may become more important for home users as convergence devices become popular.

- Bluetooth does not provide the bandwidth and range required for wireless LAN applications but instead is suited for desktop cable replacement and ad-hoc networking for both voice and data within the narrow 30-foot range of a piconet.

Wireless LAN technology is continually improving. The IEEE 802.11b Standard developers seek to improve encryption (IEEE 802.11i) and make the standard more multimedia friendly (IEEE 802.11e). Dozens of vendors are shipping IEEE 802.11b products, and the standard's proliferation in corporate and public environments is a distinct advantage. An office worker who already has an IEEE 802.11b–equipped notebook will not likely want to invest in a different network for the home.

Furthermore, the multimedia and telephony applications HomeRF advocates tout have not yet arrived to make the technology a compelling choice. Although HomeRF currently beats IEEE 802.11b in terms of security, this is not a big issue in the home. For these and other reasons, industry analysts predict that IEEE 802.11b will soon overtake HomeRF in the consumer marketplace, especially since the price difference between the two has just about reached parity.

Summary

Once expensive, slow, and proprietary, wireless LAN products are now reasonably fast, standardized, and priced for mainstream business and consumer use. Wireless LAN configurations range from simple peer-to-peer topologies to complex networks offering distributed data connectivity and roaming. To solve problems of vendor interoperability, the Wi-Fi Alliance offers a certification program that tests vendor-submitted products. Those that pass WECA's battery of tests receive the right to bear the Wireless Fidelity (Wi-Fi) logo of interoperability.

See also

Bluetooth

Digital Enhanced Cordless Telecommunication

Home RF

Infrared Networking

Spread-Spectrum Radio

Wireless Fidelity

Wireless Management Tools

WIRELESS LOCAL LOOPS

A wireless local loop (WLL) is a generic term for an access system that uses wireless links rather than conventional copper wires to connect subscribers to the local telephone company's switch. Wireless local loop—also known as "fixed wireless access" (FWA) or "simply fixed radio"—entails the use of analog or digital radio technology to provide telephone, facsimile, and data services to business and residential subscribers.

WLL systems provide rapid deployment of basic phone service in areas where the terrain or telecommunications development makes installation of traditional wireline service too expensive. WLL systems can be easily integrated into the wireline Public Switched Telephone Network (PSTN) and usually can be deployed within a month of equipment delivery, far more quickly than traditional wireline installations, which can take several months for initial deployment and years to grow capacity to meet the continually growing demand for communication services.

WLL solutions include analog systems for medium- to low-density and rural applications. For high-density, high-growth urban and suburban locations, there are WLL solutions based on Code Division Multiple Access (CDMA). Time Division Multiple Access (TDMA) and Global System for Mobile (GSM) telecommunications) systems are also offered. In addition to being able to provide higher voice quality than analog systems, digital WLL systems are able to support higher-speed fax and data services.

WLL technology is also generally compatible with existing operations support systems (OSS), as well as existing transmission and distribution systems. WLL systems are scalable, enabling operators to leverage their previous infrastructure investments as the system grows.

WLL subscribers receive phone service through a radio unit linked to the PSTN via a local base station. The radio unit consists of a transceiver, power supply, and antenna. It operates off ac or dc power and may be mounted indoors or outdoors, and it usually includes battery backup for use during line power outages. On the customer side, the radio unit connects to the premises wiring, enabling the customer to use existing phones, modems, fax machines, and answering devices (Figure W-14).

The WLL subscriber has access to all the usual voice and data features, such as caller ID, call forwarding, call waiting, three-way calling, and distinctive ringing. Some radio units provide multiple channels, which are equivalent to having multiple lines. The radio unit offers service operators

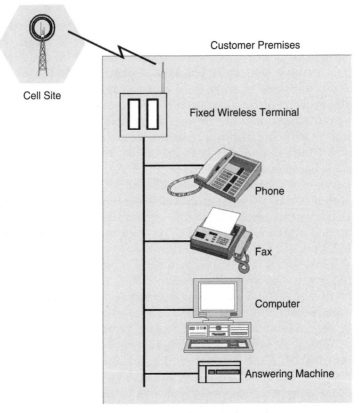

Figure W-14 The fixed wireless terminal is installed at the customer location. It connects several standard terminal devices (telephone, answering machine, fax, computer) to the nearest cell site base transceiver station (BTS).

the advantage of over-the-air programming and activation to minimize service calls and network management costs.

The radio unit contains a coding and decoding unit that converts conventional speech into a digital format during voice transmission and back into a nondigital format for reception. Many TDMA-based WLL systems use the 8-kbps Enhanced Variable Rate Coder (EVRC), a Telecommunications Industry Association (TIA) standard (IS-127). EVRC provides benefits to both network operators and subscribers.

For operators, the high-quality voice reproduction of the EVRC does not sacrifice the capacity of a network or the coverage area of a cell site. An 8-kbps EVRC system, using the same number of cell sites, provides network operators with greater than 100 percent additional capacity than the 13-kbps voice coders that are deployed in CDMA-based WLL systems. In fact, an 8-kbps EVRC system requires at least 50 percent fewer cell sites than a comparable 13-kbps system to provide similar coverage and in-building penetration.

For subscribers, the 8-kbps EVRC uses a state-of-the-art background noise suppression algorithm to improve the quality of speech in noisy environments typical of urban streets, where there is heavy pedestrian and vehicular traffic. This also is an advantage compared with traditional landline phone systems, which do not have equivalent noise suppression capabilities.

Depending on vendor, the radio unit also may include special processors to enhance call privacy on analog WLL systems. Voice privacy is enhanced through the use of a digital signal processor (DSP)–based speech coder, an echo canceler, a data encryption algorithm, and an error-detection/correction mechanism. To prevent eavesdropping, the low-bit-rate encoded speech data are encrypted using a private key algorithm, which is randomly generated during a call. The key is used by the DSPs at both ends of the communications link to decrypt the received signal. The use of DSPs in the radio units of analog WLL systems also improves fax and data transmission.

WLL Architectures

WLL systems come in several architectures: a PSTN-based direct connect network, a Mobile Telephone Switching Office/Mobile Switching Center (MTSO/MSC)–based network, and proprietary networks.

PSTN-Based Direct Connect

There are several key components of the PSTN direct connect network:

- The PSTN-to-radio interconnect system, which provides the concentration interface between the WLL and the wireline network.

- The system controller (SC), which provides radio channel control functions and serves as a performance monitoring concentration point for all cell sites.

- The base transceiver station (BTS), which is the cell-site equipment that performs the radio transmit and receive functions.

- The fixed wireless terminal (FWT), which is a fixed radio telephone unit that interfaces to a standard telephone set, acting as the transmitter and receiver between the telephone and the base station.

- The operations and maintenance center (OMC), which is responsible for the daily management of the radio network and provides the database and statistics for network management and planning.

MTSO/MSC

An MTSO/MSC-based network contains virtually the same components as the PSTN direct connect network, except that the MTSO/MSC replaces the PSTN-to-radio interconnect system. The key components of an MTSO/MSC-based network are

- Mobile telephone switching office/mobile switching center (MTSO/MSC), which performs the billing and database functions and provides a T1/E1 interface to the PSTN.

- Cell-site equipment, including the base transceiver station (BTS).

- Fixed wireless terminal (FWT).

- Operations and maintenance center (OMC).

For digital systems such as GSM and CDMA, the radio control function is performed at the base station controller (BSC) for GSM or the centralized base site controller (CBSC) for CDMA.

In GSM systems, there is a base station system controller (BSSC), which includes the base station controller (BSC) and the transcoder. The BSC manages a group of BTSs, acts as the digital processing interface between the BTSs and the MTSO/MSC, and performs GSM-defined call processing.

In CDMA systems, there is a centralized base site controller (CBSC), which consists of the mobility manager (MM) and the transcoder subsystems. The MM provides both mobile and fixed call processing control and performance monitoring for all cell sites as well as subscriber data to the switch.

As in PSTN-based networks, the FWT in MTSO/MSC-based networks is a fixed radio telephone unit that interfaces to a standard telephone set acting as the transmitter and receiver between the telephone and the base station.

Operations and maintenance functions are performed at the OMC. As in PSTN-based networks, the OMC in MTSO/MSC-based networks is responsible for the day-to-day management of the radio network and provides the database and statistics for network management and planning.

The PSTN direct connect network is appropriate when there is capacity on the existing local or central office switch. In this case, the switch continues to provide the billing and database functions, the numbering plan, and progress tones. The MTSO/MSC architecture is appropriate for adding a fixed subscriber capability to an already existing cellular mobile network or for offering both fixed and mobile services over the same network.

Proprietary Networks

While MTSO/MSC-based and PSTN direct connect networks are implemented using existing cellular technologies, proprietary WLL solutions are designed specifically as a replacement for wireline local loops. One of these proprietary solutions is Nortel's Proximity I, which is used in the United Kingdom to provide wireline-equivalent services in the 3.5-GHz band. The TDMA-based system was designed in conjunction with U.K. public operator Ionica, which is the source of the I designation. The I series provides telecommunications service from any host network switch, providing toll-quality voice, data, and fax services. The system is switch-independent and is transparent to dual-tone multifrequency (DTMF) tones and switch features.

The Proximity I system architecture consists of the following main elements:

- Residential service system (RSS), which is installed at the customer premises and provides a wireless link to the base station.

- Base station, which provides the connection between the customer's RSS and the PSTN.

- Operations, administration, and maintenance system, which provides such functions as radio link performance management and billing.

Residential Service System (RSS)

The RSS offers two lines, which can be assigned for both residential and home office use or for two customers in the same 2-kilometer area. Once an RSS is installed, the performance of the wireless link is virtually indistinguishable from a traditional wired link. The wireless link is able to handle high-speed fax and data via standard modems, as well as voice. The system supports subscriber features such as call transfer, intercom, conference call, and call pickup.

The RSS has several components: a transceiver unit, residential junction unit (RJU), network interface unit, and power supply. The transceiver unit consists of an integral 30-

centimeter octogonal array antenna with a radio transceiver encased within a weatherproof enclosure. The enclosure is mounted on the customer premises and points toward the local base station.

The RJU goes inside the house, where it interfaces with existing wiring and telephone equipment. The Proximity I system supports two 32-kbps links for every house, enabling subscribers to have a voice conversation and data connection for fax or Internet access at the same time. At this writing, work is under way to develop systems that can handle ISDN speeds of 64 kbps and beyond. Further developments will result in RSSs that can handle more lines per unit for medium-sized businesses or apartment blocks.

The network interface unit (NIU), mounted internally or externally, is a cable junction box that accepts connections from customer premises wiring. The unit also provides access for service provider diagnostics and contains lightening protection circuitry.

The power unit is usually mounted internally and connects to the local power supply (110/220 V ac). The power unit provides the dc supply to the transceiver unit. A rechargeable battery takes over in the event of a power failure and is capable of providing 12 hours of standby and 30 minutes of talk time.

Base Station The base station contains the radio frequency (RF) equipment for the microwave link between the customer's RSS and the PSTN, along with subsystems for call signal processing, frequency reference, and network management. This connection is via radio to the RSS and by microwave radio, optical fiber, or wireline to the local exchange. The base station is modular and can be configured to meet a range of subscriber densities and traffic requirements. The base station has several components: transceiver microwave unit, cabinet, power supply, and network management module.

The base station's dual antenna transceiver microwave unit provides frequency conversion and amplification functions.

Each unit provides three RF channels, the frequency of which can be set remotely. The unit can be configured for a maximum of 18 RF channels. The antennas are available in omnidirectional or sectored configurations, depending on population densities and geographic coverage. An omnidirectional system can support 600 or more customers, while a trisectored antenna can serve more than 2000 customers. Base stations in rural areas can be sited up to 20 kilometers from a subscriber's premises.

The base station can be configured with either an internal or external cabinet. The internal cabinet is for location in an equipment room, while the external cabinet is weather-sealed and vandal-proofed for outside locations. Both types of cabinets house the integrated transceiver system, transmission equipment, optional power system, and batteries. A separate power cabinet provides dc power to the base station from the local 110/220 V ac source. This cabinet may include battery backup, with battery management capability and a power distribution panel that provides power for technicians' test equipment.

The network management module is the base station polls individual RSS units to flag potential service degradation. Reports include link bit error rate (BER), signal-to-noise ratio, power supply failure, and the status of the customer standby battery.

The connection from the base stations to the local exchange on the PSTN is via the V5.2 open-standard interface. In addition to facilitating interconnections between multivendor systems, this interface enables operators to take full advantage of Proximity I's ability to maximize spectrum utilization through allocation of finite spectrum on a dynamic per-call basis rather than on a per-customer basis. Concentration allows the same finite spectrum to be shared across a much larger number of customers, producing large savings in infrastructure, installation, and operations costs for the network operator.

Operations, Administration and Maintenance OA&M functions are implemented through an element manager accessed through a field engineering terminal. In Nortel's Proximity I, the element manager is built around Hewlett-Packard's OpenView. Communications with the network of base stations and customer equipment is done through the Airside Management Protocol, which is based on the OSI Common Management Information Protocol (CMIP). The field engineering terminal can operate in a remote operations center but is intended primarily for use by on-site maintenance engineers who are responsible for the proper operation of the base stations.

All the applications software in the customer premises equipment is downloadable from the element manager. This software provides the algorithms that convert analog voice signals into 32-kbps digital ADPCM, which provides toll quality voice transmission. Other applications software includes algorithms for controlling the draw of battery-delivered power in the event of a 110/220 V ac power failure.

Via the Air Interface Protocol, the customer equipment is able to provide the element manager with information about its current status and performance, the most useful of which are measurements taken during the transmission of speech. This allows the management system to flag performance degradation for corrective action.

Summary

Wireless local loops eliminate the need for laying cables and hard-wired connections between the local switch and the subscriber's premises, resulting in faster service startup and lower installation and maintenance costs. And because the subscriber locations are fixed and not mobile, the initial deployment of radio base stations need only provide coverage to areas where immediate demand for service is apparent. Once the WLL system is in place, new customers can be added quickly and easily. Such systems support standard analog as well as digital services and provide the capability to support the evolution to new and enhanced services as the needs of the market evolve.

See also

   Fixed Wireless Access
   Wireless Centrex

WIRELESS LOCAL NUMBER PORTABILITY

Local number portability (LNP) refers to the ability of individuals, businesses, and organizations to retain their existing telephone number(s)—and the same quality of service—when switching to a new local service provider. The concept applies to cellular and other wireless services as well.

The provision of number portability is an obligation that the Telecommunications Act of 1996 imposes on all Local Exchange Carriers (LECs) as a means of fostering a procompetitive, deregulatory national policy framework. By enabling customers to switch to a new service provider without forcing them to change telephone numbers, LNP permits consumers to select a local telephone company based on service, quality, and price rather than on their desire to keep a particular telephone number. Accordingly, all wireline carriers were required to support LNP in 1999.

However, cellular and other wireless carriers were not required to provide telephone number portability at the same time as wireline service providers because of additional technical and competitive burdens but were required to do so by November 2002. But in mid-2002, the FCC extended for 1 year the compliance deadline for wireless carriers to achieve LNP to November 2003.

The decision was prompted by a request from Verizon Wireless to the FCC to eliminate the requirement altogether. Much of the wireless industry supported the petition. Carriers had argued that offering number portability would cost them $1 billion in 2003 and that customers are not wedded to their telephone number, citing the FCC's own statistics, which showed a 30 percent turnover rate among cellular subscribers. Four of the five FCC commissioners denied the petition of Verizon Wireless, voting instead to extend the compliance deadline by another year.

Summary

Wireless LNP is the ability of a user to keep the same phone number when changing service providers. It is generally available now when a customer switches wireline carriers. The FCC and state regulatory agencies contend that allowing consumers to keep their telephone numbers helps mitigate another big problem facing U.S. consumers—the shortage of telephone numbers. LNP is viewed as a way to protect this dwindling resource without having to add more digits to dial.

See also

Federal Communications Commission

WIRELESS MANAGEMENT TOOLS

The setup and management of wireless LANs is typically done with Windows-based tools. ORiNOCO Software Tools from Agere Systems, for example, is a Windows-based site survey tool to facilitate remote management, configuration, and diagnosis of spread-spectrum wireless LANs, specifically Wi-Fi–certified access points and adapters that operate in the 2.4-GHz frequency band.

The tool suite makes it easy for system administrators to monitor the quality of communications at multiple locations in a wireless network (Figure W-15), including access points and clients. It also can be used to verify building coverage, identify coverage patterns, select alternate frequencies, locate and tune around RF interference, and customize network access security. The tool suite offers the following basic functions:

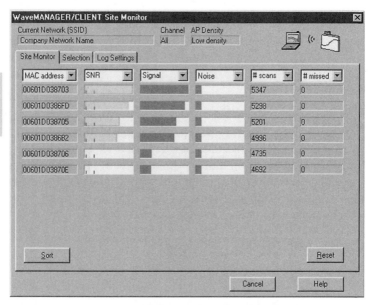

Figure W-15 ORiNOCO Software Tools from Agere Systems, formerly Lucent's WaveMANAGER, provides an administrative graphical user interface through which wireless LANs can be configured, managed, and troubleshooted.

- *Communications indicator* located on the Windows taskbar and providing mobile users graphical, real-time information on the level of communication quality between a wireless station (client) and the nearest access point.

- *Link test diagnostics* the verify the communications path between neighboring client stations as well as between a client station and access points within a wireless cell. With this feature, signal quality, signal-to-noise ratio, and the number of successfully received packets can be displayed.

- *Site monitor* that ensures optimal placement of access points. While carrying a wireless-equipped notebook computer through the facility, the site monitor graphically displays changing communication quality levels with the various access points installed in the building. This tool makes it easy to locate radio dead spots or sources of interference.

- *Frequency select* manages RF channel selection. It enables the user to choose from up to eight different channels in the 2.4-GHz frequency band.

- *Access control table manager* enables the system administrator to provide extra levels of security by restricting access to individual computers in a facility.

- *Authentication* is a MAC address–based authentication scheme using a centrally-maintained RADIUS-based database that records the users who are permitted to have access to the network. Using a MAC address–based authentication also can help prevent unauthorized access of the corporate network by visitors with their own portable wireless devices. If a laptop or desktop computer is stolen or misplaced, the unit's MAC address can be deleted easily and quickly.

For centralized management, the ORiNOCO management tools can be integrated with an HP OpenView-based enterprise network management system. Unlike most other wireless networks that require each access point's parameters to be individually updated, a single ORiNOCO HPOV management station can configure an entire wireless network.

In addition to Windows-based tools for managing wireless LANs, the Simple Network Management Protocol (SNMP) is available for managing the wireless internetwork end to end. The same SNMP tools used to manage the rest of the enterprise network can be used to support management of the bridges and routers on the wireless WAN. The result is a manageable network with reach extending to metropolitan, suburban, rural, remote, and isolated areas.

SNMP is usually implemented using a proxy agent supplied by the vendor. This is an application that continually polls the managed devices for changes in alarm and status information and updates its locally stored management information base (MIB). Events such as major and minor alarms cause the device to generate enterprise specific traps directed to the network management system (NMS). General alarm and status information stored in the MIB is made available to the NMS in response to SNMP's Get and GetNext requests.

Summary

As enterprises and service providers build out their wireless LAN networks and systems, they require tools that help IT managers recognize and troubleshoot network problems before they are reported. This can be accomplished with wireless LAN management tools that provide real-time monitoring of an entire WLAN network spread out over multiple facilities and subnets and enable an IT manager to control, update, and configure the entire wireless network from a single user interface.

See also

Access Points

Wireless Fidelity

Wireless Security

WIRELESS MEDICAL TELEMETRY SERVICE

Wireless Medical Telemetry Service (WMTS) allows operation of potentially lifesaving equipment over the air on an interference-protected basis. Medical transmissions can be unidirectional or bidirectional, but they cannot include a voice or video component. In June 2000, the FCC allocated spectrum and established rules for this service. To minimize regulatory procedures and facilitate deployment, WMTS operators are licensed by rule, like Citizens' Band radio operators. Although the FCC does not issue a formal license, operators must adhere to the rules for this service.

Medical telemetry equipment is used in hospitals and health care facilities to transmit patient measurement data to a nearby receiver, which may permit greater patient mobility and increased comfort. Examples of medical telemetry equipment include heart, blood pressure, and respiration monitors. The use of these devices can allow patients to move around early in their recovery while still being monitored for adverse symptoms. With such devices, one health care worker can monitor several patients remotely, which could reduce health care costs.

The frequency allocation for WMTS provides spectrum where the equipment can operate on a primary basis, increasing the reliability of this important service. The FCC allocated 14 MHz of spectrum for use by medical telemetry equipment in the 608- to 614-MHz, 1395- to 1400-MHz, and 1429- to 1432-MHz frequency bands. This allocation was based on a needs assessment conducted by the American Hospital Association (AHA).

The 608- to 614-MHz band, which corresponds to TV Channel 37, had been reserved for radioastronomy uses. With its action in mid-2000, the FCC elevated medical telemetry to a coprimary status with radioastronomy in this band. The 1395- to 1400-MHz and 1429- to 1432-MHz bands are former government bands reallocated for nongovernment use by the Omnibus Budget Reconciliation Act of 1993. Allocating two separate bands facilitates two-way communications and gives medical telemetry greater flexibility.

Despite existing constraints in these bands—primarily that the entire allocation is unlikely to be available in any individual market—this allocation is flexible enough to allow spectrum to be available for medical telemetry services in all locations while protecting radioastronomy and government operations currently operating in the allocated spectrum. The FCC believed, however, that the benefits of a primary allocation dedicated to this service compensates for the reduced availability of spectrum.

Summary

Medical telemetry devices once operated on a secondary basis to TV broadcasting; that is, they had to tolerate any interference that may have been caused by local television stations' broadcast signals. Users and manufacturers of medical telemetry devices had been able to avoid interference by using TV channels that were vacant locally. In other words, the medical devices used frequencies that local TV stations did not. As these vacant channels started to become used and medical telemetry services expanded, the risk of interference increased. With its own frequency allocations from the FCC, medical telemetry services and equipment now operate more reliably without the risk of interference.

See also

Citizens' Band Radio Service

Telemetry

Wireless E911

WIRELESS MESSAGING

Although "beepers" had been in existence since the 1960s and pagers since the 1980s, true wireless messaging made its debut in 1992 with EMBARC (Electronic Mail Broadcast to A Roaming Computer), built by Motorola as a nationwide one-way message service aimed at mobile executives and field-office workers who required access to e-mail. EMBARC, now defunct, was based on the 931-MHz paging technology Motorola acquired when it purchased Contemporary Communications, Inc., in 1990. At its zenith, EMBARC provided coverage spanning 250 cities in the United States and Canada.

Users were able to send messages from any e-mail system that had an EMBARC interface. Typically, the user accessed

EMBARC on a PC and addressed a message to one or more recipients. The message was then sent through an X.400 gateway to a central switch, which stored and translated it for satellite transmission. The message traveled from the satellite to one or more regional transmission sites that rebroadcast it at 931 MHz to a Motorola NewsStream receiver, which interfaced with a mobile computer via the standard RS-232C serial port. The data transmission rate was 300 bps.

EMBARC was designed from the start as an advanced messaging network, but after pouring tens of millions of dollars into EMBARC, in mid-1996 Motorola sold the service to ProNet, a large regional paging company. ProNet then merged with Teletouch Communications, another regional paging carrier.

E-mail over Paging Nets

Wireless messaging took another leap forward in 1996 with the ability of pagers to accept e-mail messages. Not only could traditional paging services support e-mail, the messages also could originate from Internet e-mail clients on desktop computers. An Internet gateway allows customers to receive e-mail messages sent across the Internet. Senders use an address consisting of the Pager Identification Number (PIN) assigned to the intended text pager followed by the domain of the paging service, such as SkyTel. For the PIN 1234567, for example, the e-mail address of a SkyTel subscriber would be

1234567@skytel.com

SkyTel's support for paging from public X.400 networks and the connected corporate e-mail systems means that e-mail users around the world can send a message to SkyTel text pagers in the same manner they would to other e-mail recipients. Most paging service providers have Web pages that allow anyone to send e-mail messages of not more than 260 characters to their subscribers.

Integrated Applications

A newer messaging solution integrates various messaging formats and various communications technologies into a single package that can be especially useful to mobile professionals. Depending on service plan, the following capabilities are available:

- Send and receive e-mail worldwide via the Internet
- Send faxes worldwide
- Convert text to speech to send telephone messages from a computer
- Receive text messages from a 24-hour, operator-assisted message center
- Send messages to any alphanumeric pager in the United States
- Receive notification of incoming messages to a computer on an alphanumeric pager in the United States
- Filter and automatically forward messages to another e-mail address
- Send and receive file attachments
- Customize faxes from a personal library of electronically stored documents

For the ultimate in messaging, there is even software available that enables users to send e-mail messages out over the Internet for receipt on all of a person's communications devices simultaneously, including cell phone, pager, wireless PDA, and notebook computer.

Summary

Not too many years ago, e-mail was considered a fad. Now there is great appreciation for e-mail and its role in supporting daily business operations. Wireless services extend the reach of e-mail even further, since they do not rely as much on the existing wireline infrastructure. In turn, this has given the distributed workforce the highest degree of mobility, enabling them to conduct business and stay in touch with coworkers without regard for location, distance from the office, or proximity to a telephone. The evolution of wireless messaging continues, with Short Messaging Services (SMS) available with most cellular services today, rendering paging virtually obsolete. And with third-generation (3G) networks coming online, there will be enough bandwidth to support multimedia messaging.

See also

Cellular Telephones

Paging

Short Messaging Services

WIRELESS PBX

With office workers spending increasing amounts of time away from their desks—supervising various projects, working at temporary assignments, attending meetings, and just walking corridors—there is a growing need for wireless technology to help them stay in touch with colleagues, customers, and suppliers. The idea behind the wireless PBX is to facilitate communication within the office environment, enabling employees to be as productive with a wireless handset as they would if they were sitting at their desk.

Applications

Typically, a wireless switch connects directly to an existing PBX, key telephone system, or Centrex service, converting an office building into an intracompany microcellular system.

This arrangement provides wireless telephone, paging, and e-mail services to mobile employees within the workplace through the use of pocket-sized portable phones, similar to those used for cellular service. Almost any organization can benefit from improved communications offered by a wireless PBX system, including

- *Manufacturing* Roving plant managers or factory supervisors do not have to leave their inspection or supervisory tasks to take important calls.

- *Retail* Customers can contact in-store managers directly, eliminating noisy paging systems.

- *Hospitality* Hotel event staff can stay informed of guest's needs and respond immediately.

- *Security* Guards can relay emergency information quickly and clearly, directly to the control room or police department, without trying to reach a desktop phone.

- *Business* Visiting vendors or customers have immediate use of preassigned phones without having to take over employee offices to use a phone.

- *Government* In-demand office managers can be available at all times for instant decision making.

A wireless PBX is especially appropriate in such areas as education and health care or any operation with multiple buildings in a campus environment. In the health care industry, for example, a typical environment for a wireless PBX would be a hospital where staff members typically are away from their workstations one-third of the workday.

System Components

Many of the wireless office systems on the market today are actually adjunct systems that interfaces to an existing PBX that provides user features and access to wireline telephones and outside trunk carrier facilities. The advantages of this approach include cost savings in terms of hardware, space requirements, and power. A wireless PBX typically consists of several discrete components (Figure W-16).

Adjunct Switch The adjunct contains the CPU and control logic. Its function is to manage the calls sent and received between the base stations. The adjunct is a standalone unit that can be wall-mounted for easy installation and maintenance. It can be collocated with the PBX or connected to the PBX via twisted-pair or optical fiber from several thousand feet away. Optional battery backup is usually available, permitting uninterrupted operation should a power failure occur. System control, management, and administration

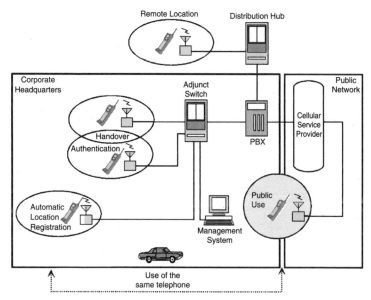

Figure W-16 A typical wireless PBX system in the corporate environment. In this case, workers can roam between the office and home using the same handset. When the handset moves within range of the local cellular service provider, the signal is handed off from the wireless PBX to the cellular carrier's nearest base station.

functions are provided through an attached terminal that is password-protected to guard against unauthorized access.

As portable telephones and base stations are added to accommodate growth, line cards are added to the PBX, and radio cards are added to the adjunct switch to handle the increasing traffic load. Each adjunct is capable of supporting several hundred portable telephones. Additional adjuncts can be added as necessary to support future growth.

Base Stations The antenna-equipped base stations, which are about the size of smoke detectors, are typically mounted on the ceiling and are connected by twisted-pair wiring to the wireless PBX. They send and receive calls between the portable telephones and the adjunct unit. As the user moves from one cell to another, the base station hands off the call to the nearest base station with an idle channel. When the next base station grabs the signal, the channel of the former base station becomes idle and is free to handle another call.

To facilitate the handoff process, each base station may be equipped with dual antennas (antenna diversity). This improves signal detection, enabling the handoff to occur in a timely manner. This is accomplished by the base station sampling the reception on each of its antennas and switching to the one that offers the best reception. This process is continuous, ensuring the best voice quality throughout the duration of the call. Some vendors offer optional external antennas for outdoor coverage or directional coverage indoors.

Wireless PBX systems are easily expanded—portable telephones and base stations are added as needed. Substantial savings can accrue over time through the elimination of traditional phone moves, adds, and changes. There also is significant savings in cabling, since there is less need to rewire offices and other locations for desktop telephones.

Telephone Handset Each portable phone has a unique identification number that must be registered with the adjunct switch. This allows only authorized users to access the communications system. The portable phone can be configured to have the same number as the user's desk phone so that when a call comes in, both phones ring. The user can even start a conversation on one phone and switch to the other. If the portable and desktop phones have different numbers, each can be programmed by the user to forward incoming calls to the other.

Since the adjunct switch becomes an integrated part of the company's existing telephone system, users have access to all its features through their portable phones. Users can even set up conference calls, forward calls, and transfer calls. If the handset is equipped with an LCD, the unit also can be used to retrieve e-mail messages, faxes, and pages. An alphanumeric display shows the name and number of the person or company calling.

The portable phone offers a number of other features, including

- Private directory of stored phone numbers for quick dialing
- Multilevel last number redial
- Audio volume, ring volume, and ring tone control
- Visual message-waiting indicator
- Silent vibrating alert
- Electronic lock for preventing outgoing calls
- In-range/out-of-range notifications
- Low battery notification

When the portable phone is not being used, a desktop unit houses the phone and charges both an internal and a spare battery. A LED indicates when the battery is fully charged. Recharging takes only a few hours and varies according to the type of battery used: Nickel cadmium (NiCd) takes about 2½ hours, while nickel metal hydride (NiMH) and the newer lithium ion (Li-Ion) batteries take about 1½ hours. Li-Ion batteries offer longer life and are lighter weight than NiCd and NiMH batteries. Some vendors offer an intelligent battery charging capability that protects the battery from overcharging.

Distribution Hub Distribution hubs are used in large installations to extend and manage communications among the base units in remote locations that are ordinarily out of range of the adjunct unit. They also allow high-traffic locations to be divided into smaller cells, called "microcells," with each cell containing multiple base stations. This arrangement makes more channels available to handle more calls.

The distribution units are connected to the adjunct unit with twisted-pair wiring or optical fiber. Optical fiber is an ideal medium for an in-building wireless network because its low attenuation over distance (approximately 1 decibel per kilometer) allows high-quality coverage even in large buildings and campus environments. Fiber is also immune to electromagnetic interference, allowing it to work effectively alongside other electronic equipment in installations such as factories and warehouses.

Frequency Bands Wireless PBXs operate in a variety of frequency bands, including the unlicensed 1910- to 1930-MHz Personal Communications Services (PCS) band. The term unlicensed refers to the spectrum that is used with equipment that can be bought and deployed without FCC approval because it is not part of the public radio spectrum. In other words, since wireless PBX operates over a dedicated frequency band for communications within a very narrow geographic area, it has little chance of interfering with other wireless services in the surrounding area. The individual channels supported by the wireless PBX system are spaced far enough apart to prevent interference with one another.

Standards

The Telecommunications Industry Association (TIA) and the American National Standards Institute (ANSI) have defined a North American standard that ensures interoperability between portable phones and wireless PBXs from different vendors. The TIA TR41.6.1 subcommittee based its development of the standard, called Personal Wireless Telecommunications (PWT), on the Digital European Cordless Telecommunications (DECT) standard. Portable phones that support PWT, formerly known as the Wireless Customer Premises Equipment standard, will interoperate with PWT-compliant wireless PBXs from any vendor.

For a wireless handset to communicate with any wireless PBX, manufacturers of both devices must agree on how the signal should be handled. As part of the PWT standard, the Customer Premises Access Profile defines the features that each side of the air interface must support to provide full, multivendor interoperability for voice services. As with most standards, vendors can add proprietary extensions to support additional features and differentiate their products.

The air interface is a layered protocol, similar to the International Organization for Standardization's (ISO's) Open Systems Interconnection (OSI) architecture. Accordingly, the air interface is composed of four protocol layers:

- *Physical Layer* Specifies radio characteristics such as channel frequencies and widths, the modulation scheme, and power and sensitivity levels. This layer also specifies the framing, so each handset can translate the bits it receives.

- *Media Access Control (MAC) Layer* Specifies the procedures by which the portable phone and the base station, or antenna, negotiate the selection of the radio channels.

- *Data Link Control (DLC) Layer* Specifies how frames are transmitted and sequenced between the handset and the base station.

- *Network Layer* Specifies messages that identify and authenticate the handset to the wireless PBX.

Call Handoff Scenario

By examining the handoff from one base station to another, the operation of these protocols can be illustrated. A handoff occurs when the mobile user walks out of the range of one base station and into the zone or cell of another base station. When the handset detects a change of signal strength from strong to weak, it will attempt to get acceptable signal strength from another channel offered by the same base station. If there is a better channel available, an exchange of messages at the MAC level occurs, which allows the conversation to continue without interruption. This channel change takes place without notification to the DLC layer.

If an acceptable channel is not available to the current base station, the handset searches for another base station. An exchange of messages at the DLC and MAC layers secures a data link via a radio channel to the new base station while the call through the original base station continues. When the data link to the second base station is established, the handset drops the old channel and begins processing the frames received through the new one. This process occurs without the network layer being notified. This means the caller and the wireless PBX are not aware that a handoff has happened.

Summary

Businesses everywhere have put a high priority on increasing the productivity of their workforce, even while they continue to cut back on staff. In order to improve profitability, serve customers, and grow market share, organizations must find ways to do more—cheaper, faster, and better. Most companies are focused on increasing efficiency and productivity while reducing time to market and improving customer service. This puts workers between a rock and a hard place: They must be mobile, away from their desks and offices, but not far from their telephones. Wireless PBX technology meets both demands.

See also

Wireless Centrex

Wireless LANs

WIRELESS TELECOMMUNICATIONS BUREAU

The Wireless Telecommunications Bureau (WTB) handles all domestic wireless telecommunications programs and policies for the FCC—except those involving satellite communications or broadcasting—including licensing, enforcement, and regulatory functions. Wireless communications services under the purview of the WTB include Amateur, Cellular, Paging, and Broadband PCS.

The amateur and amateur-satellite services are for qualified persons of any age who are interested in radio technique solely with a personal aim and without pecuniary interest. These services present an opportunity for self-training, intercommunication, and technical investigations.

Radiotelephone service, commonly referred to as "cellular" or "mobile telephone service," uses spectrum to provide a mobile telecommunications service for hire to the general public using cellular systems. Cellular licensees may operate using either analog or digital networks or both. Cellular licensees that operate digital networks also may offer advanced two-way data services.

Commercial paging is provided for profit, interconnected to the public switched network, and available to the public. Commercial paging may operate in the 35-, 43-, 152-, 158-, 454-, 929- (exclusive channels only), and 931-MHz bands. Response paging channels allow paging operators to provide two-way or response paging services.

Paging systems are traditionally one-way signaling systems. Paging services, grouped by output, include tone, tone/voice, numeric, and alphanumeric. Present systems are of two basic types: a wide-area general-use type providing subscription service to the public and in-building, private paging systems limited to a commercial building or the general area of a manufacturing plant. Currently, neither of these paging systems can initiate an answer without calling through a landline telephone.

Personal Communications Service (PCS) spectrum is used for a variety of mobile and fixed radio services, also called "wireless services." Mobile broadband PCS services include both voice and advanced two-way data capabilities that are generally available on small, mobile multifunction devices. Many broadband PCS licensees offer these services in competition with existing cellular and Specialized Mobile Radio (SMR) licensees. Examples of service providers holding a significant amount of broadband PCS spectrum include AT&T Wireless and Sprint PCS.

The goals of the Wireless Telecommunications Bureau are to

- Foster competition among different services
- Promote universal service and service to individuals with disabilities
- Maximize efficient use of spectrum
- Develop a framework for analyzing market conditions for wireless services
- Minimize regulation where appropriate
- Facilitate innovative service and product offerings, particularly by small businesses and new entrants
- Serve WTB customers efficiently, including improving licensing, eliminating backlogs, disseminating information, and making staff accessible.
- Enhance consumer outreach and protection and improve the enforcement process

In addition, the WTB is responsible for implementing the competitive bidding process for spectrum auctions, authority for which was given to the FCC by the 1993 Omnibus Budget Reconciliation Act.

Summary

The WTB also had responsibility for public safety radio communications, but following the September 11 terrorist attacks of 2001, the FCC established the National Coordination Committee (NCC) to satisfy public safety communications needs into the twenty-first century and provide the capability for a nationwide public safety interoperability communications system.

See also

Federal Communications Commission

Spectrum Auctions

Spectrum Planning

WIRELESS TELECOMMUNICATIONS INVESTMENT FRAUD

The FCC offers guidelines to help protect potential investors in wireless telecommunications services from being defrauded by unscrupulous promoters. According to the FCC, there has been a surge in the number of reported cases of fraud in emerging telecommunications services such as Paging, Specialized Mobile Radio (SMR), Wireless Cable, Interactive Video and Data Services (IVDS), Personal Communications Service (PCS), and Cellular Radio Telephone in underserved areas.

As the FCC warns, telecommunications is an undeniably alluring, fast-paced, multi-billion dollar industry. "Combine that with brain-numbingly complex technologies, and it creates the perfect environment for scam artists. The problem for the potential investor is differentiating opportunity from opportunism."

Separating truth from fiction in these sophisticated scams is not easy. However, the experience gained by the Federal

Trade Commission (FTC) and other agencies prosecuting these types of scams suggests some common warning signs that should alert consumers that the investment proposal might not be legitimate.

Common Warning Signs

- *Cold calls and infomercials* If the first contact with the telemarketer is an unsolicited call from a salesperson, be skeptical. Another favorite tactic for luring investors is via television or radio infomercials.

- *High profits, low risk* Scam artists are clever liars. If the sales representative promises high profits with little or no risk, chances are the deal is phony.

- *Urgency* Beware of promoters who say that it is urgent to invest now. Swindlers do not want people to have time to think things over. Some may even apply pressure to promptly send money by courier or wireless transfer.

- *IRA funds* Many scam artists claim that their investments have been approved for use in Investment Retirement Accounts (IRAs). In reality, there is no formal government approval process for certifying the appropriateness of funds for IRA status.

Avoiding Fraud

- *Be skeptical* The best protection against being scammed is skepticism. Investigate thoroughly any representations from salespersons before sending any money because, if duped, it is unlikely that any of it will be recovered. Make sure to fully understand the telemarketers' answers to questions. Continually assess whether their answers are reasonable.

- *High risk* A venture that seeks to obtain immediate, short-term profits from ownership of an FCC-licensed or -authorized service is a high-risk enterprise. Investors can and do lose money. In evaluating risk tolerance, would-be investors should consider whether they can afford to lose the entire investment.

- *Ask questions* Do not rely solely on the representations of the salesperson. Obtain advice from others: friends and family, an attorney, an accountant, an investment advisor, appropriate industry trade groups, the Better Business Bureau, state commission on corporations, state attorney general, or other business professionals. Be suspicious of representations that the investment is not subject to registration with or regulated by such federal agencies as the Security and Exchange Commission (SEC) or the FTC.

- *Administrative costs* Fraudulent telemarketers often take most of the money they solicit from consumers as commissions, promotions, and management and administrative costs. A shockingly small amount of anyone's investment actually goes into ownership interest.

- *Auctions* If the license is subject to an auction, be sure that
  - The auction is scheduled by the FCC.
  - The telecommunications service of interest is included in that auction.
  - The investment will be maintained pending the auction.
  - The venture is a "qualified bidder" in the auction.
  - If the venture wins a license, there will be funds and expertise available to the venture to construct and operate the system.
  - If the venture does not win a license, know how much, if any, of the investment will be returned.

- *License speculation* Be skeptical of any representation that the venture will quickly sell the license or authorization to another company for a huge profit. Sales of systems typically involve a long and complex process. The FCC has regulations that prohibit some services from being sold until the system has been constructed and is operating.

- *Construction requirements* Construction of advance telecommunications systems may cost tens of thousands or even millions of dollars beyond the initial investment. Know exactly where that money is coming from. Ultimately, the responsibility to construct and operate the system falls on the licensee. Some licenses are subject to construction deadlines. Failure to comply with these deadlines may result in the automatic cancellation of the license.

Summary

The burden is entirely on the investor to investigate the technology, the potential of wireless licenses to produce revenues, and the condition of the marketplace. License holders are responsible for being familiar with FCC rules and regulations. The FCC does not approve any individual investment proposal, nor does it provide a warranty with respect to any authorization. Receiving an authorization for a wireless license from the FCC is not a guarantee of success in the marketplace.

See also

Federal Communications Commission
Wireless Telecommunications Bureau

ACRONYMS

A

AAL	ATM Adaptation Layer
AAS	Automated Auction System
ABATS	Automated Bit Access Test System
ABM	Accunet Bandwidth Manager (AT&T)
ABR	Available Bit Rate
AC	Access Control
AC	Address Copied
ac	Alternating Current
AC	Authentication Center
ACD	Automatic Call Distributor
ACELP	Algebraic Code Excited Linear Predictive
ACP	Access Control Point
ACL	Access Control List
ACL	Asynchronous Connectionless (link)
ACS	Asynchronous Communications Server
ACTS	Advanced Communication Technology Satellite (NASA)
ADA	Americans with Disabilities Act

ADCR	Alternate Destination Call Routing (AT&T)
ADM	Add-Drop Multiplexer
ADN	Advanced Digital Network (Pacific Bell)
ADPCM	Adaptive Differential Pulse Code Modulation
ADSL	Asymmetric Digital Subscriber Line
AES	Advanced Encryption Algorithm
AFP	Apple File Protocol
AGC	Automatic Gain Control
Aglet	Agent Applet (IBM Corp.)
AGRAS	Air-Ground Radiotelephone Automated Service
AIOD	Automatic Identification of Outward Dialed Calls
AIN	Advanced Intelligent Network
ALI	Automatic Location Identification or Information
ALM	Antilocking Mechanism
AM	Amplitude Modulation
AMPS	Advanced Mobile Phone Service
AMTS	Automated Marine Telecommunications System
ANI	Automatic Number Identification
ANR	Automatic Network Routing (IBM Corp.)
ANSI	American National Standards Institute
ANT	ADSL Network Terminator
AOL	America Online
AP	Access Point
APC	Access Protection Capability (AT&T)
API	Application Programming Interface
APPC	Advanced Program-to-Program Communications (IBM Corp.)

APPN	Advanced Peer-to-Peer Network (IBM Corp.)
APS	Automatic Protection Switching
ARCnet	Attached Resource Computer Network (Datapoint Corp.)
ARB	Adaptive Rate Based (IBM Corp.)
ARIB	Association of Radio Industries and Businesses (Japan)
ARP	Address Resolution Protocol
ARPA	Advanced Research Projects Agency
ARQ	Automatic Repeat Request
ARS	Action Request System (Remedy Systems, Inc.)
AS	Autonomous System
ASCII	American Standard Code for Information Interchange
ASIC	Application-Specific Integrated Circuit
ASN.1	Abstract Syntax Notation 1
ASTN	Alternate Signaling Transport Network (AT&T)
AT&T	American Telephone & Telegraph
ATDMA	Asynchronous Time Division Multiplexing Access
ATIS	Alliance for Telecommunications Industry Solutions (formerly, ECSA)
ATM	Asynchronous Transfer Mode
ATM	Automated Teller Machine
ATSC	Advanced Television Systems Committee
ATT	Auction Tracking Tool
AUI	Attachment Unit Interface
AVM	Administrative View Module
AWG	American Wire Gauge
AWT	Abstract Window Toolkit

B

B8ZS	Binary Eight Zero Substitution
BACP	Bandwidth Allocation Control Protocol
BBS	Bulletin Board System
BCCH	Broadcast Control Channel
BDCS	Broadband Digital Cross-Connect System
BECN	Backward Explicit Congestion Notification
Bellcore	Bell Communications Research, Inc.
BER	Bit Error Rate
BERT	Bit Error Rate Tester
BETRS	Basic Exchange Telephone Radio Service
BFT	Binary File Transfer
BGP	Border Gateway Protocol
BHCA	Busy Hour Call Attempts
BHCR	Busy Hour Call Rate
BIB	Backward Indicator Bit
BIOS	Basic Input-Output System
BMC	Block Multiplexer Channel (IBM Corp.)
BMS-E	Bandwidth Management Service–Extended (AT&T)
BOC	Bell Operating Company
BONDING	Bandwidth on Demand Interoperability Group
BootP	Boot Protocol
BPDU	Bridge Protocol Data Unit
bps	Bits per Second
BPV	Bipolar Violation
BREW	Binary Runtime Environment for Wireless (Qualcomm)
BRI	Basic Rate Interface (ISDN)
BSA	Basis Serving Arrangement

BSC	Base Station Controller
BSC	Binary Synchronous Communications
BSE	Basic Service Element
BSN	Backward Sequence Number
BSSC	Base Station System Controller
BST	Base Station Transceiver
BT	Both-Way Trunk
BTA	Business Trading Area
BTS	Base Transceiver Station

C

CA	Communications Assistant
CAD	Computer-Aided Design
CAI	Common Air Interface
CAM	Computer-Aided Manufacturing
CAN	Campus Area Network
CAP	Carrierless Amplitude/Phase (Modulation)
CAP	Competitive Access Provider
CARS	Cable Antenna Relay Services
CAS	Customer Activation System
CASE	Computer-Aided Software Engineering
CAT	Computed Axial Tomography
CATV	Cable Television
CB	Cell Broadcast
CB	Citizens' Band
CBR	Constant Bit Rate
CBSC	Centralized Base Site Controller
CCC	Clear Channel Capability
CCCH	Common Control Channel

CCF	Call Control Function
CCITT	Consultative Committee for International Telegraphy and Telephony
CCR	Customer-Controlled Reconfiguration
CCS	Common Channel Signaling
CCSNC	Common Channel Signaling Network Controller
CCSS 6	Common Channel Signaling System 6
CD	Compact Disc
CD-R	Compact Disc–Recordable
CD-ROM	Compact Disc–Read Only Memory
CDCS	Continuous Dynamic Channel Selection
CDDI	Copper Distributed Data Interface
CDG	CDMA Development Group
CDMA	Code Division Multiple Access
CDMP	Cellular Digital Messaging Protocol
CDO	Community Dial Office
CDPD	Cellular Digital Packet Data
CDR	Call Detail Record or Recording
CEA	Component Economic Area
CEI	Comparably Efficient Interconnection
Centrex	Central Office Exchange
CEP	Circular Error Probable
CEPT	Conference of European Posts and Telegraphs
CGI	Common Gateway Interface
CGSA	Cellular Geographic Servicing Areas
CHAP	Challenge Handshake Authentication Protocol
CIF	Common Intermediate Format
CIM	Common Information Model

CIMD	Computer Interface to Message Distribution (Nokia Telecommunications)
CIR	Committed Information Rate
CLAN	Cordless Local Area Network
CLASS	Custom Local Area Signaling Services
CLEC	Competitive Local Exchange Carrier
CLI	Calling Line Identification
CLNP	Connectionless Network Protocol
CLP	Cell Loss Priority
CM-ES	Circuit-Switched Mobile-End System
CMD-IS	Circuit-Switched Mobile Data Intermediate System
CMI	Cable Microcell Integrator
CMIP	Common Management Information Protocol
CMIS	Common Management Information Services
CMRS	Commercial Mobile Radio Service
CNCS	Clearlink Network Control System (AT&T Tridom)
CNIR	Calling Number Identification Presentation
CNR	Customer Network Reconfiguration
CNS	Complementary Network Service
CO	Central Office
COM	Component Object Model (Microsoft Corp.)
COMSAT	Communications Satellite Corp.
CON	Concentrator
CORBA	Common Object Request Broker Architecture
CoS	Class of Service
COT	Central Office Terminal

CP	Coordination Processor
CPE	Customer Premises Equipment
CPODA	Contention, Priority-Oriented Demand Allocation
cps	Cycles per Second (Hertz)
CPU	Central Processing Unit
CR	Carriage Return
CRC	Cyclic Redundancy Check
CRM	Cisco Resource Manager (Cisco Systems)
CRM	Customer Relationship Management
CS	Cell Station
CS-CDPD	Circuit-Switched Cellular Digital Packet Data
CSA	Carrier Serving Area
CSACELP	Conjugate Structure Algebraic Code Excited Linear Prediction
CSC	Circuit-Switched Cellular
CSCCP	Circuit-Switched CDPD Control Protocol
CSD	Circuit-Switched Data
CSM	Communications Services Management
CSMA/CA	Carrier Sense Multiple Access with Collision Avoidance
CSMA/CD	Carrier Sense Multiple Access with Collision Detection
CSR	Customer Service Representative
CSU	Channel Service Unit
CT	Cordless Telecommunications
CTI	Computer-Telephony Integration
CUG	Closed User Groups
CVSD	Continuously Variable Slope Delta (Modulation)

D

D-AMPS	Digital Advanced Mobile Phone Service
DA	Destination Address
DAC	Digital-to-Analog Converter
DACS	Digital Access and Cross-Connect System (AT&T)
DAP	Demand Access Protocol
DAS	Dual Attached Station
DASD	Direct Access Storage Device (IBM Corp.)
DAT	Digital Audio Tape
dB	Decibel
DBMS	Database Management System
DBS	Direct Broadcast Satellite
DBU	Dial Backup Unit
dc	Direct Current
DCC	Digital Control Channel
DCCH	Digital Control Channel
DCE	Data Communications Equipment
DCE	Distributed Computing Environment
DCF	Data Communication Function
DCS	Digital Cross-Connect System
DDF	Digital Data Fast (Motorola)
DDS	Digital Data Services
DDS/SC	Digital Data Service with Secondary Channel
D/E	Debt-Equity (Ratio)
DECT	Digital Enhanced (formerly, European) Cordless Telecommunication
DES	Data Encryption Standard
DFSMS	Data Facility Storage Management Subsystem (IBM Corp.)

DHCP	Dynamic Host Configuration Protocol
DID	Direct Inward Dialing
DIF	Digital Interface Frame
DISCO	Domestic and International Satellite Consolidation Orders (FCC)
DLCS	Digital Loop Carrier System
DLL	Data Link Layer
DLL	Dynamic Link Library (Microsoft Corp.)
DLSw	Data Link Switching (IBM Corp.)
DLU	Digital Line Unit
DM	Distributed Management
DME	Distributed Management Environment
DMI	Desktop Management Interface
DMT	Discrete Multitone
DMTF	Desktop Management Task Force
DNS	Domain Name Service
DoD	Department of Defense (U.S.)
DOD	Direct Outward Dialing
DOS	Disk Operating System
DOT	Department of Transportation (U.S.)
DOV	Data over Voice
DQDB	Distributed Queue Dual Bus
DQPSK	Differential Quadrature Phase-Shift Keying
DRAM	Dynamic Random Access Memory
DS0	Digital Signal–Level 0 (64 kbps)
DS1	Digital Signal–Level 1 (1.544 Mbps)
DS1C	Digital Signal–Level 1C (3.152 Mbps)
DS2	Digital Signal–Level 2 (6.312 Mbps)

DS3	Digital Signal–Level 3 (44.736 Mbps)
DS4	Digital Signal–Level 4 (274.176 Mbps)
DSI	Digital Speech Interpolation
DSL	Digital Subscriber Line
DSMA	Digital Sense Multiple Access
DSN	Defense Switched Network
DSP	Digital Signal Processor
DSS	Decision Support System
DSSS	Direct Sequence Spread Spectrum
DSU	Data Service Unit
DSX1	Digital Systems Cross-Connect 1
DTE	Data Terminal Equipment
DTMF	Dual Tone Multifrequency
DTR	Dedicated Token Ring
DTU	Data Transfer Unit
DTV	Digital Television
DTX	Discontinuous Transmission
DWMT	Discrete Wavelet Multitone
DXI	Data Exchange Interface

E

E-AMPS	Enhanced Advanced Mobile Phone Service
E-Mail	Electronic Mail
E-TDMA	Expanded Time Division Multiple Access
ECC	Enterprise Command Center (Bay Networks)
ECG	Electrocardiogram
ECSA	Exchange Carriers Standards Association
ED	Ending Delimiter
EDGE	Enhanced Data Rates for Global Evolution

EDI	Electronic Data Interchange
EDRO	Enhanced Diversity Routing Option (AT&T)
EEPROM	Electronically Erasable Programmable Read-Only Memory
EEROM	Electronically Erasable Read-Only Memory
EFRC	Enhanced Full Rate Codec
EFT	Electronic Funds Transfer
EGP	External Gateway Protocol
EHF	Extremely High Frequency (More than 30 GHz)
EIA	Electronic Industries Association
EIR	Equipment Identity Register
EISA	Extended Industry Standard Architecture
EMA	Electronic Messaging Association
EMBARC	Electronic Mail Broadcast to a Roaming Computer
EMI	Electromechanical Inteference
EMS	Element Management System
EOC	Embedded Overhead Channel
EOT	End of Transmission
EP	Extension Point
EPFD	Equivalent Power Flux Density
ERMES	Enhanced Radio Messaging System
ERP	Effective Radiated Power
ERP	Enterprise Resource Planning
ESCON	Enterprise System Connection (IBM Corp.)
ESD	Electronic Software Distribution
ESF	Extended Super Frame
ESMR	Enhanced Specialized Mobile Radio

ESMS	Enhanced Short Message Service
ESMTP	Extended Simple Mail Transfer Protocol
ESN	Electronic Serial Number
ESQ	End System Query
ETC	Enhanced Cellular Throughput
ETSI	European Telecommunication Standards Institute
EVRC	Enhanced Variable Rate Coder

F

4GL	Fourth-Generation Language
FA	Foreign Agent
FACCH	Fast Associated Control Channel
FASB	Financial Accounting Standards Board
FASC	Fraud Analysis and Surveillance Center (AT&T)
FAT	File Allocation Table
FC	Frame Control
FC	Fibre Channel
FCC	Federal Communications Commission (U.S.)
FCS	Frame Check Sequence
FDDI	Fiber Distributed Data Interface
FDIC	Federal Deposit Insurance Corporation
FDL	Facilities Data Link
FECN	Forward Explicit Congestion Notification
FEP	Front-End Processor
FHSS	Frequency Hopping Spread Spectrum
FIB	Forward Indicator Bit
FIB	Forwarding Information Base
FIR	Fast Infrared

FITL	Fiber in the Loop
FM	Frequency Modulation
FOCC	Forward Control Channel
FOD	Fax on Demand
FPF	Fraud Protection Feature
FPGA	Field Programmable Gate Array
FRAD	Frame Relay Access Device
FRS	Family Radio Service
FS	Frame Status
FSN	Forward Sequence Number
FSTN	Film Super-Twisted Nematic
FTAM	File Transfer, Access, and Management
FT1	Fractional T1
FTP	File Transfer Protocol
FTS	Federal Telecommunications System
FTTB	Fiber to the Building
FTTC	Fiber to the Curb
FTTH	Fiber to the Home
FWA	Fixed Wireless Access
FWT	Fixed Wireless Terminal
FX	Foreign Exchange (Line)

G

GAO	General Accounting Office (U.S.)
GATT	General Agreement on Tariffs and Trade
GCS	Ground Control Station
GDS	Generic Digital Services
GEO	Geostationary Earth Orbit
GFC	Generic Flow Control

GGSN	Gateway GPRS Support Node
GHz	Gigahertz (Billions of Cycles per Second)
GIS	Geographic Information Systems
GMDSS	Global Maritime Distress and Safety System
GMRS	General Mobile Radio Service
GMSK	Gaussian Minimum Shift Keying
GPI	General Purpose Interface
GPRS	General Packet Radio Services
GPS	Global Positioning System
GSA	General Services Administration
GSM	Global System for Mobile Telecommunications (formerly Groupe Spéciale Mobile)
GUI	Graphical User Interface

H

H0	High-Capacity ISDN Channel Operating at 384 kbps
H11	High-Capacity ISDN Channel Operating at 1.536 Mbps
HA	Home Agent
HDLC	High-Level Data Link Control
HDML	Handheld Device Markup Language
HDSL	High-Bit-Rate Digital Subscriber Line
HDTV	High Definition Television
HEC	Header Error Check
HF	High Frequency (1.8–30 MHz)
HFC	Hybrid Fiber/Coax
HIC	Head-End Interface Converter
HLR	Home Location Register

HMMP	HyperMedia Management Protocol
HMMS	HyperMedia Management Schema
HMOM	HyperMedia Object Manager
HSCSD	High-Speed Circuit-Switched Data
HTML	HyperText Markup Language
HTTP	HyperText Transfer Protocol
HVAC	Heating, Ventilation, and Air Conditioning
Hz	Hertz (Cycles per Second)

I

I/O	Input-Output
IAB	Internet Architecture Board
IANA	Internet Assigned Numbers Authority
IAPP	Inter Access Point Protocol
IC	Integrated Circuit
IC	Integrity Check
ICI	Interexchange Carrier Interface
ICMP	Internet Control Message Protocol
ICP	Integrated Communications Provider
ICR	Intelligent Call Routing
ICS	Intelligent Calling System
ID	Identification
IDDD	International Direct Dialing Designator
iDEN	Integrated Digital Enhanced Network
IDPR	Inter-Domain Policy Routing
IEC	International Electrotechnical Commission
IEEE	Institute of Electrical and Electronic Engineers

IESG	Internet Engineering Steering Group
IETF	Internet Engineering Task Force
IF	Intermediate Frequency
IGP	Interior Gateway Protocol
IIG	Intercast Industry Group
IIOP	Internet Inter-ORB Protocol
ILEC	Incumbent Local Exchange Carrier
IM	Instant Messaging
IMAP	Internet Mail Access Protocol
IMEI	International Mobile Equipment Identity
IMO	International Maritime Organization
IMS/VS	Information Management System/Virtual Storage (IBM Corp.)
IMSI	International Mobile Subscriber Identity
IMTS	Improved Mobile Telephone Service
IN	Intelligent Network
INMARSAT	International Maritime Satellite Organization
INMS	Integrated Network Management System
ITV	Interactive Television
IOC	Interoffice Channel
IP	Internet Protocol
IPv4	Internet Protocol Version 4 (Current)
IPv6	Internet Protocol Version 6 (Future)
IPH	Integrated Packet Handler
IPI	Intelligent Peripheral Interface
IPN	Intelligent Peripheral Node
IPsec	Internet Protocol Secure
IPX	Internet Packet Exchange
IR	Infrared

IRAC	Interdepartment Radio Advisory Committee
IRC	Internet Relay Chat
IrDA	Infrared Data Association
IrDA-SIR	Infrared Data Association Serial Infrared (Standard)
IrFM	Infrared Financial Messaging
IrLAN	Infrared Local Area Network
IrLAP	Infrared Link Access Protocol
IrLMP	Infrared Link Management Protocol
IrPL	Infrared Physical Layer
IrTTP	Infrared Tiny Transport Protocol
IRQ	Interrupt Request
IrTTP	Infrared Transport Protocol
IS	Information System
IS	Industry Standard
IS-IS	Intraautonomous System to Intraautonomous System
ISA	Industry Standard Architecture
ISDL	ISDN Subscriber Digital Line
ISDN	Integrated Services Digital Network
ISM	Industrial, Scientific, and Medical (Frequency Bands)
ISO	International Organization for Standardization
ISOC	Internet Society
ISP	Internet Service Provider
ISSI	Inter-Switching Systems Interface
IT	Information Technology
ITFS	Instructional Television Fixed Service
ITR	Intelligent Text Retrieval

ITS	Internet Telephony Server (Lucent Technologies)
ITU-TSS	International Telecommunications Union–Telecommunications Standardization Sector (formerly CCITT)
IV	Initialization Vector
IVDS	Interactive Video and Data Service
IVR	Interactive Voice Response
IXC	Interexchange Carrier

J

J2ME	Java 2 Micro Edition
JDBC	Java Database Connectivity
JDC	Japanese Digital Cellular
JDK	Java Development Kit
JEPI	Joint Electronic Payments Initiative
JIT	Just in Time
JMAPI	Java Management Application Programming Interface
JPEG	Joint Photographic Experts Group
JTC	Joint Technical Committee
JVM	Java Virtual Machine

K

k (kilo)	One Thousand (e.g., kbps)
kB	Kilobyte
KSU	Key Service Unit
KTS	Key Telephone System
kHz	Kilohertz (Thousands of Cycles per Second)

L

L2F	Layer 2 Forwarding
L2TP	Layer 2 Tunneling Protocol
LAN	Local Area Network
LAPB	Link Access Procedure–Balanced
LAPD	Link Access Procedure on the D-Channel
LAT	Local Area Transport (Digital Equipment Corp.)
LATA	Local Access and Transport Area
LBO	Line Build Out
LCD	Liquid-Crystal Display
LCN	Local Channel Number
LCP	Link Control Protocol
LD	Laser Diode
LDAP	Lightweight Directory Access Protocol
LDCELP	Low Delay Code Excited Linear Prediction
LEC	Local Exchange Carrier
LED	Light-Emitting Diode
LEO	Low Earth Orbit
LF	Line Feed
LF	Low Frequency (30–300 kHz)
LI	Length Indicator
Li-Ion	Lithium Ion
LLC	Logical Link Control
LMDS	Local Multipoint Distribution System
LMS	Location and Monitoring Service
LNP	Local Number Portability
LPFM	Low-Power Frequency-Modulated
LPRS	Low-Power Radio Service
LSI	Large-Scale Integration

LSM	Limited-Size Messaging
LTG	Line Trunk Group
LU	Logical Unit (IBM Corp.)

M

M (Mega)	One Million (e.g., Mbps)
M-ES	Mobile End System
MAC	Media Access Control
MAC	Moves, Adds, Changes
MAE	Major Economic Area
MAN	Metropolitan Area Network
MAPI	Messaging Applications Programming Interface (Microsft Corp.)
MAU	Multistation Access Unit
MB	Megabyte
MD	Mediation Device
MD-IS	Mobile Data–Intermediate System
MDBS	Mobile Database System
MDI	Mobile Data Initiative
MDLP	Mobile Data-Link Layer Protocol
MDS	Multipoint Distribution Service
MEO	Middle Earth Orbit
MES	Master Earth Station
MF	Mediation Function
MH	Mobile Host
MHz	Megahertz (Millions of Cycles per Second)
MIB	Management Information Base
MIN	Mobile Identification Number
MIPS	Millions of Instructions per Second
MIS	Management Information Services

MM	Mobility Manager
MMDS	Multichannel, Multipoint Distribution Service
MMS	Maritime Mobile Service
MMS	Multimedia Messaging Service
MNLP	Mobile Network Location Protocol
MNRP	Mobile Network Registration Protocol
MO	Mobile Originating
Modem	Modulation/demodulation
MPEG	Moving Pictures Experts Group
MPLS	Multiprotocol Label Switching
MPMLQ	Multipulse Maximum Likelihood Quantization
MR	Message Register
ms	Millisecond (Thousandths of a Second)
MS	Mobile Station
MSC	Mobile Switching Center
MSG	Message
MSRN	Mobile Station Roaming Number
MSS	Mobile Satellite Service
MT	Mobile Terminating
MTBF	Mean Time Between Failure
MTSO	Mobile Telephone Switching Office
MTSO	Mobile Transport Serving Office
MVC	Multicast Virtual Circuit
MVDS	Microwave Video Distribution System
MVDDS	Multichannel Video Distribution and Data Service
MVPD	Multichannel Video Program Distribution
mW	Milliwatt
MXU	Mobile Exchange Unit

N

N-AMPS	Narrowband Advanced Mobile Phone Service
N-PCS	Narrowband Personal Communications Service
NAM	Numeric Assignment Module
NAP	Network Access Point
NASA	National Aeronautics and Space Administration (U.S.)
NAT	Network Address Translation
NAVSTAR	Navigation System with Timing and Ranging
NCC	National Coordination Committee
NE	Network Element
NEI	Network Equipment Identifier
NetBIOS	Network Basic Input-Output System
NFS	Network File System (or Server)
NIC	Network Interface Card
NiCd	Nickel Cadmium
NIF	Network Interconnect Facility (Metricom, Inc.)
NiMH	Nickel-Metal Hydride
NIST	National Institute of Standards and Technology
NIU	Network Interface Unit
nm	Nanometer
NM	Network Manager
NMS	Network Management System
NMT	Nordic Mobile Telephone (Ericsson)
NOC	Network Operations Center
NOS	Network Operating System
NPCS	Narrowband Personal Communication Services

NSA	National Security Agency (U.S.)
NSF	National Science Foundation (U.S.)
NTIA	National Telecommunications and Information Administration (U.S.)
NTSC	National Television Standards Committee

O

OAM	Operations, Administration, Management
OAM&P	Operations, Administration, Maintenance, and Provisioning
OBEX	Object Exchange
OC	Optical Carrier
OEM	Original Equipment Manufacturer
OET	Office of Engineering and Technology (FCC)
OFDM	Orthogonal Frequency Division Multiplexing
OMC	Operations and Maintenance Center
OS	Operating System
OSI	Open Systems Interconnection
OSPF	Open Shortest Path First

P

PA	Preamble
PACS	Personal Access Communications System
pACT	Personal Air Communications Technology
PAP	Password Authentication Protocol
PAT	Port Address Translation
PBX	Private Branch Exchange
PC	Personal Computer
PCB	Printed Circuit Board

PCH	Paging Channel
PCM	Pulse Code Modulation
PCMCIA	Personal Computer Memory Card International Association
PCN	Personal Communications Networks
PCS	Personal Communications Services
PCT	Private Communication Technology
PDA	Personal Digital Assistant
PDBS	pACT Database Station
PDIS	pACT Data Intermediate System
PDN	Packet Data Network
PDSN	Packet Data Serving Node
PDU	Payload Data Unit
PGP	Pretty Good Privacy
PHS	Personal Handyphone System
PHY	Physical Layer
PIAF	PHS Internet Access Forum (Japan)
PIM	Personal Information Manager
PIN	Personal Identification Number
PLMRS	Private Land Mobile Radio Services
POCSAG	Post Office Code Standardization Advisory Group
POP	Point of Presence
POTS	Plain Old Telephone Service
PPP	Point-to-Point Protocol
PPS	Packets per Second
PRI	Primary Rate Interface (ISDN)
PSAP	Public Safety Answering Point
PSN	Packet Switched Network
PSTN	Public Switched Telephone Network

PT	Payload Type
PTT	Post Telephone & Telegraph
PUC	Public Utility Commission
PVC	Permanent Virtual Circuit
PWT	Personal Wireless Telecommunications

Q

QA	Quality Assurance
QoS	Quality of Service
QPSK	Quadrature Phase Shift Keying

R

RACH	Random Access Channel
RAD	Remote Antenna Driver
RADIUS	Remote Authentication Dial-In User Service
RAM	Random Access Memory
RASP	Remote Antenna Signal Processor
RDSS	Radio Determination Satellite Service
REAG	Regional Economic Area Grouping
RECC	Reverse Control Channel
RF	Radiofrequency
RF	Routing Field
RFI	Radio frequency Interference
RIP	Routing Information Protocol
RMON	Remote Monitoring
ROI	Return on Investment
RRM	Radio Resource Management
RSS	Residential Service System (Nortel)
RT	Remote Terminal

RTCMSC	Radio Technical Commission for Maritime Services Special Committee
RX	Receive

S

SA	Source Address
SACCH	Slow Associated Control Channel
SAN	Satellite Access Node
SC	System Controller
SCC	Satellite Control Center
SCF	Service Control Function
SCO	Synchronous Connection Oriented
SD	Starting Delimiter
SDF	Service Data Function
SDSL	Symmetric Digital Subscriber Line
SET	Secure Electronic Transaction
SFD	Start Frame Delimiter
SHF	Super High Frequency (3–30 GHz)
SHVIA	Satellite Home Viewer Improvement Act
SHTTP	Secure HyperText Transfer Protocol
SIG	Special Interest Group
SID	System Identification
SSID	Service Set Identifier
SIM	Subscriber Identification Module
SMR	Specialized Mobile Radio
SMS	Service Management System
SMS	Short Message Service
SMS-SC	Short Message Service Service Center
SMS/CB	Short Message Service/Cell Broadcast
SMS/PP	Short Message Service/Point-to-Point

SMTP	Simple Mail Transfer Protocol
SNMP	Simple Network Management Protocol
SOHO	Small Office/Home Office
SOLAS	Safety of Life at Sea
SQL	Structured Query Language
SS	Switching System
SS7	Signaling System No. 7
SSF	Service Switching Function
SSL	Secure Sockets Layer

T

T1	Transmission Service at the DS1 Rate of 1.544 Mbps
T3	Transmission Service at the DS3 Rate of 44.736 Mbps
TA	Technical Advisor
TA	Technical Advisory
TACS	Total Access Communications System
TAP	Telocator Alphanumeric Protocol
TASI	Time Assigned Speech Interpolation
TB	Terabyte (Trillion Bytes)
Tbps	Terabits per Second
TCO	Total Cost of Ownership
TCP	Transmission Control Protocol
TDD	Time Division Duplexing
TDM	Time Division Multiplexer
TDMA	Time Division Multiple Access
TDMA/TDD	Time Division Multiple Access with Time Division Duplexing
TDOA	Time Difference of Arrival

TDR	Time Domain Reflectometer
TEM	Transcend Enterprise Manager (3Com Corp.)
TFTP	Trivial File Transfer Protocol
TIA	Telecommunications Industry Association
TIB	Tag Information Base
TMN	Telecommunications Management Network
TNPP	Telocator Network Paging Protocol
TV	Television
TX	Transmit

U

UAPROF	User Agent Profile
UART	Universal Asynchronous Receiver/Transmitter
UBR	Unspecified Bit Rate
UDP	User Datagram Protocol
UDP/IP	User Datagram Protocol/Internet Protocol
UHF	Ultra High Frequency (238 MHz to 1.3 GHz)
ULS	Universal Licensing System
UMS	Universal Messaging System
UMTS	Universal Mobile Telephone Service
UMTS	Universal Mobile Telecommunications System
UN	United Nations
UNI	User-Network Interface
UPS	Uninterruptible Power Supply
URL	Uniform Resource Locator
USAT	Ultra Small Aperture Terminal
UTP	Unshielded Twisted-Pair
UWB	Ultra Wide Band

V

VC	Virtual Circuit
VEC	Volunteer Examiner-Coordinator
VF	Voice Frequency
VFIR	Very Fast Infrared
VHF	Very High Frequency (50–146 MHz)
VLF	Very Low Frequency (Less than 30 kHz)
VLR	Visitor Location Register
VLSI	Very Large Scale Integration
VOD	Video on Demand
VPN	Virtual Private Network
VSAT	Very Small Aperture Terminal

W

W3C	World Wide Web Consortium
WACS	Wireless Access Communications System (Bellcore)
WAE	Wireless Application Environment
WAN	Wide Area Network
WAP	Wireless Access Point
WAP	Wireless Application Protocol
WASP	Wireless Application Service Provider
WBM	Web-Based Management
WCDMA	Wideband Code Division Multiple Access
WCS	Wireless Communications Service
WDCT	Worldwide Digital Cordless Telephone
WECA	Wireless Ethernet Compatibility Alliance
WEP	Wired Equivalent Privacy
WFQ	Weighted Fair Queuing (Cisco Systems)

WGS	Worldwide Geodetic System
Wi-Fi	Wireless Fidelity
WISP	Wireless Internet Service Provider
WLAN	Wireless Local Area Network
WLL	Wireless Local Loop
WML	Wireless Markup Language
WMTS	Wireless Medical Telemetry Service
WRC	World Radio Conference
WSP	Wireless Session Protocol
WTA	Wireless Telephony Application
WTB	Wireless Telecommunications Bureau
WWW	World Wide Web

X

XOR	Exclusive Or

INDEX

Boldface page range indicates a main entry.

About the Author

Nathan J. Muller is an independent consultant specializing in telecommunications technology marketing, research, and education. A resident of Sterling, VA, he serves on the Editorial Board of the *International Journal of Network Management* and the Advisory Panel of Faulkner Information Services. Among the 24 books he has authored are *The Desktop Encyclopedia of Telecommunications* and *Network Manager's Handbook*.